Human and Animal Sensitivity

Human and Animal Sensitivity: How Stock-People and Consumer Perception Can Affect Animal Welfare

Special Issue Editor

Fabio Napolitano

MDPI • Basel • Beijing • Wuhan • Barcelona • Belgrade

MDPI

Special Issue Editor
Fabio Napolitano
Universita degli Studi della Basilicata,
Italy

Editorial Office
MDPI
St. Alban-Anlage 66
4052 Basel, Switzerland

This is a reprint of articles from the Special Issue published online in the open access journal *Animals* (ISSN 2076-2615) from 2018 to 2019 (available at: https://www.mdpi.com/journal/animals/special_issues/Human_and_Animal_Sensitivity_How_Stock-People_and_Consumer_Perception_Can_Affect_Animal_Welfare)

For citation purposes, cite each article independently as indicated on the article page online and as indicated below:

LastName, A.A.; LastName, B.B.; LastName, C.C. Article Title. *Journal Name* **Year**, *Article Number, Page Range.*

ISBN 978-3-03921-261-3 (Pbk)
ISBN 978-3-03921-262-0 (PDF)

Contents

About the Special Issue Editor

Fabio Napolitano, Associate Professor, after achieving a Ph.D. in animal science in 1995, was appointed to a researcher position and, in 2006, became an associate professor at the School of Agriculture, Food, Forestry and Environment (University of Basilicata, Italy). His research is particularly centered on the study of animal behavior, animal welfare, and the effect of information about animal welfare on consumer perception of product quality. His teaching is conducted at the master and Ph.D. level. He has been included in the list of experts in animal welfare and, between 2009 and 2011, was nominated as a member of the scientific committee of external reviewers by the European Food Safety Authority (EFSA). He is the author of over 125 peer-reviewed articles and several book chapters.

Preface to "Human and Animal Sensitivity: How Stock-People and Consumer Perception Can Affect Animal Welfare"

This book is based on an oxymoron. Although animal welfare is all about how the animals perceive the surrounding environment, the actual welfare of the animals is dependent on how the stakeholders perceive and weigh animal welfare. The stakeholders can, either directly (i.e. through stock-people interacting with the animals) or indirectly (e.g. when retailers and consumers are willing to pay more for high welfare animal-based products), affect the way animals are kept and handled at the farm, and while transported and at slaughter. "How to improve stock-people attitude and behavior towards the animals? How to increase consumer sensitivity to animal welfare issues?" These are some of the questions addressed in this book. The entire food chain is at stake, from stock-people and farmers to veterinarians, retailers, consumers and citizens in general, including the children and young adults who represent the future perceptions of animal welfare. Therefore, the book presents cross-discipline studies that cover aspects ranging from animal science to social/consumer sciences and psychology with the aim of disseminating information suitable to promote the enhancement of animal welfare by improving stakeholders' perception of animal welfare.

Fabio Napolitano
Special Issue Editor

animals

MDPI

Article

The Human-Animal Relationship in Australian Caged Laying Hens

Lauren E. Edwards [1,*], Grahame J. Coleman [1], Kym L. Butler [1,2] and Paul H. Hemsworth [1]

[1] Animal Welfare Science Centre, Faculty of Veterinary and Agricultural Sciences, University of Melbourne, Parkville, Victoria 3010, Australia; grahame.coleman@unimelb.edu.au (G.J.C.); kym.butler@unimelb.edu.au (K.L.B.); phh@unimelb.edu.au (P.H.H.)

[2] Department of Jobs, Precincts and Regions, Agriculture Research Victoria, Hamilton, Victoria 3300, Australia

[*] Correspondence: ledwards@unimelb.edu.au

Received: 9 April 2019; Accepted: 29 April 2019; Published: 2 May 2019

Simple Summary: Stockperson behaviour can influence fear of humans and welfare of farm animals. This study observed the human-animal relationship (HAR) in 19 Australian caged laying hen flocks to determine whether stockperson behaviour was associated with behavioural indicators of fear of humans and stress in caged laying hens. The average avoidance response of each flock toward an approaching human was assessed using two behavioural tests, and stress was measured using the concentration of corticosterone in an egg sample collected immediately prior to these observations. Stockperson behaviour was observed for 2 days in each flock and compared to hen fear. Unexpectedly, no relationships were found between the observed stockperson behaviour and avoidance of humans in the hens, but flocks were more productive when they showed less avoidance of humans, and when stockpeople made less noise in the laying house. This suggests that stockperson behaviour and hen fear may influence productivity, but there was no evidence that any effect of fear on productivity was caused by stockperson behaviour. Unexpectedly, the most fearful flocks also had the lowest stress levels. These results clearly need further research to be fully understood but could not confirm the existence of the HAR on caged egg farms in Australia.

Abstract: Studies on farm animals have shown relationships between stockperson attitudes and behaviour and farm animal fear, stress and productivity. This study investigated how the avoidance behaviour of Australian commercial caged laying hens may be related to stockperson behaviour, albumen corticosterone, and the number of weeks producing within 5% of peak egg production. Nineteen laying houses were assessed over 3 days. Fear of humans in hens, based on their avoidance response to an unfamiliar human, was assessed using two behavioural tests. Albumen corticosterone concentrations were measured from egg samples collected immediately prior to behavioural testing. Stockperson attitudes were assessed using a questionnaire and stockperson behaviour was observed over 2 days. Productivity records for each laying house were also obtained. The duration of peak production was negatively related to both noise made by the stockperson and hen avoidance. No relationship between stockperson behaviour or attitudes and hen avoidance was found, but stockpeople with negative attitudes made more noise. In conclusion, this study could not confirm a relationship between stockperson behaviour and hen avoidance behaviour for Australian caged laying hens. However, this study did confirm a relationship between hen avoidance behaviour, albumen corticosterone concentration, and the duration of peak egg production.

Keywords: human-animal relationship; fear; laying hen; stockpeople attitudes; stockperson behaviour; egg farm; albumen corticosterone; welfare

1. Introduction

There is increasing evidence that the interactions between stockpeople and their animals can have a substantial effect on the behaviour, welfare and productivity of farm animals by causing fear of humans and fear-related stress in these animals [1,2]. The nature and extent of these human-animal interactions will determine the quality of the human-animal relationship (HAR), and are determined by stockperson attitudes towards their animals and their work, their beliefs about other people's expectations of them, and their beliefs about the extent to which they have control over their ability to appropriately interact with the animals [1,2]. For example, the frequent use of aversive behaviours by stockpeople increases fear of humans in both sows [3] and dairy cattle [4–6], and the reduced productivity observed in these animals has been attributed to fear-induced stress [1].

Similar relationships have also been demonstrated in laying hens. Under experimental conditions, Barnett et al. (1994) [7] demonstrated that exposing cage-housed hens to additional human contact of a positive nature reduced their fear of humans, based on their avoidance behaviour in an approaching human test and corticosterone response to handling. The additional human contact also improved their immune function, based on a cell-mediated response following injection with a mitogen, and hen day production in comparison to hens that received minimal human contact. Under commercial free-range conditions, Waiblinger et al. (2018) [8] found significant relationships between stockperson attitudes, their subsequent interactions with the laying hens, the number of hens that could be approached and touched in the flock, and the feather damage and mortality rates of those hens. Stockpeople with positive attitudes toward hen care had flocks with lower avoidance behaviour, while stockpeople with negative general attitudes had flocks with more feather damage and higher mortality. Several other studies have shown relationships between human behaviour and fear of humans in hens [9–11], and between fear of humans and hen productivity [12]. These studies suggest that stockperson behaviour may be an important determinant of the avoidance behaviour, stress physiology and productivity of commercial laying hens.

This paper investigates how one aspect of hen productivity, namely the number of production weeks within 5% of peak egg production, may, through fear of humans, be related to attitudes and behaviours of stockpersons in commercial cage-housing systems within Australia.

2. Materials and Methods

This research had ethics approval from the Department of Primary Industries Animal Ethics Committee (AEC Code No. 2627) and the Melbourne School of Land and Environment Human Ethics Advisory Group (HREC Project No. 050041).

2.1. Subjects, Location and Summary of Data Collection

Data were collected at 19 laying houses from 10 farms, representative of commercial caged egg enterprises that were located in the states of Victoria or New South Wales, Australia. The study was conducted over 15 months. A summary of the laying houses visited is presented in Table 1. Each laying house was considered a separate unit if a separate team of stockpeople maintained it, and thus more than one laying house could be studied at the same farm. Multiple houses were studied at four farms. In each laying house a sample of focal cages was selected for closer study.

Where possible, the productivity records for the entire laying life of the flock in each study house were obtained. Data were collected at each laying house over a 3-day period, with 1 day spent conducting avoidance behaviour tests on the hens and the following 2 days spent observing the stockperson behaviour that occurred in the laying house. At the end of the third day a written attitude questionnaire was administered to the main stockperson that worked in each house.

Table 1. A description of each of the laying houses used in the study, highlighting variation in design and genetics.

Farm	Laying House within Farm	Flock Size (in '000s)	No. of Stock People per House	No. of Tiers	Av. No. of Birds per Cage	Aisle Length (m)	Age of Birds (weeks)	Strain of Birds	No. of Focal Cages per House
1	a	14	5	3	2.9	86	50	IB	77
2	a	12.3	4	3	3.0	84	57	HB	90
3	a	24	6	4	4.7	74	57	IB	61
4	a	2.5	2	1	2.9	30	48	IB	52
5	a	1.3	2	1	3.0	39	48	HB	34
6	a	22	6	4	3.7	85	55	IB	87
6	b	68.3	9	8	6.2	83	52	IB	74
6	c	22	4	4	3.9	84	90 [†]	IB	87
7	a	22	4	4	5.8	54	36	IH	58
8	a	29.4	1	5	8.9	76	48	HB	88
8	b	21.9	4	4	5.6	74	51	HB	62
9	a	108.1	5	6 [*]	6.2	129	69	IB	159
9	b	122	6	6 [*]	4.7	129	66	IB	107
9	c	116.4	5	6 [*]	6.8	129	62	IB	80
9	d	116.4	4	6 [*]	6.5	129	60	IB	80
10	a	67.5	5	6 [*]	5.3	93	65	IB	91
10	b	67.5	10	6 [*]	6.4	93	63	HB	90
10	c	67.5	5	6 [*]	7.0	93	42	HB	91
10	d	67.5	8	6 [*]	6.8	93	43	HB	91

[*] Denotes laying houses with six tiers of cages that had a wire mesh mezzanine floor installed between tiers 3 and 4. Stockpeople walked on the ground floor to monitor tiers 1–3, and on the mezzanine floor to monitor tiers 4–6. Husbandry observations were only made on the ground floor as this was the location of the focal birds. [†] Denotes a 90 wk old flock that had been moulted to extend the productive life of the hens. The number of birds in each focal cage varied due to mortalities. IB = ISA Brown, HB = Hyline Brown, IH = Ingham Hisex.

2.2. Physical Features of the Laying House

The following measurements were made of the laying houses: the total number of stockpeople that entered the laying house (No. of Stockpeople); the age of the flock in weeks (Age of Birds); the length of the aisle in m (Aisle Length); the height, depth and width of the cages in cm (Cage Height, Cage Depth, Cage Width); the average number of birds in each focal cage (Av. No. of Birds per Cage); the number of birds in the laying house (Flock Size); the average space allowance in cm^2 per bird in each focal cage (Space Allowance), the light level at the door of each focal cage in lux (Lux) and the strain of laying hen used (Strain). Lux was logarithmically transformed before being used in statistical analysis.

It should be noted that the conventional cages used for laying hens in Australia do not contain enrichments such as perches or nest boxes. Legislation exists in regard to the minimum space allowance (550 cm^2/hen) and certain aspects of cage design (e.g., floor slope $\leq 8°$, 10 cm trough space per hen).

2.3. Egg Sample

A sample of eggs was collected from each laying house at approximately 07:00–08:00 h, prior to commercial egg collection commencing. Four eggs were collected from one side of each aisle (one from each end and two from the middle), prior to commercial egg collection starting. Each egg was separated and the albumen frozen and stored at −20 °C until it could be analysed for corticosterone content by a third party using competitive protein binding radioimmunoassay [13]. The average value of albumen corticosterone content (ng/g) of all eggs in each sample was calculated to create a single value for each laying house (Corticosterone).

2.4. Behavioural Testing of Laying Hens

All behavioural tests were conducted in the morning following egg sample collection. The time taken to conduct the tests varied with the size of the flock but was approximately 2–3 h.

2.4.1. Approaching Human Test

The Approaching Human Test (AHT) was adapted from a test used by Hemsworth and Barnett (1993) [10], and assessed the avoidance response of caged laying hens to an unfamiliar human approaching their cage in a standard manner. Every tenth cage on the second tier on the right-hand side of each aisle was designated as a focal cage. The second tier was approximately 1 m from the ground, although the exact height varied between farms. The same person (L.E.E.) was used as the human stimulus in all behavioural tests. This researcher was always dressed in a standard manner (blue overalls and a white dustmask), and her hands remained in her pockets. This attire was visually distinct from the clothing worn by stockpeople, who did not wear dustmasks. An assistant was dressed in a similar manner to the researcher (overalls), but held a stopwatch and clipboard and did not wear a white dustmask. The assistant remained at a distance of at least 2 m from the focal cage but was still visible to the hens being tested.

Each focal cage was tested once, and the testing occurred sequentially along each aisle. Each test consisted of five stages, and each stage was 5 s in duration. The first stage was a familiarisation stage, and allowed the hens in the focal cage to make visual contact with the researcher prior to the test commencing, avoiding a startle response. During this familiarisation stage the researcher stood on the opposite side of the aisle (approximately 1 m) to the cage that was adjacent to the focal cage to be tested. The researcher remained in this position for a period of 5 s, after which she stepped sideways so that she was directly in front of the focal cage and the Approaching Human Test began. After 5 s had elapsed, the researcher stepped toward the cage so that her torso was in contact with the feed trough at the front of the cage (approximate 20 cm from the cage front). The researcher waited in this position, observing the birds, for 5 s before stepping back to the opposite side of the aisle. The researcher stood in the original starting position for 5 s, after which she again stepped forward to the front of the focal cage and waited in this position for a further 5 s. Thus, the entire test for each cage took 20 s, and consisted of four movements made at 5-s intervals, depicted in Figure 1.

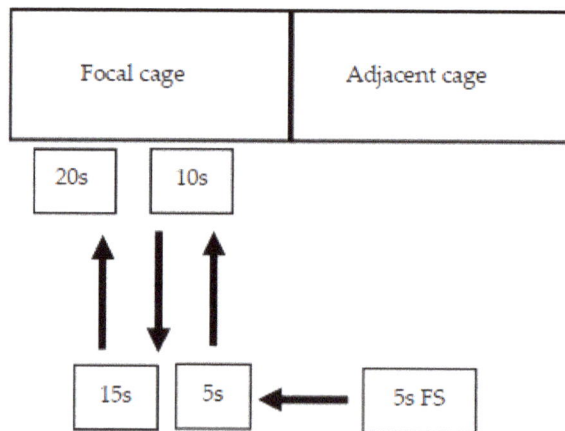

Figure 1. Movement of the researcher during the four stages of the Approaching Human Test. 5s FS = 5-second familiarization stage prior to the test commencing.

Whilst the researcher was standing stationary during each 5-s interval she made the following observations for each focal cage, and verbally relayed these to the assistant standing 2 m away:

i.   The maximum proportion of hens that simultaneously placed their beaks in the front 5 cm of the cage at any point during the 5-s period (variable labelled 'Maximum Heads');

ii.   The proportion of hens with their beaks in the front 5 cm of the cage at the end of the 5-s period (variable labelled 'Point Count');

iii.   The maximum proportion of hens that simultaneously placed their heads out of the cage at any point during the 5-s period (variable labelled 'Heads Out').

This resulted in 12 behavioural measurements for each focal cage (three variables measured four times each). A principal components analysis was used to reduce the number of variables. Two components of behaviour accounted for 82% of the variation in avoidance response. The first component was composed of the variables 'Maximum Heads' at 5, 10, 15 and 20 s, the 'Point Count' at 5, 10, 15 and 20 s, and 'Heads Out' at 5 s. These nine variables were combined into a single score that represented the first behavioural component and was labelled the 'Forward Score' (Cronbach's alpha = 0.94). The second component was composed of the variable 'Heads Out' at 10, 15 and 20 s. These variables were combined into a single score that represented the second behavioural component, labelled the 'Heads Out Score' (Cronbach's alpha = 0.86). A high value for either score indicates a greater proportion of birds at the cage front (Forward Score) or with their heads out (Heads Out Score) during the Approaching Human Test and thus indicating less avoidance of the researcher.

2.4.2. Stroll Test

The Stroll Test was adapted from a similar test used by Cransberg et al. (2000) [14], and assessed the avoidance response of caged laying hens to an unfamiliar human walking along the aisles. The procedure involved the researcher walking through the laying house at a standard speed of one step/second and filming the response of the birds in the second tier with a hand-held video camera. The test was administered in one movement through the laying house, filming the same side of each aisle that the Approaching Human Test had been administered. The angle of the camera was held such that the distance to the first bird with its head out of the cage was able to be recorded, with a field of view of at least four cages ahead of the researcher.

During subsequent video analysis the footage was paused at 5-s intervals and the withdrawal distance was measured. The withdrawal distance was classified as the distance (cm) between the experimenter and the first bird with its head out of the cage, calculated by counting the number of cages between the two and multiplying by the cage width in cm (Withdrawal Distance). This variable measured the distance at which the hens would let the researcher approach before withdrawing their head into the cage. A greater 'Withdrawal Distance' indicates greater avoidance of the researcher.

The second variable measured was the number of birds with their heads extended through the front of the four cages directly ahead of the researcher. Only four cages were counted due to low visibility beyond this distance. The total number of birds with their heads out of the four cages was summed, converted to a proportion of the total number of birds in those four cages, and divided by the total length of the four cages in m. This allowed the proportion of birds with their heads out per m to be calculated, creating the variable 'Prop Heads Out/m'. A greater value for the 'Prop Heads Out/m' indicates less avoidance of the researcher. This standardised unit of measurement was calculated to account for the between-farm differences in cage widths and stocking densities.

2.5. Stockperson Behaviour Observations

The behaviour of all stockpeople that entered the laying house was observed over a 2-day period. These observations are referred to as 'All SP' and represent the total human behaviours that the hens were exposed to. An ethogram of all stockperson behaviours recorded and how they were categorised is presented in Table 2. A subset of these observations are the behaviours performed by the focal stockperson only, referred to as 'Main SP'. The focal stockperson was the main person that was responsible for the maintenance and husbandry within each house, and spent more time in the laying house than other farm staff. The focal stockperson behaviours were used to assess the relationships between stockperson attitudes and behaviours.

Table 2. An ethogram of the specific stockperson behaviours that were recorded, and how they were grouped into each behavioural category.

General Behaviour Category	Definition of Behaviour	Specific Stockperson Behaviours
Visual	Any behaviour that was performed in front of the cages, but did not involve noise or approaching within 20 cm of the cage.	• Manipulate clothing • Manipulate object • Pause • Push object • Carry object • Kick object • Sweep • Crouch • Turn around • Bend down • Drop object
Noise	Any behaviour that generated noise.	• Noise (e.g., knock broom on cage) • Yell • Loud noise (e.g., leaf blower)
Approach	Any behaviour that involved a body part or an object held by the stockperson to approach within 20 cm of the cage without contacting the cage.	• Close visual inspection of cage • Throw object onto manure belt • Hand approaches cage • Torch use • Face over feeder • Object approaches cage • Object over cage • Object under cage
Contact	Any behaviour that involved a body part or object held by the stockperson to contact the cage, feed trough, egg belt or rail.	• Contact feeder • Lean on feeder • Manipulate egg • Object on egg belt • Object in feeder • Hand in feeder • Hand under feeder • Hand on rail • Stand on feeder • Stand on rail • Stand on cage • Hand on cage • Manipulate cage • Bang cage • Kick egg belt • Kick feeder • Contact cage • Contact feeder with object • Contact cage with object
Entry	Any behaviour that involved inserting a hand or an object into the cage.	• Insert object into cage • Insert one hand into cage • Insert both hands into cage
Handle	Any behaviour that involved touching or handling the hens.	• Touch bird • Handle bird
Near Cage		• A composite variable made by summing the behavioural categories 'Approach', 'Contact', 'Entry', 'Handle'

The researcher (L.E.E.) followed the stockperson at a short distance (10–15 m) and verbally recorded observations on a small handheld tape recorder in real time. The frequency of each stockperson behaviour was recorded when analysing the cassette tapes at a later date, and a criterion of 5-s was used to record the frequency of behaviours that were performed for >5-s. For example, a behaviour that occurred for 8-s would be classified as occurring twice. The stockpeople were informed that their activities in the house were being observed, and were instructed to continue their work as if the researcher was not present. Due to the large number of different stockperson behaviours observed, all behaviours were grouped on the basis of their type of interaction with the birds, resulting in seven categories of stockperson behaviour (Visual, Noise, Approach, Contact, Entry, Handle, Near Cage).

In addition, the researcher also recorded the amount of time (seconds) that each stockperson spent in the aisles (Time in Aisle); at either the start or end of the laying house (Time SOH); in total for both the aisles and the start of the laying house (Total Time); at the end of each aisle without walking the entire length of the aisle (such as when maintaining equipment, Time Ends Aisles); and the time spent standing stationary in the aisles (Stationary). Speed of movement was calculated by dividing the time spent walking along an aisle by the length of the aisle (Av SOM, Min SOM, Max SOM, m/seconds).

The total numbers of occurrences of behaviours in each category were summed for each laying house over the 2-day observation period, divided by two to calculate the average number of each behaviour per day, and then divided by the total number of cages in the laying house to provide a standard measure of stockperson behaviour per cage. This conversion allowed the behavioural observations to be converted into a standard unit (average behaviours/cage/day) for comparison between laying houses of different sizes. The duration of time that individual stockpeople spent in each area of the laying house (such as the aisles or ends of the laying house) was also converted into the units 'seconds in each area per cage'. Prior to inclusion in statistical analysis, several of these measurements were transformed to reduce skewness in the measurement. The transformations used for each variable are included in the results tables.

2.6. Stockperson Attitude Assessment

Stockperson attitudes were assessed using a written questionnaire. The questionnaire was administered to the focal stockperson working in each house at the end of the 2-day behaviour observation period. Complete questionnaires were obtained for 14 stockpeople.

The questionnaire consisted of three sections: the first section contained statements about working with laying hens and assessed stockperson beliefs about the behaviours that they performed when working in the laying houses (behavioural beliefs); the second section contained statements about the characteristics of laying hens and assessed general beliefs about laying hens; and the third section contained statements about interacting with laying hens and assessed behavioural beliefs about interacting with laying hens. Stockpeople were asked to indicate the extent to which they agreed or disagreed with these statements using a five-point Likert scale, ranging from 'Strongly agree' to 'Strongly disagree' with 'Neither agree or disagree' as the middle option. Each stockperson was given a score from 1–5 for each answer, with 'Strongly disagree' receiving a score of one and 'Strongly agree' receiving a score of five.

Initial questionnaire items were developed from focus group discussions with stockpeople working in the egg industry. The large number of initial questionnaire items could not be reduced using a principal components analysis due to the small sample size, and so the questionnaire items were manually sorted into subscales with similar content using item-total correlations to exclude non-correlated items in each subscale. This resulted in 10 subscales, and the scores for the items in each subscale were summed to create a single value for each subscale for each stockperson. All of the items in a subscale were significantly correlated with the subscale total, and all subscales had a Cronbach's alpha score above 0.7 indicating that each total was a reliable measure of each subscale. Each subscale is described below.

Statements relating to the stockperson's general beliefs about the characteristics of laying hens were divided into positive and negative statements about laying hens. The positive statements formed the first attitude subscale 'Pos Gen Atts' (Cronbach's alpha = 0.83), which consisted of nine statements such as 'Laying hens are entertaining to watch' and 'Laying hens are intelligent animals', and stockpeople who agreed with these statements were considered to have a positive general attitude toward laying hens. Conversely, negative statements about the characteristics of laying hens were grouped to form the second attitude subscale 'Neg Gen Atts' (Cronbach's alpha = 0.81) and consisted of 13 statements such as 'Laying hens are dirty animals' and 'Laying hens have an ugly appearance'. Stockpeople who agreed with these statements were considered to have a negative general attitude toward laying hens.

Statements relating to the stockperson's behavioural beliefs about interacting with laying hens were also grouped into subscales. The third subscale was labelled 'Insensitivity of SP' (Cronbach's alpha = 0.75), and consisted of 13 statements such as 'Yelling at the birds quietens them down' and 'Laying hens aren't affected by the way they are treated'. Stockpeople who agreed with these statements were considered to be unaware of or unconcerned with the effects that their behaviour had on the hens. Conversely, the fourth subscale labelled 'Sensitivity of SP' (Cronbach's alpha = 0.74) consisted of five statements such as 'I should act carefully around laying hens so as not to scare them' and 'I notice differences in the way laying hens respond to me'. Stockpeople who agreed with these statements were considered to be aware of the impact of their behaviour on the hens. The fifth attitude subscale was labelled 'Unpleasantness of Job' (Cronbach's alpha = 0.87) and consisted of 13 statements such as 'Laying hens are frustrating to work with' and 'I find the laying house too noisy'. Stockpeople who agreed with these statements were considered to find their work unpleasant, or dislike certain aspects of their job. The sixth attitude subscale was labelled 'SP Enjoys Job' (Cronbach's alpha = 0.79), and consisted of statements such as 'I don't mind the dust in the laying house' and 'I am happy with the amount of walking that I have to do in the laying house'. Stockpeople who agreed with these statements were considered to enjoy their work, or find some aspects of their work pleasant. The seventh attitude subscale was labelled 'No Job Control' (Cronbach's alpha = 0.73) and consisted of six statements such as 'I don't have much control over what I do in the laying house' and 'I have no input into how my job is done'. Stockpeople who agreed with these statements were considered to have little perceived control over the work that they do, and felt unvalued. The eighth attitude subscale was labelled 'Job Control' (Cronbach's alpha = 0.70) and consisted of two statements, 'Management listens to my suggestions' and 'I am an important member of a team'. Stockpeople who agreed with these statements were considered to feel valued, and able to contribute to how well his or her job was done. The ninth attitude subscale was labelled 'Diligence' (Cronbach's alpha = 0.76) and consisted of eight statements such as ''I try to make the hens as comfortable as possible' and 'I am very thorough in my work'. Stockpeople who agreed with these statements were considered to value their work, and place emphasis on doing their work thoroughly and correctly.

The final subscale related to the empathic capabilities of the stockperson rather than their attitudes. However, for the sake of brevity, the empathy subscale will be included in the term 'attitude subscales' throughout this paper. Labelled 'Empathy' (Cronbach's alpha = 0.81), this subscale consisted of four statements such as 'It is kinder to handle the birds gently' and 'I feel bad if the hens go without food or water'. Stockpeople who agreed with these statements were considered capable of vicariously feeling a similar subjective state to that of the hens, and were concerned with the comfort of the hens.

2.7. Productivity Records

The production records were obtained from the farm managers at the conclusion of the study. The values for peak hen day production (PHDP), age at peak hen day production (Age at PHDP) and rate of lay in the week of testing (hen day production, HDP) were determined for each flock. The number of production weeks within 5% of peak egg production were calculated as a measure of persistency of lay [15]. Due to the age-related variation in hen productivity, the productivity variables for each flock were standardised by comparing them to the appropriate production standard [16,17] for each age (week) for each strain of hen, rather than between flocks of different ages. The standard value for each productivity variable was subtracted from the actual value for each variable. Thus, a negative value for these productivity measures meant that the flock had a lower productivity than expected at that age for that strain, and a positive value meant that the flock was producing better than expected at that age. The cumulative mortality rate on the week of testing was also determined for each flock and the production standard value subtracted to create the standardised variable 'Mortality'.

A list of all the variables collected in this study along with their descriptive statistics are presented in Table 3.

Table 3. Descriptive statistics of laying houses, for all variables included in this study.

Variables Measured	N	Mean	SD	Minimum	Maximum
Hen avoidance behaviour					
Forward Score	19	0.52	0.66	−0.38	1.87
Heads Out Score	19	0.16	0.63	−0.34	2.22
Withdrawal distance (cm)	19	101	36	46	170
Proportion heads out/m	19	0.25	0.13	0.06	0.46
Baseline stress physiology					
Albumen corticosterone concentration (ng/g)	19	1.56	0.27	1.10	2.18
Laying house features					
No. of stockpeople in house	19	5	2	1	10
Age of birds (weeks)	19	56	12	36	90
Flock size	19	51,193	41,163	1320	121,968
Length of aisle (m)	19	87	28	30	129
Cage width (cm)	19	54	12	31	70
Av. no. of birds/cage	19	5.3	1.7	2.9	8.9
Space allowance (cm^2/bird)	19	538	84	382	689
Cage height (cm)	16	48.2	5.6	39.0	55.0
Cage depth (cm)	18	51.4	6.9	38.0	59.0
Av. lux at cage front	19	107.2	317.6	0.03	1248.7
All stockperson (SP) behaviours					
All SP Visual/cage/day	19	0.36	0.41	0.01	1.65
All SP Noise/cage/day	19	0.15	0.17	0.00	0.51
All SP Approach/cage/day	19	0.11	0.06	0.00	0.22
All SP Contact/cage/day	19	0.22	0.34	0.02	1.21
All SP Entry/cage/day	19	0.01	0.01	0.00	0.04
All SP Handle/cage/day	19	0.005	0.01	0.00	0.04
All SP Near Cage/cage/day	19	0.34	0.34	0.07	1.25
All SP Time SOH/cage/day (second)	19	1.03	0.49	0.20	1.93
All SP Time in aisles/cage/day (second)	19	3.78	1.81	1.45	8.89
All SP Total time/cage/day (second)	19	3.82	1.99	0.85	8.82
All SP Time Ends Aisles/cage/day (second)	19	0.38	0.32	0.00	1.02
All SP Stationary/cage/day (second)	19	0.24	0.22	0.00	0.84
All SP Av SOM (m/second)	19	0.95	0.34	0.40	1.81
All SP Max SOM (m/second)	19	2.76	1.15	1.01	4.93
All SP Min SOM (m/second)	19	0.22	0.24	0.04	1.01
Main stockperson (SP) behaviours					
Main SP Visual/cage/day	14	0.17	0.15	0.01	0.46
Main SP Noise/cage/day	14	0.08	0.10	0.00	0.25
Main SP Approach/cage/day	14	0.06	0.04	0.01	0.16
Main SP Contact/cage/day	14	0.08	0.09	0.01	0.35
Main SP Entry/cage/day	14	0.004	0.004	0.00	0.02
Main SP Handle/cage/day	14	0.001	0.002	0.00	0.01
Main SP Near cage/cage/day	14	0.16	0.13	0.04	0.49
Main SP Time SOH/cage/day (second)	14	0.78	0.37	0.06	1.47
Main SP Time Ends Aisles/cage/day (second)	14	0.12	0.12	0.00	0.38
Main SP Stationary/cage/day (second)	14	0.11	0.09	0.00	0.33
Main SP Time In Aisles/cage/day (second)	14	1.27	0.70	0.21	2.49
Main SP Total time/cage/day (second)	14	1.81	0.87	0.51	3.46
Main SP Av SOM (m/second)	14	0.99	0.18	0.71	1.23
Main SP Min SOM (m/second)	14	0.21	0.27	0.04	1.01
Main SP Max SOM (m/second)	14	2.56	0.88	1.01	4.43

Table 3. Cont.

Variables Measured	N	Mean	SD	Minimum	Maximum
Main stockperson attitudes					
Empathy	14	14.2	3.3	6.0	17.0
Sensitivity of stockperson	14	35.7	5.7	25.0	43.0
Insensitivity of stockperson	14	41.7	6.3	32.0	51.0
Job control	14	7.6	1.8	2.0	10.0
No job control	14	16.5	4.1	9.0	25.0
Stockperson enjoys job	14	26.7	6.5	12.0	35.0
Diligence	14	33.6	6.1	20.0	41.0
Unpleasantness of job	14	31.7	9.8	17.0	51.0
Neg general attitudes	14	38.8	9.2	25.0	55.0
Pos general attitudes	14	35.3	7.0	20.0	44.0
Flock productivity					
Peak Hen Day Production (standardized %)	16	0.73	1.77	−2.60	3.88
Hen Day Production (standardized %)	15	3.37	1.78	0.80	7.12
Cumulative mortality (standardized %)	15	−1.47	2.09	−4.35	3.90
Age at Peak Hen Day Production (weeks)	16	9	7	1	21
Weeks at 5% of peak egg production (weeks)	16	16	7	1	27

All stockperson behavior units are frequencies/cage/day in the laying house except where followed by (second), in which case they are durations/cage/day. SP = stockperson, SOH = start of laying house, SOM = speed of movement, SD = standard deviation. Hen Day Production refers to the average % of hens laying an egg on each day over a 1-week period. All production data have been subtracted from the expected production values for flocks of the same strain and age to account for age differences in production between flocks. Thus, a positive value indicates a higher value than expected when compared to the breed standard at that age, while a minus value indicates a lower value than expected when compared to the breed standard at that age.

2.8. Statistical Analysis

The general approach was to select an outcome variate related to productivity (5% peak egg production duration) that had a likelihood of being directly or indirectly affected by stockperson behaviour and hen fear and stress. This variable was chosen based on the results of preliminary analyses (unpresented) that revealed few likely relationships between other production variables and hen behaviour variables. For example, an analysis relating peak hen day production (PHDP) to albumen corticosterone, laying house parameters, hen avoidance behaviours, stockperson behaviours and stockperson attitudes was attempted. The parsimonious model for the logarithm of PHDP only included laying house terms. In particular, PHDP increased as the size (in terms of flock size) of the laying house increased and the light level at the cage front (Lux) increased. These responses do not appear to be directly related to a stockperson-animal relationship, and thus the detail of this analysis is not reported in this paper.

On a between laying house basis, a parsimonious statistical model was developed relating 5% peak egg production duration to albumen corticosterone, laying house parameters (e.g., bird strain, number of birds, cage design etc.), hen avoidance behaviours, stockperson behaviours and stockperson attitudes. There were only three predictors in the parsimonious model (see results for detail), namely:

- A predictor associated with cage design, namely cage width;
- A predictor often associated with animal fear, namely average withdrawal distance during the Stroll Test (AvWD);
- A predictor related to stockperson behaviour, namely the square root of the frequency of noise behaviours of all stockpeople (AllSPNoise).

To understand the HAR further, the average withdrawal distance of each laying house was related to the physical features of the laying house, all stockperson behaviour measurements, main stockperson behaviour measurements and main stockperson attitude measurements. As the results of the AHT were not predictive of the number of weeks that the hens spent within 5% of peak egg production, these results were not analysed further. If a measurement is a predictor of the average withdrawal distance, without being a predictor of the outcome variate in the parsimonious model (i.e., no effect on the outcome variate after adjusting for the effect of average withdrawal distance), then the measurement is a candidate for mediating or affecting productivity via the mechanism of affecting the average withdrawal distance.

To understand the factors that influence the amount of noise made by the main stockperson, the square root of the frequency of noise behaviours of the main stockperson (square root of MainSPNoise) for each laying house was related to physical features of the laying house, all stockperson behaviour measurements other than noise behaviour, main stockperson behaviour measurements other than noise behaviour and main stockperson attitude measurements. The main stockperson noise behaviour, rather than all stockperson noise behaviour, was used as the dependent variate because attitude measurements were only available for the main stockperson. Stockperson behaviour measurements, other than noise behaviour, were included as predictor variates for MainSPNoise because, while noise behaviour is a stockperson behaviour, it is partly determined by management decisions like the cleaning routines in the laying house.

Finally, since fear-related stress is considered to be an important component of the HAR, the logarithm of albumen corticosterone was related to physical features of the laying house, hen avoidance behaviours, stockperson behaviours and stockperson attitudes. Cage width was not considered further because its effect on the outcome variates was considered likely to be caused by farm management decisions, and thus not influenced by the stockperson.

Using residual maximum likelihood (REML) mixed model analysis, a parsimonious model was developed to relate 5% peak egg production duration, of each laying house, to fixed effects for the logarithm of albumen corticosterone, features of the laying house, avoidance behaviour test results, all stockperson behaviour measurements, main stockperson behaviour measurements and main stockperson attitude measurements and random effects associated with farm identity. Prior to examination of some fixed effects that had skewed distributions, those effects were transformed to reduce the skewness of the distribution. Fixed effects were included or excluded from the models using Wald F tests, and random effects were included or excluded using χ^2 change in deviance tests. Random effects for farm are likely to be necessary because animal productivity is likely to be affected by unmeasured factors outside the HAR paradigm (such as nutrition and genetics), and these factors are likely to be determined on a farm basis. This is likely to induce correlation between different laying houses on the same farm, which in turn induces correlation in the errors of the model. Confidence intervals for predictors in the parsimonious model were calculated on the logarithmically transformed scale, using the asymptotic normal distribution, before back-transforming to the original albumen concentration scale.

For each of the logarithm of albumen corticosterone, average withdrawal distance and the square root of MainSPNoise, a parsimonious general linear model was developed to relate the measurement to the sets of predictor variates in the paragraphs above. Terms were included and excluded from the models using standard F tests for linear models. To examine whether systematic variation between farm needed to be included in the models, for each of the logarithm of albumen corticosterone, average withdrawal distance and the square root of MainSPNoise, an attempt was made to fit a residual maximum likelihood (REML) mixed model that included the parsimonious model terms plus an extra random term for farm identity. However, in each case the best estimate of the variance for the farm identity effect was negative. Since it is implausible that the true variance is negative, although a negative estimate often occurs when the variance is zero, a random effect for farm was not included in these models. Confidence intervals for predictors in the model for MainSPNoise were calculated on the square root scale, using the t distribution, before squaring to obtain a confidence interval on the original MainSPNoise scale.

The unit of all analysis was a single laying house. Statistical analyses were carried out using the general linear model component, and the REML analysis of mixed models component, in GenStat 17 [18].

3. Results

The sheds were representative of the commercial caged egg industry in south east Australia (Table 1). There was almost a 100-fold difference in shed size, with sheds varying from 1300 birds to 116,000 birds

3.1. The Number of Weeks within 5% of Peak Egg Production

The most parsimonious model for 5% peak egg production duration included a random term for farm (χ^2 change in deviance = 12.69 on 1 degrees of freedom; p = 0.00089) and additive fixed effect terms for cage width, a linear response to average withdrawal distance in the Stroll Test and the noise behaviour of all stockpeople (Table 4). The cage width effect could be described as having two groupings, namely narrower cages (≤50 cm) and wider cages (>60 cm), with no further effect of cage width (p = 0.32 for additional linear response to cage width, Table 4). Chi-square change in deviance tests were also carried out, using random coefficient regression models and variance component models, to examine whether any of the fixed effects varied randomly with farm but, in all cases, either the models were not statistically significant (p > 0.1) or did not numerically converge.

Table 4. Tests for including and excluding fixed effect terms in the model relating 5% peak egg production duration to physical features of the laying house, hen avoidance behaviours and all stockperson behaviours. Tests for the inclusion of terms for main stockperson behaviours and attitudes of main stockperson had p > 0.05, and thus are not presented. All Wald F tests have 1 numerator degrees of freedom. A transformation in parentheses after a term indicates that the variable was examined for inclusion/exclusion in the model after the variable had been transformed, so as to reduce skewness of the variable.

Terms Included	F Value	Denominator Degrees of Freedom	p-Value
Cage Width Grouping (≤50 cm or >60 cm)	118.66	7.0	0.000012
Average withdrawal distance in Stroll Test (AvWD)	10.28	7.7	0.010
Noise (square root) behaviour for all stockpersons (AllSPNoise)	12.66	7.5	0.0082

Terms Excluded	F Value	Denominator Degrees of Freedom	p-Value
Extension of terms in model			
Cage width (included as a variate)	1.08	9.7	0.32
Square of AvWD	0.11	9.9	0.75
Square of Noise	0.12	6.7	0.74
AvWD response differs with cage width grouping	0.43	11.0	0.53
Noise response differs with cage width grouping	0.66	6.9	0.44
Product of AvWD and Noise	0.35	11.0	0.57
Stress physiology			
Albumen corticosterone (logarithm)	0.30	7.2	0.60
Shed features			
Strain of bird	0.00	7.1	0.97
No. of stock people	0.56	6.3	0.48
Age of birds	0.04	7.2	0.84
Flock size	0.12	10.1	0.74
Aisle length	0.26	5.3	0.63
Av. No. of Birds per cage	0.28	7.3	0.62
Space allowance	0.43	7.1	0.53
Cage height	0.85	6.6	0.39
Cage depth	0.01	5.8	0.92
Lux (log(y+1) transformed)	0.03	5.8	0.86
Hen avoidance behaviours			
Proportion of heads out/m in Stroll Test	0.06	6.7	0.82
Forward score in AHT	6.22	5.5	0.051
Heads out score in AHT	0.04	6.6	0.85

Table 4. Cont.

Terms Included	F Value	Denominator Degrees of Freedom	*p*-Value
All stockperson behaviours			
Approach (square root)	0.06	5.9	0.82
Av SOM	1.73	10.6	0.22
Contact (square root)	0.46	10.9	0.51
Entry (square root)	0.09	5.6	0.78
Handle (square root)	0.02	5.8	0.88
Max SOM	2.43	6.8	0.16
Min SOM	3.14	5.4	0.10
Near Cage (square root)	0.49	10.4	0.50
Stationary (square root)	0.12	5.2	0.74
Time Ends Aisles	0.01	6.3	0.92
Time in Aisle (logarithm)	4.30	8.9	0.068
Time SOH	0.03	6.0	0.86
Total Time (square root)	2.54	9.2	0.15
Visual (square root)	0.82	8.6	0.39

AvWD = Average withdrawal distance, SOM = speed of movement, SOH = start of laying house, AHT = Approaching Human Test. *p*-values less than 0.05 are indicated in bold.

The largest effect was the effect of cage width grouping, with the wider cages (>60 cm) having at least a 10 week greater 5% peak egg production duration than narrow cages (≤50 cm), for laying houses with the same average withdrawal distance in the Stroll Test and noise behaviour made by all stockpeople (Figure 2). Despite the wide confidence limits in the predicted means in Figure 2a,b, the marginal slope for 5% peak egg production duration (weeks) on the square root of noise made by all stockpeople (behaviours/cage/day) and the marginal slope of 5% peak egg production duration slope per day on average withdrawal distance (cm) are greater than three (slope = −11.7 (SE = 3.29) for square root of noise frequency; slope = −0.062 (SE = 0.0192) for average withdrawal distance), and thus there is reasonably good precision for the estimation of these slopes. For laying houses with the same width cages, the number of weeks spent within 5% of the peak in egg production decreased by about 5 weeks from the laying houses with the least noise made by all stockpeople (AllSPNoise) to the laying houses with the most noise made by all stockpeople. There was also a 5-week difference between the laying houses with the smallest average withdrawal distance to the laying houses with the greatest average withdrawal distance.

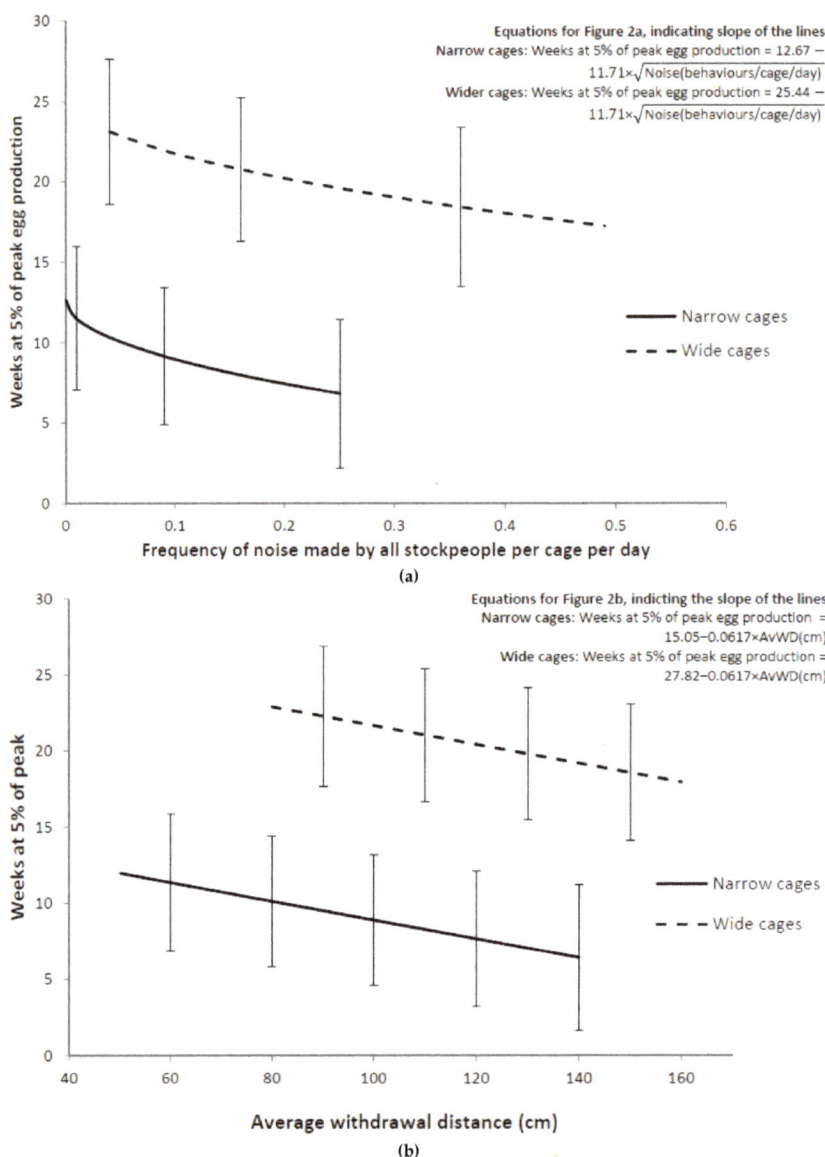

Figure 2. (a) Relationship between weeks at 5% of peak egg production and all stockperson noise, after adjusting for average withdrawal distance (AvWD), for the two cage width groupings. The noise responses, for each grouping, are presented at the between shed mean of average withdrawal distance, for sheds in that grouping. (b) Relationship between weeks at 5% of peak egg production and AvWD, after adjusting for the square root of all stockperson noise, for the two cage width groupings. The AvWD responses, for each grouping, are presented at the between shed mean for all stockperson noise, for sheds in that grouping. Error bars are 95% confidence intervals using the asymptotic normal approximation.

3.2. Albumen Corticosterone Concentration

The most parsimonious model for the logarithm of albumen corticosterone concentration only included a negative linear response to the average withdrawal distance in the Stroll Test (i.e., a single linear regression on average withdrawal distance, Table 5). The average withdrawal distance of hens was strongly related, in a negative direction, to albumen corticosterone concentration (Figure 3).

Table 5. Tests for including and excluding fixed effect terms in the model relating the logarithm of albumen corticosterone concentration to physical features of the laying house, hen avoidance behaviours and all stockperson behaviours. Tests for the inclusion of terms for the main stockperson behaviours and attitudes of main stockperson had $p > 0.05$, and thus are not presented. Except where indicated, all F tests have 1 numerator degrees of freedom. A transformation in parentheses after a term indicates that the variable was examined for inclusion/exclusion in the model after the variable had been transformed, so as to reduce skewness of the variable.

Terms Included	F Value	Denominator Degrees of Freedom	*p*-Value
Average withdrawal distance in Stroll Test (AvWD)	13.81	17	**0.0017**

Terms Excluded	F Value	Denominator Degrees of Freedom	*p*-Value
Extension of terms in model			
Square of AvWD	0.07	16	0.80
Shed features			
Strain of bird (2 denominator degrees of freedom)	0.11	15	0.90
No. of stock people	0.17	16	0.69
Age of birds	1.48	16	0.24
Flock size	0.19	16	0.67
Aisle length	0.65	16	0.43
Cage width	1.11	16	0.31
Av. No. of Birds per cage	0.67	16	0.43
Space allowance	0.21	16	0.66
Cage height	0.03	13	0.87
Cage depth	0.36	15	0.56
Lux (log(y+1) transformed)	0.56	16	0.46
Hen avoidance behaviours			
Proportion of heads out/m in Stroll Test	0.76	16	0.40
Forward score in AHT	0.11	16	0.75
Heads out score in AHT	0.97	16	0.34
All stockperson behaviours			
Approach (square root)	0.25	16	0.62
Av SOM	0.06	16	0.81
Contact (square root)	0.66	16	0.43
Entry (square root)	0.00	16	0.98
Handle (square root)	1.40	16	0.26
Max SOM	0.00	16	0.97
Min SOM	0.01	16	0.93
Near Cage (square root)	0.30	16	0.59
Stationary (square root)	0.22	16	0.64
Time Ends Aisles	0.48	16	0.50
Time in Aisle (logarithm)	0.02	16	0.89
Time SOH	0.10	16	0.75
Total Time (square root)	0.03	16	0.86
Noise (square root)	3.32	16	0.09
Visual (square root)	0.05	16	0.83

AvWD = Average withdrawal distance, SOM = speed of movement, SOH = start of laying house, AHT = Approaching Human Test. *p*-values less than 0.05 are indicated in bold.

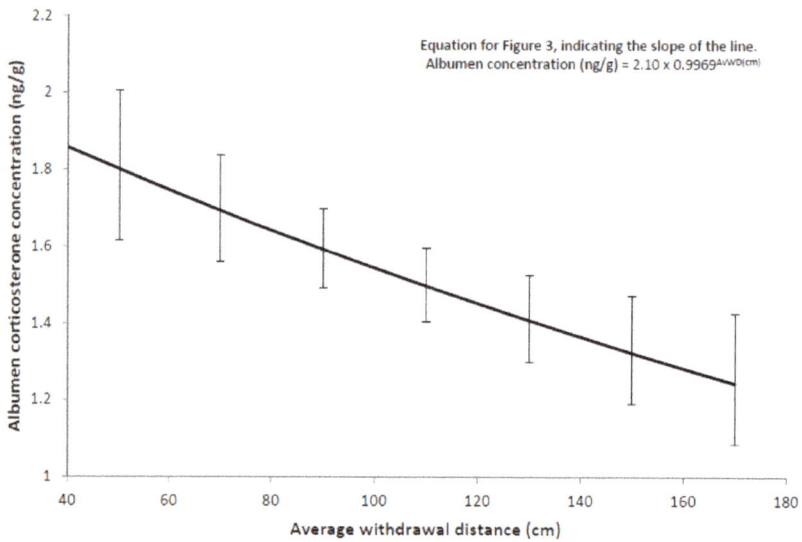

Equation for Figure 3, indicating the slope of the line.
Albumen concentration (ng/g) = 2.10 x 0.9969^{AWD(cm)}

Figure 3. Relationship between albumen corticosterone concentration and the average withdrawal distance in the Stroll Test. Error bars represent 95% confidence intervals.

3.3. Average Withdrawal Distance during the Stroll Test

The average withdrawal distance during the Stroll Test was the only hen behaviour variable that was associated with the 5% peak egg production duration and the albumen corticosterone concentration. For this reason, the results of the AHT are not presented. The most parsimonious model for the average withdrawal distance only included a positive linear response to cage height (i.e., a single linear regression on cage height, Table 6, Figure 4). Two stockperson behaviours (proportion of time at the end of the aisles, both for all stockpeople and for the main stockperson) had *p* values between 0.05 and 0.03 when added to the parsimonious model, but two out of 32 stockperson behaviours being statistically significant at the 5% level were judged to be insufficient to be reliable evidence of an effect, and thus these behaviours were not included in the parsimonious model. After adjusting for cage height, there was also a slight negative relationship between the average withdrawal distance and the insensitivity score for the main stockperson ($p = 0.05$, Table 6).

Table 6. Tests for including and excluding terms in the model relating average withdrawal distance in the Stroll Test to strain of bird, all stockperson behaviours except noise, and attitudes. Tests for the inclusion of terms for main stockperson behaviours gave similar results to those for all stockperson behaviours parameters, but with less residual degrees of freedom, and thus are not presented. Tests for the inclusion of terms for other layer house parameters had $p > 0.05$, and thus are not presented. Except where indicated, all F tests have 1 numerator degrees of freedom. A transformation in parentheses after a term indicates that the variable was examined for inclusion/exclusion in the model after the variable had been transformed, so as to reduce skewness of the variable.

Terms Included	F Value	Denominator Degrees of Freedom	*p*-Value
Cage Height	12.69	14	0.0031
Terms Excluded	**F Value**	**Denominator Degrees of Freedom**	***p*-Value**
Extension of terms in model Square of Cage Height	0.03	13	0.88
Strain of bird Strain of bird (2 numerator degrees of freedom)	1.46	2, 12	0.27

Table 6. *Cont.*

Terms Included	F Value	Denominator Degrees of Freedom	*p*-Value
All stockperson behaviours			
Approach (square root)	0.01	13	0.91
Av SOM	2.37	13	0.15
Contact (square root)	3.92	13	0.069
Entry (square root)	0.37	13	0.55
Handle (square root)	1.51	13	0.24
Max SOM	0.58	13	0.46
Min SOM	0.13	13	0.73
Near Cage (square root)	3.09	13	0.10
Stationary (square root)	1.98	13	0.18
Time Ends Aisles	5.22	13	**0.040**
Time in Aisle (logarithm)	0.31	13	0.59
Time SOH	0.03	13	0.88
Total Time (square root)	0.35	13	0.56
Noise (square root)	0.20	13	0.66
Visual (square root)	0.69	13	0.42
Main stockperson attitudes			
Empathy	0.18	9	0.68
Sensitivity of stockperson	0.00	9	0.95
Insensitivity of stockperson	4.98	9	0.053
Job control	1.14	9	0.31
Stockperson enjoys job	1.13	9	0.32
Diligence	0.00	9	0.96
No job control	1.14	9	0.31
Unpleasantness of job	0.47	9	0.51
Negative general attitudes	1.38	9	0.27
Positive general attitudes	1.13	9	0.32

SOM = speed of movement, SOH = start of laying house. *p*-values less than 0.05 are indicated in bold.

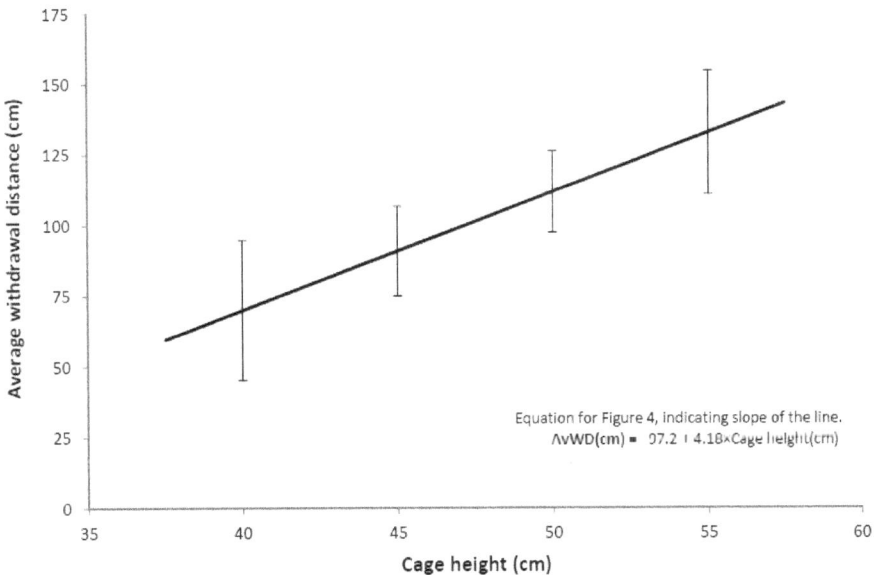

Figure 4. Relationship between average withdrawal distance in the Stroll Test and cage height. Error bars represent 95% confidence intervals.

3.4. Noise Made by the Main Stockperson

The most parsimonious model for the frequency of noise (square root transformed) made by the main stockperson included additive linear responses to the square root of all stockperson hand entries per cage, main stockperson positive general attitude score and main stockperson insensitivity score (i.e., a multiple linear regression for square root of all stockperson entries per cage, main stockperson positive general attitude score and main stockperson insensitivity attitude score, Table 7).

Table 7. Tests for including and excluding terms in the model relating main stockperson noise (square root transformed) to strain of bird, all stockperson behaviours except noise and attitudes. Tests for the inclusion of terms for other physical features of the laying house, main stockperson behaviours had $p > 0.1$, and thus are not presented. Except where indicated, all F tests have 1 numerator degrees of freedom.

Terms Included	Wald F Value	Denominator Degrees of Freedom	*p*-Value
All stockperson entry behaviour (square root) (AllSPEntry)	25.34	10	**0.00051**
Main stockperson positive attitude score (MainPositive)	14.96	10	**0.0031**
Main stockperson insensitivity attitude score (MainInsensitivity)	10.68	10	**0.0085**

Terms Included	Wald F Value	Denominator Degrees of Freedom	*p*-Value
Extension of terms in model			
Square of AllSPEntry	0.11	9	0.75
Square of MainSPPositive	0.43	9	0.53
Square of MainSPInsensitivity	3.06	9	0.11
Product of AllEntry and MainSPPositive	0.32	9	0.59
Product of AllEntry and MainInsensitivity	0.08	9	0.78
Product of MainPositive and MainInsensitivity	0.00	9	0.95
Strain of bird			
Strain of bird (2 numerator degrees of freedom)	5.57	8	**0.031**
Strain of bird (after deleting a single laying house with Ingham Hisex)	2.13	8	0.18
All stockperson behaviours			
Approach (square root)	0.87	8	0.38
Av SOM	0.87	8	0.38
Contact (square root)	0.00	8	0.95
Handle (square root)	0.02	8	0.89
Max SOM	0.24	8	0.64
Min SOM	0.59	8	0.46
Near Cage (square root)	0.04	8	0.85
Stationary (square root)	0.72	8	0.42
Time Ends Aisles	0.04	8	0.84
Time in Aisle (logarithm)	0.00	8	0.95
Time SOH	0.22	8	0.65
Total Time (square root)	0.02	8	0.90
Visual (square root)	0.07	8	0.80
Main stockperson attitudes			
Empathy	2.08	9	0.18
Sensitivity of stockperson	0.07	9	0.79
Job control	0.36	9	0.56
Stockperson enjoys job	1.27	9	0.29
Diligence	3.69	9	0.087
No job control	0.99	9	0.35
Unpleasantness of job	0.26	9	0.62
Negative general attitudes	0.01	9	0.94

SOM = speed of movement, SOH = start of laying house. *p*-values less than 0.05 are indicated in bold.

The amount of noise made by the main stockperson was greater when the number of cage entries made by all stockpeople in the shed was greater, and this was the strongest relationship with noise (Figure 5a). The main stockpeople also made more noise when they had a low score for positive general attitudes and a high score for the insensitivity attitude (Figure 5b,c).

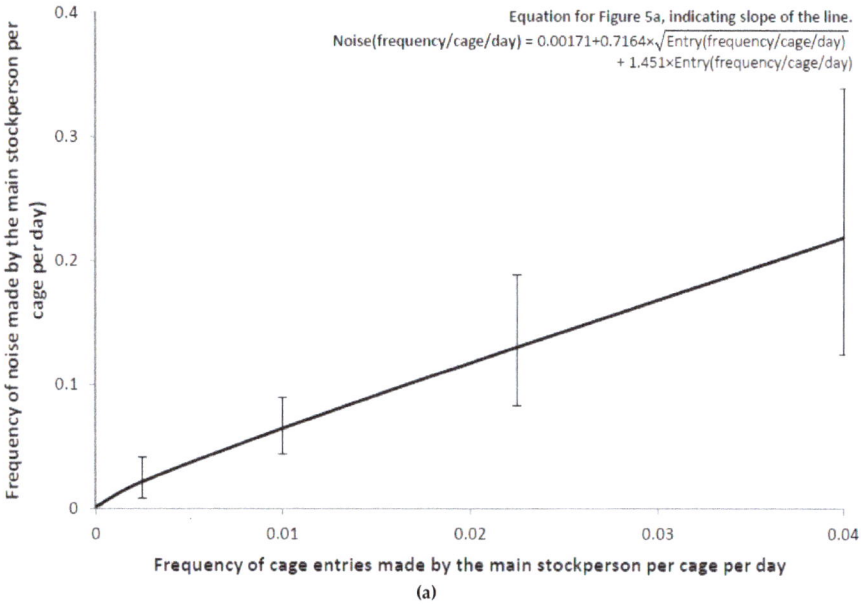

Equation for Figure 5a, indicating slope of the line.
Noise(frequency/cage/day) = 0.00171+0.7164×√Entry(frequency/cage/day) + 1.451×Entry(frequency/cage/day)

(a)

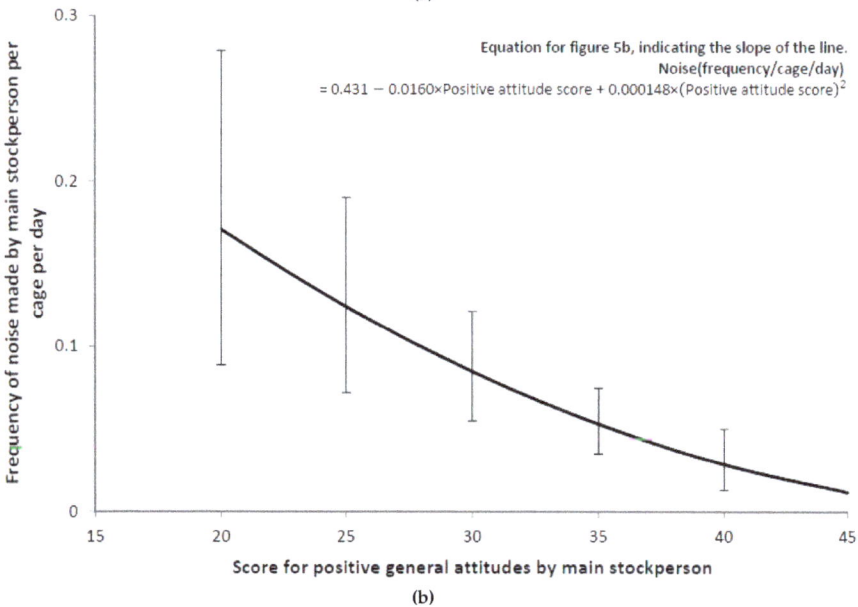

Equation for figure 5b, indicating the slope of the line.
Noise(frequency/cage/day) = 0.431 − 0.0160×Positive attitude score + 0.000148×(Positive attitude score)^2

(b)

Figure 5. *Cont.*

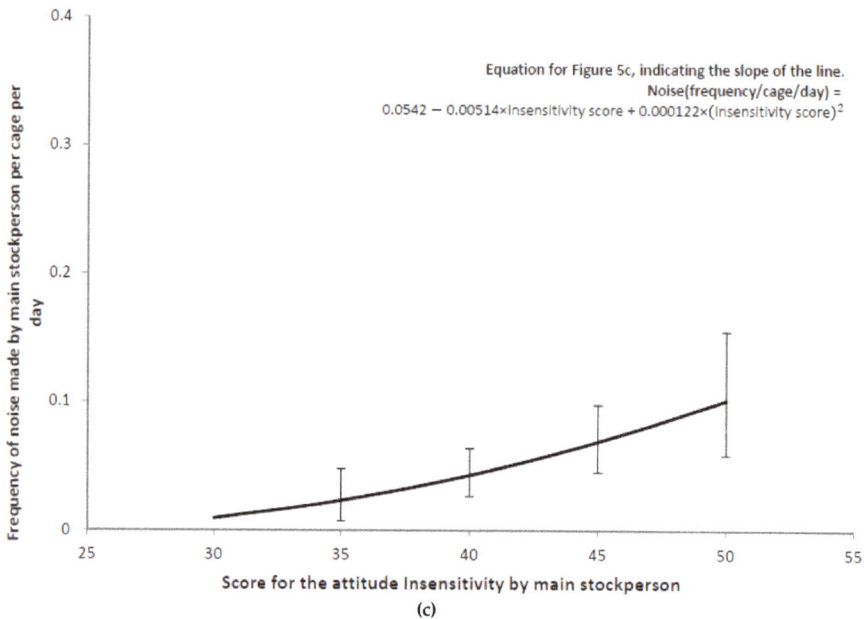

Equation for Figure 5c, indicating the slope of the line.
Noise(frequency/cage/day) =
$0.0542 - 0.00514 \times$ Insensitivity score $+ 0.000122 \times ($ Insensitivity score$)^2$

(c)

Figure 5. (a) Relationship between main stockperson noise and all stockperson entries per cage, after adjusting for stockperson positive attitude score and stockperson insensitivity attitude score. (b) Relationship between main stockperson noise and main stockperson positive attitude score, after adjusting for square root of all stockperson entries per cage and stockperson insensitivity attitude score. (c) Relationship between main stockperson noise and main stockperson insensitivity attitude score, after adjusting for square root of all stockperson entries per cage and stockperson positive attitude score. Error bars represent 95% confidence intervals.

4. Discussion

The human-animal relationship model predicts that there is a relationship between the attitudes of stockpeople and their behaviour toward animals, which in turn influences the welfare of farm animals by mediating their fear of humans and the associated stress response, with subsequent reductions in productivity. The results of the current study will be discussed in the reverse order of this model, starting with the productivity results, so that only the relevant variables linking each section of the model are presented. It needs to be recognised that while there were significant relationships found between several variables in this study, these relationships were observational and thus cannot be interpreted as causal until they are tested using controlled experiments.

The productivity of the hens (the number of weeks producing within 5% of the peak in egg production) was most closely related to whether cage width was less than 50 cm or greater than 60 cm (there were no cages between these two widths). Cages that were wider than 60 cm showed an additional 10 weeks of sustained high egg production compared to cages that were 50 cm wide or less. While it is tempting to attribute this effect to the wider feed trough decreasing competition and increasing the availability of feed to the hens [19], this explanation does not take into account the group size within the cage, which will affect the amount of feeder space available to each hen. In addition, the duration of peak egg production did not vary within each of the cage width categories, suggesting the effect is not linearly dependent on cage width. Thus, the difference could partly be due to other aspects of the design of cages that happen to be associated with width but were not included in the analyses (cage depth, cage height and group size were included). This may include variables such as the amount of light or air movement at the cage front. The conclusion from this part of the results is that cage design may be very important to productivity, rather than cage width *per se* being the important factor.

Hens with similar cages were able to maintain a higher production curve (duration within 5% of peak egg production) for longer when they allowed the researcher to approach more closely before withdrawing, and when they were exposed to quieter stockpeople. The model estimates that, for hens from laying sheds with a similar cage width grouping, an additional 5 weeks of high production can be obtained for flocks that showed the lowest withdrawal distance compared with those with the greatest withdrawal distance. Additionally, for hens from laying sheds with similar cage design, an additional 5 weeks of high production can be obtained for flocks that were exposed to the least amount of stockperson noise compared to those exposed to the greatest amount of stockperson noise. This is a promising result, as it indicates that both fear of humans and stockperson behaviour may influence egg production, and is consistent with a fear-related stress response [1,3,20]. For example, Barnett et al. (1992) [12] found that commercial caged laying hens could maintain egg production within 5% of their peak for longer when they showed less avoidance in a human approach test. In addition, O'Connor et al. (2011) [21] found that laying hens exposed to chronic noise at 80 dB produced less eggs, with no appreciable increase in physiological stress (based on plasma corticosterone concentration and heterophil:lymphocyte ratios), similar to the current study. Much of the noise made in the laying houses was created by loud cleaning machinery, such as leaf blowers and air hoses that were used to blow the dust off horizontal surfaces. This machinery could be used for long periods, and substantially increased the amount of airborne dust within the laying house during cleaning.

Negative relationships between hen avoidance behaviour and productivity are usually assumed to be due to a fear-related stress response. However, the result that the parsimonious model for 5% peak egg production duration included a term for average withdrawal distance, but no term for albumen corticosterone concentration, indicates that this mediation is not occurring.

In addition, the baseline stress level of each flock, as indicated by average albumen corticosterone concentration, was lower for flocks that showed a greater withdrawal distance toward an approaching unfamiliar human. This is the opposite to what would be expected with a fear-related explanation of the relationship between hen avoidance behaviour and productivity. While these two results are contrary to the previously established positive relationship between fear of humans and stress physiology in laying hens [22–24], a recent study with pullets has found the same relationship. Sprafke et al. (2018) [25] reported that the number of pullets in a cage-free flock that could be physically touched by the researcher was greater when the pullets had higher basal corticosterone concentrations, as indicated by the concentration of corticosterone metabolites in their droppings. No explanation for this result was provided by the authors. Albumen corticosterone concentration has been successfully correlated with plasma corticosterone concentration [13,26], (although see Engel et al. (2011) [27] for an opposing view), and while plasma corticosterone was not measured in the current study, albumen corticosterone concentration was negatively correlated with space allowance (r = −0.46, p = 0.048) as expected [28]. This indicates that the albumen corticosterone measurements are accurate, and a definitive explanation for this apparently anomalous result is not available.

The relationship between withdrawal distance and cage height may be due to the greater mobility of these birds to move backward within the cage when approached, as conventional cages have sloped floors and the available space at the back of the cage may be limited by cage height. Alternatively, due to the additive effects of cage height, hens in taller cages are likely to be closer to the researcher's face during the behavioural testing. Hens will show longer tonic immobility [29] and greater avoidance of humans [30] when the researcher's eyes are visible during testing. It is possible that closer proximity to the researcher's face may have provoked a greater avoidance response. As the exact height of the cages from the ground or the roof height at the back of the cage was not measured in this study, these effects could not be directly examined.

A major prediction of the human-animal relationship model is that the avoidance response of the laying hens would be related to the behaviour and attitudes of the stockperson. This was not the case in the current study, in which the parsimonious model for average withdrawal distance included no terms for stockperson behaviour or attitudes. If attitudes and behaviour are related to fear of humans in hens then, at least after adjusting for the effect of cage height, a relationship between average withdrawal distance and stockperson behaviour would be expected. The result that no such relationships were observed indicates that stockperson behaviours did not substantially affect fear of humans in the hens. This is in contrast to recent research on the human-animal relationship by Waiblinger et al. (2018) [8], who found relationships between stockperson attitudes, stockperson behaviour and hen avoidance behaviour on free-range egg farms.

Despite this, after adjusting for terms that might not be associated with the human-animal relationship, a shorter duration spent within 5% of peak egg production was associated with more stockperson noise, and the amount of noise produced by an individual stockperson was associated with the attitudes of that stockperson and the number of cage entries made by all stockpeople in the laying house. Stockpeople made more noise when they had negative attitudes towards the hens, indicated by scoring a low value for Positive General Attitudes, or a high value for Insensitivity. The amount of noise made by stockpeople was also greater when all people working in the laying house put their hands in the cages more often (cage entries). There is no obvious explanation for this relationship, but it may be related to farm management decisions or the design of the laying house influencing cleaning and inspection procedures.

These attitude-behaviour relationships are in the expected direction [1] and may represent the degree of volitional control that the stockperson has over their noise-related behaviours. For example, a stockperson who is unaware or insensitive to the impact of their behaviour on the hens may not moderate their noise levels as much as a stockperson who is aware of their impact. Similarly, a stockperson with few positive general attitudes toward hens may be aware of the impact that noise has on the hens but may have little motivation to reduce that impact by moderating noise production. The observed relationships between stockperson attitudes and noise behaviour are consistent with previous research on the human-animal relationship [1].

5. Conclusions

In conclusion, the major prediction of the HAR that the behavioural and physiological responses of hens to an approaching human would be related to the attitudes and behaviours of stock people did not hold, on a between laying house basis, for commercial caged hens in Australia. There was evidence of a relationship between stockperson attitudes and noise behaviour, and a high frequency of noise was associated with a reduction in one measure of egg production. If noise was affecting hen productivity it appears not to be mediated through its effects on fear of humans in hens. In addition, flocks that showed the least avoidance of the researcher had the greatest concentration of corticosterone in their egg albumen. Clearly further research is required to more thoroughly understand this human-animal relationship and its implications for hen welfare and productivity, on Australian cage-egg farms. The imperative for this is the findings by Waiblinger et al. (2018) of relationships between stockperson attitudes and behaviour and hen avoidance behaviour and productivity on free-range egg farms.

Author Contributions: This research formed part of the doctoral candidature of L.E.E. and was supervised by P.H.H. and G.J.C. Conceptualization, P.H.H. and G.J.C.; methodology, L.E.E., P.H.H., G.J.C.; formal analysis, K.L.B.; investigation, L.E.E.; writing—original draft preparation, L.E.E.; writing—review and editing, L.E.E., P.H.H., G.J.C., K.L.B.; supervision, P.H.H., G.J.C.; project administration, L.E.E.; funding acquisition, P.H.H., G.J.C.

Funding: This research was funded by the Australian Poultry Cooperative Research Centre.

Acknowledgments: The authors wish to thank the staff and students at the Animal Welfare Science Centre who assisted with data collection, the farms that participated in this research, and the two reviewers who provided helpful feedback on this manuscript.

Conflicts of Interest: The authors declare no conflicts of interest.

References

1. Hemsworth, P.H.; Coleman, G.J. *Human-Livestock Interactions: The Stockperson and the Productivity and Welfare of Intensively Farmed Animals*; CABI: Wallingford, UK, 2011.
2. Hemsworth, P.H.; Sherwen, S.L.; Coleman, G.J. Human contact. In *Animal Welfare*, 3rd ed.; Appleby, M.C., Olsson, I.A.S., Galindo, F., Eds.; CABI: Wallingford, UK, 2018; pp. 294–314.
3. Hemsworth, P.H.; Barnett, J.L.; Coleman, G.J.; Hansen, C. A study of the relationships between the attitudinal and behavioural profiles of stockpersons and the level of fear of human and reproductive performance of commercial pigs. *Appl. Anim. Behav. Sci.* **1989**, *23*, 301–314. [CrossRef]
4. Breuer, K.; Hemsworth, P.H.; Barnett, J.L.; Matthews, L.R.; Coleman, G.J. Behavioural response to humans and the productivity of commercial dairy cows. *Appl. Anim. Behav. Sci.* **2000**, *66*, 273–288. [CrossRef]
5. Waiblinger, S.; Menke, C.; Coleman, G. The relationship between attitudes, personal characteristics and behaviour of stockpeople and subsequent behaviour and production of dairy cows. *Appl. Anim. Behav. Sci.* **2002**, *79*, 195–219. [CrossRef]
6. Des Roches, A.D.B.; Veissier, I.; Boivin, X.; Gilot-Fromont, E.; Mounier, L. A prospective exploration of farm, farmer, and animal characteristics in human-animal relationships: An epidemiological survey. *J. Dairy Sci.* **2016**, *99*, 5573–5585. [CrossRef]
7. Barnett, J.L.; Hemsworth, P.H.; Hennessy, D.P.; McCallum, T.H.; Newman, E.A. The effects of modifying the amount of human contact on behavioural, physiological and production responses of laying hens. *Appl. Anim. Behav. Sci.* **1994**, *41*, 87–100. [CrossRef]
8. Waiblinger, S.; Zaludik, K.; Raubek, J.; Gruber, B.; Niebuhr, K. Human-Hen Relationship on Grower and Laying Hen Farms in Austria and Associations with Hen Welfare. In Proceedings of the 52nd Congress of the International Society of Applied Ethology, Charlottetown, PE, Canada, 30 July–3 August 2018; p. 187.
9. Graml, C.; Niebuhr, K.; Waiblinger, S. Reaction of laying hens to humans in the home or a novel environment. *Appl. Anim. Behav. Sci.* **2008**, *113*, 98–109. [CrossRef]
10. Hemsworth, P.H.; Barnett, J.L.; Jones, R.B. Situational factors that influence the level of fear of humans by laying hens. *Appl. Anim. Behav. Sci.* **1993**, *36*, 197–210. [CrossRef]
11. Jones, R.B. Reduction of the domestic chick's fear of human beings by regular handling and related treatments. *Anim. Behav.* **1993**, *46*, 991–998. [CrossRef]
12. Barnett, J.L.; Hemsworth, P.H.; Newman, E.A. Fear of humans and its relationships with productivity in laying hens at commercial farms. *Br. Poult. Sci.* **1992**, *33*, 699–710. [CrossRef] [PubMed]
13. Downing, J.A.; Bryden, W.L. Determination of corticosterone concentrations in egg albumen: A non-invasive indicator of stress in laying hens. *Phys. Behav.* **2008**, *95*, 381–387. [CrossRef] [PubMed]
14. Cransberg, P.H.; Hemsworth, P.H.; Coleman, G.J. Human factors affecting the behaviour and productivity of commercial broiler chickens. *Br. Poult. Sci.* **2000**, *41*, 272–279. [CrossRef]
15. Grossman, M.; Gossman, T.N.; Koops, W.J. A model for persistency of egg production. *Poult. Sci.* **2000**, *79*, 1715–1724. [CrossRef]
16. Hyline, I. Hyline Performance Standards Tables. 2007. Available online: https://www.hyline.com/aspx/general/dynamicpage.aspx?id=255 (accessed on 30 March 2019).
17. Hendrix-Genetics. Isa Brown Product Performance Manual. 2007. Available online: https://www.isa-poultry.com/en/product/isa-brown/ (accessed on 30 March 2019).
18. Payne, R.; Murray, D.; Harding, S.; Baird, D.; Soutar, D. *Introduction to Genstat® Forwindows™*; VSN International: Hemel Hempstead, Hertfordshire, UK, 2014.
19. Garner, J.P.; Kiess, A.S.; Mench, J.A.; Newberry, R.C.; Hester, P.Y. The effect of cage and house design on egg production and egg weight of white leghorn hens: An epidemiological study. *Poult. Sci.* **2012**, *91*, 1522–1535. [CrossRef]
20. Barnett, J.L.; Hemsworth, P.H. Fear of humans by laying hens in different tiers of a battery: Behavioural and physiological responses. *Br. Poult. Sci.* **1989**, *30*, 497–504. [CrossRef]
21. O'Connor, E.A.; Parker, M.O.; Davey, E.L.; Grist, H.; Owen, R.C.; Szladovits, B.; Demmers, T.G.M.; Wathes, C.M.; Abeyesinghe, S.M. Effect of low light and high noise on behavioural activity, physiological indicators of stress and production in laying hens. *Br. Poult. Sci.* **2011**, *52*, 666–674. [CrossRef]
22. Beuving, G.; Vonder, G.M.A. Effect of stressing factors on corticosterone levels in the plasma of laying hens. *Gen. Comp. Endocrinol.* **1978**, *35*, 153–159. [CrossRef]

23. Fraisse, F.; Cockrem, J.F. Corticosterone and fear behaviour in white and brown caged laying hens. *Br. Poult. Sci.* **2006**, *47*, 110–119. [CrossRef]

24. Hazard, D.; Couty, M.; Richard, S.; Guémené, D. Intensity and duration of corticosterone response to stressful situations in japanese quail divergently selected for tonic immobility. *Gen. Comp. Endocrinol.* **2008**, *155*, 288–297. [CrossRef]

25. Sprafke, H.; Palme, R.; Schmidt, P.; Erhard, M.; Bergmann, S. Effect of two transport options on the welfare of two genetic lines of organic free range pullets in switzerland. *Animals* **2018**, *8*, 183. [CrossRef]

26. Rettenbacher, S.; Möstl, E.; Hackl, R.; Palme, R. Corticosterone in chicken eggs. *Ann. N. Y. Acad. Sci.* **2005**, *1046*, 193–203. [CrossRef]

27. Engel, J.; Widowski, T.M.; Tilbrook, A.J.; Hemsworth, P.H. Further Investigation of Non-Invasive Measures of Stress in Laying Hens. In Proceedings of the 22nd Annual Australian Poultry Science Symposium, Sydney, Australia, 14–16 February 2011; pp. 126–129.

28. Campbell, D.L.M.; Hinch, G.N.; Downing, J.A.; Lee, C. Outdoor stocking density in free-range laying hens: Effects on behaviour and welfare. *Animal* **2017**, *11*, 1036–1045. [CrossRef] [PubMed]

29. Gallup, G.G.; Nash, R.F.; Ellison, A.L. Tonic immobility as a reaction to predation: Artificial eyes as a fear stimulus for chickens. *Psychon. Sci.* **1971**, *23*, 79–80. [CrossRef]

30. Edwards, L.; Coleman, G.; Hemsworth, P. Close human presence reduces avoidance behaviour in commercial caged laying hens to an approaching human. *Anim. Prod. Sci.* **2013**, *53*, 1276–1282. [CrossRef]

animals

MDPI

Article

Human-Animal Interactions in Dairy Buffalo Farms

Fabio Napolitano [1], Francesco Serrapica [2], Ada Braghieri [1], Felicia Masucci [2], Emilio Sabia [3] and Giuseppe De Rosa [2,*]

[1] Scuola di Scienze Agrarie, Forestali, Alimentari ed Ambientali, Università degli Studi della Basilicata, Via dell'Ateneo Lucano 10, 85100 Potenza, Italy; fabio.napolitano@unibas.it (F.N.); ada.braghieri@unibas.it (A.B.)

[2] Dipartimento di Agraria, Università degli Studi di Napoli Federico II, Via Università 133, 80055 Portici, NA, Italy; francesco.serrapica83@gmail.com (F.S.); felicia.masucci@unina.it (F.M.)

[3] Free University of Bozen-Bolzano, Faculty of Science and Technology, Piazza Università 5, 39100 Bolzano, Italy; emilio.sabia@unibz.it

* Correspondence: giuseppe.derosa@unina.it; Tel.: +39-081-2539300

Received: 15 February 2019; Accepted: 12 May 2019; Published: 16 May 2019

Simple Summary: The quality of the human-animal relationship plays a central role in determining animal welfare. In this study, we assessed the relationship between stockperson behavior and buffalo behavior. In particular, during milking, we recorded the behavior of stockpeople in terms of quality and quantity of interactions, and we recorded the behavior of animals in terms of restlessness, whereas at the feeding place, we measured the avoidance distance. Avoidance distance of an animal can be defined as the distance to which the animal will allow an unknown person to approach before moving to the side or away. We found that a high percentage of negative stockperson interactions (shouting, talking impatiently, slapping, and handling forcefully) were associated with a high avoidance distance at the feeding place and restlessness during milking. Therefore, appropriate stockpeople training should be conducted to improve the human-animal relationship with positive effects on animal welfare, productivity, and stockpeople safety.

Abstract: This study aimed to assess the relationship between stockperson behavior and buffalo behavior. The research was carried out in 27 buffalo farms. The behavior of stockpeople and animals during milking and the avoidance distance at the feeding place were recorded. Recordings were repeated within one month to assess test-retest reliability. A high degree of test-retest reliability was observed for all the variables with Spearman rank correlation coefficients (r_s) ranging from 0.578 ($p = 0.002$, $df = 25$) for the number of kicks performed during milking to 0.937 ($p < 0.001$, $df = 25$) for the percentage of animals moving when approached by ≤ 0.5 m. The number of negative stockperson interactions correlated positively with the number of kicks during milking ($r_s = 0.421$, $p < 0.028$, $df = 25$) and the percentage of animals injected with oxytocin ($r_s = 0.424$, $p < 0.027$), whereas the percentage of negative stockperson interactions correlated positively with the percentage of buffaloes moving when approached at a distance >1 m ($r_s = 0.415$, $p < 0.031$, $df = 25$). In a subsample of 14 farms, milk yield was correlated positively with the number of positive interactions ($r_s = 0.588$, $p < 0.027$, $df = 12$) and correlated negatively with the number of steps performed by the animals during milking ($r_s = -0.820$, $p < 0.001$, $df = 12$). This study showed that the quality of stockpeople interactions may affect buffalo behavior and production.

Keywords: dairy buffalo; human-animal relationship; animal behavior; test-retest reliability; avoidance distance; milk production; animal welfare

1. Introduction

Over the last few decades, farm animal welfare has become of great interest for the consumers of many different regions, including Europe and North America but also Oceania, Latin America, and Asia [1,2]. As a matter of fact, the consumers' perceptions of food quality are not only determined by its overall nature and safety but also by the welfare status of the animal from which it was produced [3]. In other words, animal welfare is an important component of an overall "food quality concept". The human-animal relationship is an important factor when considering farm animal welfare [4,5]. In intensive systems, farm animals are under human control and interact with stockpeople in several situations, including handling and milking. In dairy buffaloes, the farming system has recently become more intensive than it is in dairy cattle, causing potentially higher impacts of human interactions on the animals [6].

The quantity and the quality of these interactions can have distinct outcomes on the emotional state and the cognitive bias of farm animals [4,7]. In addition, there is a large body of evidence suggesting that negative interactions may have a detrimental effect on fertility, growth rates, milk yield, and behavior of farm animals [8–11]. In buffalo farms, when lactating animals face unfavorable conditions (e.g., stress and lack of habituation to the milking procedures), milk let-down is facilitated by using injections of exogenous oxytocin. Therefore, negative human interactions may increase the use of this practice [12]. Conversely, measures intended to improve this relationship can affect animal reaction to humans [13,14], thus reducing the risk of stockpeople injuries during farm procedures (handling, therapeutic treatments, growth control, loading and unloading from the lorry, etc.). Moreover, beneficial effects of gentling on growth of calves and veal quality have been documented [11,15].

The avoidance distance of animals to humans has been widely used as a measure of the quality of the human-animal relationship [5,10,16–18]. Avoidance distance of an animal is the distance to which the animal will allow an observer to approach before it moves to the side or away [5,19]. The rationale behind this measure is that the lower the distance between animal and observer is, the lower the level of fear towards humans is [5,16]. In cattle and buffaloes, the avoidance distance has been measured both in the barn and at the feeding place [17,18,20,21], and, at least in cattle, these two measures have proven to be highly correlated [17,21]. Therefore, due to feasibility reasons, the avoidance distance at the feeding place was used in the Welfare Quality® assessment protocol for cattle [22] and buffaloes [23]. Previously, De Rosa et al. [23] reported a high inter-observer reliability of the avoidance distance at the feeding place, with a Spearman coefficient of 0.92, whereas no information was available on test-retest reliability (i.e., consistency of the measurement when repeated within a certain period of time) of the avoidance distance at the feeding place and animal and stockpeople behaviors in the milking parlor.

In addition, while previous investigations conducted on dairy cattle [5], pigs [24], and poultry [25] found significant correlations between stockpeople behavior, animal behavior, and productive parameters such as milk yield, milk quality, and growth rate, little is known about these relationships in dairy buffaloes [12]. In particular, no information is available on the relationship between the behavior of buffaloes in the milking parlor and their reaction to an approaching person at the feeding place (i.e., avoidance distance) or on the relationship between the quality of human-animal relationships and milk production.

Therefore, the present study aimed to assess the test-retest reliability of the stockperson and buffalo behavior in the milking parlor and the test-retest reliability of the response of buffaloes when approached at the feeding place. Then, within the frame of the human-animal relationship model as set by Hemsworth [26], the relationships between human behavior in the milking parlor, animal behavior in the milking parlor, buffalo reactivity to an unknown person at the feeding place, and milk production was studied. In addition, the effect of the milking parlor design on human and buffalo behaviors was studied.

2. Methods

2.1. Farms, Animals, and Procedure

The research was carried out in 27 Italian buffalo farms located in Campania (n = 18) and Apulia (n = 9) regions. The herd size ranged from 90 to 1,400 head (number of buffalo cows ranging from 46 to 930), with lactating buffalo ranging from 20 to 300. Fourteen farms were equipped with herring-bone parlors, whereas the remaining used tandem parlors. Observations were conducted from September 2014 to February 2015 by two trained assessors. They were trained in buffalo farms not involved in the experiment for identifications of stockperson and animal behaviors through direct observations. The human-animal relationship was assessed by performing two different methods—avoidance distance at the feeding place 5 min after morning feed distribution and behavioral observations of stockperson and animals during afternoon milking. The time interval between the two tests—albeit variable, as it was dependent on farm routine practices (feeding routine and milking routine, respectively)—ranged from 5 to 7 h. These tests were repeated within one month to assess test-retest reliability; before starting the second observation session, the observers verified that no major changes had occurred in farm management. However, the animals observed in the two visits were not exactly the same. In all farms, stockperson and buffalo behaviors in the milking parlor were recorded by the same observer, whereas the stockperson's behavior when moving the animals from the waiting area to the milking parlor was always observed by the second observer. The latter also measured the avoidance distance at the feeding place. A management questionnaire aimed to gather information about herd size, milking routine, farm management, housing, and milking parlor characteristics was administered by the observer to farm owners during the first visit. In all farms, testing order was always the same: avoidance distance at the feeding place (morning,), filling questionnaire (morning), and behavioral observations during afternoon milking. Fourteen farms were enrolled in the national milk recording scheme. Therefore, for these farms, the data on milk production (expressed as kg/head/year) and milk quality (in terms of percentage of fat and protein) were collected.

2.2. Avoidance Distance at the Feeding Place

Five minutes after feed distribution, the avoidance distance at the feeding place was measured. The number of animals tested in each farm ranged from 20 to 100. The test was conducted according to the procedure reported by Waiblinger et al. [21]. The observer waited for the individual buffalo to look at him before approaching the animal. Animals were approached by the test person in a standardized way, i.e., directly from the front, starting, whenever possible, from a distance of 2 m, walking slowly (around one step per second), looking at the animal muzzle without staring at the buffalo's eyes, and keeping an arm at an angle of about 45° in front of the body. The test was ended whenever the animal withdrew (i.e., taking steps away from the observer or turning the head more than 45°). If the buffalo cow accepted the touch on the muzzle/nose, the experimenter tried to stroke the cheek of the animal for at least 1 s but not longer than 3 s. Avoidance distance was estimated at the moment of buffalo cow withdrawal as the distance between the observer's hand and the animal's head with a resolution of 10 cm. The distance was measured by counting the steps of the observer and converting into meters by measuring the length of the observer's step. In a case of withdrawing at the moment of touching the nose or muzzle, an avoidance distance of 10 cm was recorded, whereas a distance of 0 cm was assigned when the animal allowed itself to be touched and stroked. Animals were consecutively tested, but, in order to reduce the risk of influencing the neighbor's behavior, every second animal was tested. For each farm and each visit, the mean avoidance distance at the feeding place was calculated. In addition, the following variables were calculated, as reported in the Welfare Quality protocol for buffaloes [23]:

animals moving at a distance >1 m, %;
animals moving at a distance ≤1 m and ≥0.5 m, %;
animals moving at a distance ≤0.5 m, %;

animals that can be touched, %.

2.3. Stockperson and Buffalo Behavior during Milking

Stockperson's behavior was observed from moving the animals to the waiting area to the exit from the milking parlor. A total of 55 stockpeople were observed. Before starting the second observation session, the observers verified that no change in the personnel had occurred. We defined the interactions promoting social partnership as positive, the interactions showing a dominant role of the stockperson towards the animals as neutral, and those involving harsh physical or verbal interplay as negative. Therefore, the variables concerning the stockpeople behavior were classified and recorded as follows: number of positive (talk quiet, pet, touch gentle), neutral (talk dominant, hand gentle, stick gentle) and negative (shout, talk impatient, stick, slap, and handle forceful) interactions, as indicated by Waiblinger et al. [5]. The percentages of these three variables in relation to the total of interactions were also calculated. For each farm and each visit, the average number of interactions per milked buffalo was calculated. If two persons milked together, the sum of both was used to calculate this average. The occurrence of oxytocin injection at milking (number of injected animals/number of observed animals) was also recorded.

The buffalo behaviors, recorded from the entrance in the milking parlor to the removal of the milking cluster, were step (foot lifted less than 15 cm off of the ground) and kick (raised above 15 cm off of the ground, even if a clear kick was not visible). They were registered whenever the stockperson was within 0.5 m of the animals.

2.4. Statistical Analysis

Data were analyzed with the Statistical Analysis Systems Institute package [27]. The farm was used as the experimental unit. Therefore, for each visit, the values of the behavioral variables concerning both the animals and the stockpeople were averaged within farms. In order to avoid non-independent results, the number and the percentage of neutral interactions and the percentage of animals moving at a distance ≤1 m and ≥0.5 m were excluded from the analyses, as they corresponding to intermediate, thus less informative, categories. Then, the means were used to calculate the test-retest reliability of the variables concerning the behaviors of the stockperson (number of positive and negative interactions/milked buffalo, percentage of positive and negative interactions, percentage of animals injected with oxytocin) and the animals (number of steps and kicks/milked buffalo, the avoidance distance at the feeding place, percentage of animals that can be touched, percentage of animals moving at a distance ≤0.5 m, percentage of animals moving at a distance >1 m). Test–retest reliability was calculated using the Spearman rank correlation test (r_s). Limits of agreement were also calculated to test whether bias existed between visits [28]. Subsequently, these variables were averaged within the farm across the two visits. These farm averages were used to compute the correlations between stockpeople behavior variables, including the number of milked buffaloes/stockperson, and animal behavior variables using the Spearman rank correlation test. In addition, only for the 14 farms enrolled in the national milk recording scheme, we calculated the correlation between milk production and the stockpeople behavior variables as well as the correlation between milk production and the animal behavior variables using the Spearman rank correlation test.

The Kruskal-Wallis one-way ANOVA test was used to assess the effect of milking parlor design (herring-bone: n = 14 and tandem: n = 13) on the variables collected on stockpeople and animals.

3. Results and Discussion

3.1. Test-Retest Reliability

Tables 1 and 2 indicate the test-retest reliability of the variables concerning the behavior of stockpeople and animals, respectively. According to Martin and Bateson [29], a satisfactory threshold for correlation coefficients may be considered 0.7, as roughly 50% of variance in one set of observations is explained by the other set of observations. In our study, the reliability of the variables measured on the stockpeople may be considered satisfactory, with r_s value above 0.7 for most of them, whereas the number of positive interactions and the percentage of negative interactions were 0.650 and 0.677, respectively; $p < 0.001$, $df = 25$. However, the values of these coefficients were lower than those obtained for the animal-based variables, which ranged from 0.578 ($p = 0.002$, $df = 25$) for the number of kicks performed by the animals during milking to 0.937 ($p < 0.001$, $df = 25$) for the percentage of animals that moved when approached by ≤0.5 m. Limits of agreement mostly confirm the results expressed in terms of r_s. In cattle, high long-term consistency (farm visits were conducted at bimonthly intervals) was observed by Winckler et al. [30] for both avoidance distance measured in the barn and at the feeding place, whereas moderate to high test-retest reliability (recorded at 2–3 week intervals) was observed in buffaloes for avoidance distance measured in the barn [20]. Our results indicate that the avoidance distance at the feeding place and stockpeoples' and animals' behaviors during milking can be reliably used as indicators of the quality of the human-animal relationship, as also suggested by Hemsworth et al. [16] and Waiblinger et al. [21] for dairy cows.

3.2. Stockperson Behavior

The median and the range of stockperson behavioral variables are shown in Table 3. The behavior of the stockpeople was characterized by a low number of interactions with the animals. These interactions were mainly neutral, whereas negative interactions were the lowest, with a high degree of variability among farms. Although Breuer et al. [10] and Hemsworth et al. [16] used different categories to classify human-animal interactions in cattle, if the definitions are considered, these categories roughly correspond to those used in the present study (i.e., positive, negative, and very negative interactions from Breuer et al.'s [10] and Hemsworth et al.'s [16] studies corresponding to positive, neutral, and negative interactions, respectively, in our study). These authors recorded a similar total number of interactions. Conversely, Waiblinger et al. [5] and Ivemeyer et al. [31] reported a higher number of interactions. However, this discrepancy is likely due to the fact that the latter authors recorded the stockpeople behavior since the animals were moved from the barn, whereas in this and the other previously mentioned studies, the behavior of stockpeople was observed when the animals were moved from the waiting area to the milking parlor until their exit from it. Percent negative interactions showed the lowest value in this study, as in all the previous studies conducted on dairy cattle, whereas neutral and positive interactions showed the highest and the intermediate percentages, respectively. Although neutral interactions were the most represented in a study conducted by Saltalamacchia et al. [12] in a previous work on dairy buffaloes, these authors recorded a higher number of negative as compared to positive interactions. We can hypothesize that the attitude and consequently the behavior of the stockpeople working in buffalo milking parlors has improved over the last decade. Negative interactions are able to increase the level of fear of humans, whereas positive interactions can decrease it with more beneficial effects on animal welfare as compared with neutral interactions [4].

The design of the milking parlor affected only the number (median, range: 0.09, 0.03–0.64 versus 0.01, 0.00–0.37, respectively) and percentage (15.00, 4.71–87.88 versus 2.56, 0.00–32.35, respectively) of negative interactions ($x^2 = 7.26$, $p = 0.007$ and $x^2 = 5.75$, $p = 0.016$, respectively, $df = 1$) with higher levels of negative interactions in tandem parlors as compared with herring-bone parlors. Although in buffaloes, no studies about the effect of milking parlor design on human-animal interaction are available to support our hypothesis, we postulate that tandem parlors require animals to be individually handled in order to let them in and out of each stall, whereas in herring-bone parlors, animals are handled in groups, which may facilitate their entrance and exit with a reduced likelihood of negative interactions because of reduced handling and individual interactions. In addition, the type of parlor tended to influence the percentage of animals injected with oxytocin ($x^2 = 3.32$, $p = 0.067$) with lower percentages in herring-bone parlors as compared with tandem parlors (6.55, 0.00–27.81 versus 14.21, 0.00–100, respectively).

Table 1. Test–retest reliability of the variables observed on the stockpeople using Spearman rank correlation coefficient (r_s) and limits of agreement (mean of differences ±2 SD). Calculations were based on two farm visits (n = 27).

Variable	Spearman Statistics			Limits of Agreement	
	r_s	*p*-Value	Mean	Mean +2 SD	Mean −2 SD
Positive interactions/milked buffalo, n	0.650	<0.001	−0.01644	0.30416	−0.33703
Negative interactions/milked buffalo, n	0.825	<0.001	0.002949	0.33771	−0.33181
Positive interactions, %	0.771	<0.001	0.641434	34.20647	−32.9236
Negative interactions, %	0.677	<0.001	0.718693	23.07633	−21.6389
Animals injected with oxytocin, %	0.799	<0.001	2.616128	22.58423	−17.352

Table 2. Test–retest reliability of the variables observed on the animals using the Spearman rank correlation coefficient (r_s) and limits of agreement (mean of differences ±2 SD). Calculations were based on two farm visits (n = 27).

Variable	Spearman Statistics			Limits of Agreement	
	r_s	*p*-Value	Mean	Mean +2 SD	Mean −2 SD
Steps/buffalo, n	0.900	<0.001	0.079351	1.385843	−1.22714
Kicks/buffalo, n	0.578	0.002	0.007591	0.682179	−0.667
Animals moving at a distance >1 m, %	0.821	<0.001	−0.52111	6.873861	−7.91608
Animals moving at a distance ≤0.5 m, %	0.937	<0.001	0.051073	16.34284	−16.2407
Animals that can be touched, %	0.903	<0.001	−0.68902	16.08882	−17.4669
Median avoidance distance at the feeding place, m	0.923	<0.001	0.013462	0.135306	−0.10838

Table 3. Median and range of the variables observed on the stockpeople (n = 27).

Variable	Median	Range
Positive interactions/milked buffalo, n	0.15	0–0.79
Neutral interactions/milked buffalo, n	0.36	0.03–3.28
Negative interactions/milked buffalo, n	0.04	0–0.64
Positive interactions, %	19.64	0–83.88
Neutral interactions, %	61.29	9.09–93.88
Negative interactions, %	7.61	0–87.88
Animals injected with oxytocin, %	9.75	0–100
Milked buffalo/stockperson, n	55.50	19.12–100

3.3. Buffalo Behavior and Production

The median and the range of buffalo behavioral variables are shown in Table 4. Animal restlessness at milking may be caused by many different factors, such as pushing of adjacent cows (only in herringbone parlors), lameness, presence of hematophage insects, poor maintenance of milking machine, etc. However, it is widely accepted that at least a component of these behavioral expressions is related to the quality of the human-animal relationship. In this study, the number for stepping was in line with that reported in some studies conducted on dairy cows [16,31] but lower than that reported in other studies conducted in dairy buffaloes [12] and cattle [5,10]. Conversely, the number for kicking was higher than that reported in all the previously cited articles on cattle, albeit it was lower than that observed in buffaloes [12]. This may be due to a higher sensitivity and reactivity of buffaloes to the milking routine, as also suggested by the high number of animals injected with oxytocin observed in this study and in previous studies conducted at farms [12] and individual levels [32] as compared with cattle [33].

Table 4. Median and range of the variables observed in the animals (n = 27).

Variable	Median	Range
Steps/milked buffalo, n	0.88	0.11–6.62
Kicks/milked buffalo, n	0.22	0–0.84
Animals moving at a distance >1 m, %	0	0–52.78
Animals moving at a distance ≤1 m and ≥0.5 m, %	7.32	0–46.30
Animals moving at a distance ≤0.5 m, %	49.31	0.93–100
Animals that can be touched, %	34.17	0–77.38
Median avoidance distance at the feeding place, m	0.23	0.04–1.17
Milk production [1], kg/head/year	1995	1593–2540

[1] This variable was measured only in 14 farms enrolled in the national milk recording scheme.

In this study, the avoidance distance at the feeding place was lower than that reported by Shahin et al. in dairy cattle (mean = 52 cm) [34], higher than those measured in fattening bulls (mean = 12–15 cm) [18] and in dairy cattle (median = 8–10 cm) [35], and comparable to those reported by De Rosa et al. [23] for buffaloes (median = 20 cm) and by Windschnurer et al. [17] (median = 18.0 cm) and Battini et al. in dairy cattle (mean = 25) [36]. A previous study also reported a lower avoidance distance in buffaloes as compared with cattle kept in the same management and housing conditions [20].

Only the percentage of buffaloes moving when approached at a distance >1 m was affected by the design of the milking parlor, with tandem milking parlors showing a percentage of buffaloes moving when approached at a distance >1 m higher than herring-bone milking parlors (χ^2 = 5.06, p = 0.024, df = 1, median = 2.67, range = 0.00–52.78 versus 0.00, 0.00–11.90, respectively). These results soundly match those on stockperson behavior, where the tandem milking parlor induced a higher level of negative interactions and tended to increase the number of animals injected with oxytocin.

3.4. Correlating Stockperson and Buffalo Behaviors

Negative stockperson interactions—both in terms of absolute number and percentage—correlated positively with the number of kicks during milking ($r_s = 0.421$, $p = 0.028$ and $r_s = 0.430$, $p = 0.025$, respectively; $df = 25$), whereas only the number of negative interactions correlated positively with the percentage of animals injected with oxytocin ($r_s = 0.424$, $p = 0.027$, $df = 25$). These results are consistent with the hypothesis that the stockperson behavior can influence animal behavior during milking [5,16] and indicate that a negative behavior expressed by stockpeople at the time of milking can have a detrimental effect on buffalo cows. In our study, buffaloes, concomitant to a negative human approach, displayed higher levels of restlessness in terms of number of kicks, with an increased number of animals injected with oxytocin due to either impaired milk let down or stockpeople willing to speed up the milking routine. The number of negative stockperson interactions tended to be correlated positively with the percentage of buffaloes moving when approached at a distance >1 m ($r_s = 0.355$, $p = 0.069$, $df = 25$), whereas the percentage of negative stockperson interactions was correlated positively with the percentage of buffaloes moving when approached at a distance >1 m ($r_s = 0.415$, $p = 0.031$). This result suggests that buffaloes, as with other farm animals, are able to generalize their response to humans, and if they perceive negative stimuli from the stockpeople in the parlor, they also tend to increase their avoidance response to an unknown person approaching them at the feeding place. The percentage of buffaloes moving when approached at a distance >1 m also tended to be correlated positively with the number of lactating animals per stockperson ($r_s = 0.345$, $p = 0.078$, $df = 25$), which may indicate that a high animals to milkers ratio may impair the establishment of a positive human-animal relationship and reduce the confidence of the animals towards humans. A previous study primarily conducted on dairy cattle found no association between the number of animals per farm and the level of animal welfare [37] and suggested that efforts should concentrate on the improvement of animal welfare independently from the size of the farms. However, in the present study, the effect of the ratio of animal to milker was investigated (rather than the effect of the number of animals per farm), and a high ratio, while increasing the work load for stockpeople, may have potentially negative consequences on the quality of the human-animal relationship. The percentage of positive stockperson interactions tended to be correlated negatively with the percentage of animals injected with oxytocin ($r_s = -0.343$, $p = 0.080$, $df = 25$).

The milk production per animal per year was correlated negatively with the number of steps ($r_s = 0.820$, $p < 0.001$, $df = 12$) and correlated positively with the number of positive stockperson interactions ($r_s = 0.588$, $p = 0.027$, $df = 12$). This latter finding suggests that positive stockperson interactions may improve the quality of human-animal relationships and increase the welfare of the animals with beneficial effects on milk production, as also reported by other authors in previous studies on dairy cattle (e.g., [16]). However, these results should be taken with caution, as they are based on a limited number of farms (n = 14, i.e., only those adhering to the official Italian recording system).

4. Conclusions

The present study confirmed that the test-retest reliability of the variables used to assess the human-animal relationship in buffaloes was high. Also relevant are the findings showing the relationship between negative stockperson behavior and the reaction of buffalo cows in terms of restlessness and, possibly, consequent impaired milk let down. Therefore, appropriate stockpeople training should be conducted to improve human-animal relationships with positive effects on animal welfare, productivity, and stockpeople safety. In addition, the correlation between the percentage of negative stockperson interactions at milking and the percentage of animals moving when approached at a distance >1 m showed that buffalo cows are able to generalize their responses to humans, and if they perceive negative stimuli from the stockpeople in the parlor, they also tend to increase their avoidance response to an unknown person approaching them at the feeding place. Therefore, the avoidance distance of buffaloes at the feeding place is a promising variable to be used for the assessment of the quality of human-animal relationships, as demonstrated in other animal species. Although based on a limited number of farms, also relevant were the results displaying a correlation between stockperson positive interactions and milk production.

Stockpeople behavior was also affected by milking parlor design, with higher negative interactions in tandem parlors than in herring-bone parlors, possibly due to the fact that animals had to be individually handled to let them in and out of each stall. In farms equipped with tandem parlors, the percentage of buffaloes reacting at more than 1 m to an approaching human was higher, and the percentage of animals injected with oxytocin tended to increase.

Author Contributions: Conceptualization, F.N. and G.D.R.; data curation, F.S., F.M. and G.D.R.; formal analysis, F.M. and G.D.R.; funding acquisition, A.B.; investigation, F.S. and E.S.; methodology, F.N. and G.D.R.; project administration, A.B.; supervision, A.B.; visualization, F.S. and E.S.; writing—original draft, F.N. and G.D.R.; writing—review & editing, F.N.

Funding: The research was funded by Scuola di Scienze Agrarie, Forestali, Alimentari ed Ambientali, Università degli Studi della Basilicata (Potenza, Italy).

Acknowledgments: Thanks are due to Amelia M. Riviezzi and Giovanni Migliori for expert technical assistance.

Conflicts of Interest: The authors declare no conflicts of interest.

References

1. Schnettler, B.; Vidal, R.; Silva, R.; Vallejos, L.; Sepulveda, N. Consumer perception of animal welfare and livestock production in the Araucania Region, Chile. *Chil. J. Agric. Res.* **2008**, *68*, 80–93. [CrossRef]
2. Murray, G.; Ashley, K.; Kolesar, R. Drivers for animal welfare policies in Asia, the Far East and Oceania. *Rev. Sci. Tech. Rev. Off. Int. Epiz.* **2014**, *33*, 77–83. [CrossRef]
3. Napolitano, F.; Girolami, A.; Braghieri, A. Consumer liking and willingness to pay high welfare animal-based products. *Trends Food Sci. Technol.* **2010**, *21*, 537–543. [CrossRef]
4. Hemsworth, P.H.; Coleman, G.J. *Human—Livestock Interactions*, 2nd ed.; CABI Head Office Nosworthy Way: Wallingford, UK, 2011; pp. 1–168.
5. Waiblinger, S.; Menke, C.; Coleman, G. The relationship between attitudes, personal characteristics and behavior of stockpeople and subsequent behavior and production of dairy cows. *Appl. Anim. Behav. Sci.* **2002**, *79*, 195–219. [CrossRef]
6. Napolitano, F.; Pacelli, C.; Grasso, F.; Braghieri, A.; De Rosa, G. The behavior and welfare of buffaloes (*Bubalus bubalis*) in modern dairy enterprises. *Animal* **2013**, *7*, 1704–1713. [CrossRef] [PubMed]
7. Mendl, M.; Burman, O.H.P.; Paul, E.S. An integrative and functional framework for the study of animal emotion and mood. *Proc. R. Soc. B* **2010**, *277*, 2895–2904. [CrossRef] [PubMed]
8. Hemsworth, P.H.; Pedersen, V.; Cox, M.; Cronin, G.M.; Coleman, G.J. A note on the relationship between the behavioral response of lactating sows to humans and the survival of their piglets. *Appl. Anim. Behav. Sci.* **1999**, *65*, 43–52. [CrossRef]
9. Rushen, J.; De Passille, A.M.; Munksgaard, L. Fear of people by cows and effects on milk yield, behavior and heart rate at milking. *J. Dairy Sci.* **1999**, *82*, 720–727. [CrossRef]
10. Breuer, K.; Hemsworth, P.H.; Barnett, J.L.; Matthews, L.R.; Coleman, G. Behavioural response to humans and the productivity of commercial dairy cows. *Appl. Anim. Behav. Sci.* **2000**, *66*, 273–288. [CrossRef]
11. Lensink, J.; Fernandez, X.; Cozzi, G.; Florand, L.; Veissier, I. The influence of farmers' behavior on calves' reactions to transport and quality of veal meat. *J. Anim. Sci.* **2001**, *79*, 642–652. [CrossRef]
12. Saltalamacchia, F.; Tripaldi, C.; Castellano, A.; Napolitano, F.; Musto, M.; De Rosa, G. Human and animal behavior in dairy buffalo at milking. *Anim. Welf.* **2007**, *16*, 139–142.
13. Lensink, J.; Fernandez, X.; Boivin, X.; Pradel, P.; LeNeindre, P.; Veissier, I. The impact of gentle contacts on ease of handling, welfare, and growth of calves and on quality of veal meat. *J. Anim. Sci.* **2000**, *78*, 1219–1226. [CrossRef]
14. Grandin, T. How to improve livestock handling and reduce stress. In *Improving Animal Welfare: A Practical Approach*; Grandin, T., Ed.; CABI International: Wallingford, UK, 2010; pp. 64–87, ISBN 978-1-84593-541-2.

15. Lürzel, S.; Münsch, C.; Windschnurer, I.; Futschik, A.; Palmed, R.; Waiblinger, S. The influence of gentle interactions on avoidance distance towards humans, weight gain and physiological parameters in group-housed dairy calves. *Appl. Anim. Behav. Sci.* **2015**, *172*, 9–16. [CrossRef]

16. Hemsworth, P.H.; Coleman, G.J.; Barnett, J.L.; Borg, S. Relationships between human-animal interactions and productivity of commercial dairy cows. *J. Anim. Sci.* **2000**, *78*, 2821–2831. [CrossRef]

17. Windschnurer, I.; Schmied, C.; Boivin, X.; Waiblinger, S. Reliability and inter-test relationship of tests for on-farm assessment of dairy cows' relationship to humans. *Appl. Anim. Behav. Sci.* **2008**, *114*, 37–53. [CrossRef]

18. Windschnurer, I.; Boivin, X.; Waiblinger, S. Reliability of an avoidance distance test for the assessment of animals responsiveness to humans and a preliminary investigation of its association with farmers' attitudes on bull fattening farms. *Appl. Anim. Behav. Sci.* **2009**, *117*, 117–127. [CrossRef]

19. Fisher, A.D.; Morris, C.A.; Matthews, L.R. Cattle behavior: Comparison of measures of temperament in beef cattle. *Proc. N. Z. Soc. Anim.* **2000**, *60*, 214–217.

20. Napolitano, F.; Grasso, F.; Bordi, A.; Tripaldi, C.; Pacelli, C.; Saltalamacchia, F.; De Rosa, G. On-farm welfare assessment in dairy cattle and buffaloes: Evaluation of some animal-based parameters. *Ital. J. Anim. Sci.* **2005**, *4*, 223–231. [CrossRef]

21. Waiblinger, S.; Menke, C.; Fölsch, D.W. Influences on the avoidance and approach behavior of dairy cows towards humans on 35 farms. *Appl. Anim. Behav. Sci.* **2003**, *84*, 23–29. [CrossRef]

22. Welfare Quality®. *Welfare Quality®Assessment Protocol for Cattle*; Welfare Quality®Consortium: Lelystad, The Netherlands, 2009; ISBN 978-90-78240-04-4.

23. De Rosa, G.; Grasso, F.; Winckler, C.; Bilancione, A.; Pacelli, C.; Masucci, F.; Napolitano, F. Application of the Welfare Quality protocol to dairy buffalo farms: Prevalence and reliability of selected measures. *J. Dairy Sci.* **2015**, *98*, 6886–6896. [CrossRef]

24. Coleman, G.; Hemsworth, P.H.; Cox, H.M. Modifying stockperson attitudes and behavior towards pigs at a large commercial farm. *Appl. Anim. Behav. Sci.* **2000**, *66*, 11–20. [CrossRef]

25. Barnett, J.L.; Hemsworth, P.H.; Hennessy, D.P.; McCallum, T.M.; Newman, E.A. The effects of modifying the amount of human contact on the behavioral, physiological and production responses of laying hens. *Appl. Anim. Behav. Sci.* **1994**, *41*, 87–100. [CrossRef]

26. Hemsworth, P.H. Human–animal interactions in livestock production. *Appl. Anim. Behav. Sci.* **2003**, *81*, 185–198. [CrossRef]

27. SAS Institute. *User's Guide: Statistics*; SAS Inst. Inc.: Cary, NC, USA, 1990.

28. Bland, J.M.; Altman, D.G. Statistical methods for assessing agreement between two methods of clinical measurement. *Lancet* **1986**, *327*, 307–310. [CrossRef]

29. Martin, P.; Bateson, P. *Measuring Behavior: An Introductory Guide*, 3rd ed.; Cambridge University Press: Cambridge, UK, 2007; ISBN 978-0-521-53563-2.

30. Winckler, C.; Brinkmann, J.; Glatz, J. Long-term consistency of selected animal-related welfare parameters in dairy farms. *Anim. Welf.* **2007**, *16*, 197–199.

31. Ivemeyer, S.; Knierim, U.; Waiblinger, S. Effect of human-animal relationship and management on udder health in Swiss dairy herds. *J. Dairy Sci.* **2011**, *94*, 5890–5902. [CrossRef] [PubMed]

32. Polikarpus, A.; Grasso, F.; Pacelli, C.; Napolitano, F.; De Rosa, G. Milking behavior of buffalo cows: Entrance order and side preference in the milking parlor. *J. Dairy Res.* **2014**, *81*, 24–29. [CrossRef]

33. Bruckmaier, R.M. Normal and disturbed milk ejection in dairy cows. *Domest. Anim. Endocrinol.* **2005**, *28*, 268–273. [CrossRef] [PubMed]

34. Shahin, M. The effects of positive human contact by tactile stimulation on dairy cows with different personalities. *Appl. Anim. Behav Sci.* **2018**, *204*, 23–28. [CrossRef]

35. Lürzel, S.; Barth, K.; Windschnurer, I.; Futschik, A.; Waiblinger, S. The influence of gentle interactions with an experimenter during milking on dairy cows' avoidance distance and milk yield, flow and composition. *Animal* **2018**, *12*, 340–349. [CrossRef] [PubMed]

36. Battini, M.; Andreoli, E.; Barbieri, S.; Mattiello, S. Long-term stability of Avoidance Distance tests for on-farm assessment of dairy cow relationship to humans in alpine traditional husbandry systems. *Appl. Anim. Behav. Sci.* **2011**, *135*, 267–270. [CrossRef]

37. Robbins, J.; Von Keyserlingk, M.; Fraser, D.; Weary, D. Farm size and animal welfare. *J. Anim. Sci.* **2016**, *94*, 5439–5455. [CrossRef] [PubMed]

Article

Farmer Perceptions of Pig Aggression Compared to Animal-Based Measures of Fight Outcome

Rachel S. E. Peden [1],*, Irene Camerlink [2], Laura A. Boyle [3], Faical Akaichi [4] and Simon P. Turner [1]

[1] Animal Behaviour & Welfare, Animal and Veterinary Sciences Research Group, Scotland's Rural College (SRUC), West Mains Rd., Edinburgh EH9 3JG, UK; Simon.Turner@SRUC.ac.uk

[2] Institute of Animal Welfare Science, University of Veterinary Medicine, Veterinärplatz 1, 1210 Vienna, Austria; Irene.Camerlink@vetmeduni.ac.at

[3] Teagasc, Pig Development Department, Animal & Grassland Research and Innovation Centre, Moorepark, Fermoy Co., Cork P61 C997, Ireland; Laura.Boyle@teagasc.ie

[4] Land Economy Environment and Society Research Group, Scotland's Rural College (SRUC), West Mains Rd., Edinburgh EH9 3JG, UK; Faical.Akaichi@SRUC.ac.uk

* Correspondence: Rachel.Peden@SRUC.ac.uk; Tel.: +44-0131-6519360

Received: 20 November 2018; Accepted: 7 January 2019; Published: 10 January 2019

Simple Summary: Aggression between pigs is a major animal welfare issue in commercial farming, however only a minority of farmers believe that aggression is a problem that needs to be addressed. We investigated whether the farmers' reluctance to reduce aggression is linked to desensitization as a result of their frequent exposure to the behavior. We showed farmers video clips of pigs during and immediately after a fight and they judged through a questionnaire the severity of what they saw. These judgments were compared to (a) animal-based measures of injury (skin lesions) and exhaustion (blood lactate), and (b) human observers with and without experience of working with pigs. Farmers perceived fights as severe and were motivated to prevent them continuing. They were not desensitized to aggression as their judgments were similar to those of participants who had never worked with pigs. When farmers (and comparison groups) did not see the fight occurring, they judged exhaustion and injuries to be lower than indicated by the animal-based measures. Farmers could benefit from information on how to better assess the impact of aggression by scoring lesions and from evidence of the economic and welfare impact of these lesions.

Abstract: Several animal welfare issues persist in practice despite extensive research which has been linked to the unwillingness of stakeholders to make changes. For example, most farmers do not perceive pig aggression to be a problem that requires action despite the fact that stress and injuries are common, and that several solutions exist. Frequent exposure to animal suffering could affect farmer responses to distressed animals. This study investigated for the first time whether this occurs, using pig aggression as a focus. Using video clips, 90 pig farmers judged the severity of aggression, level of pig exhaustion and the strength of their own emotional response. Their judgments were compared to objective measures of severity (pigs' skin lesions and blood lactate), and against control groups with similar pig experience (10 pig veterinarians) and without experience (26 agricultural students; 24 animal science students). Famers did not show desensitization to aggression. However, all groups underestimated the outcome of aggression when they did not see the fight occurring as compared to witnessing a fight in progress. We suggest that farmers be provided with evidence of the economic and welfare impact of aggression as indicated by lesions and that they be advised to score lesions on affected animals.

Keywords: aggression; animal welfare; desensitization; perception; pigs

---

1. Introduction

Farmers are frequently exposed to a range of animal welfare issues, yet they are often unwilling to implement recommendations to improve animal welfare [1–3]. It is known that frequently witnessing human suffering can disrupt human emotional, cognitive and behavioral responses to witnessing distress [4,5]. The current study investigates for the first time whether exposure to animal suffering disrupts farmer responses to animal suffering, using pig aggression as a case study.

Desensitization is a well-established defense mechanism which occurs automatically and unconsciously [4,5]. For example, regular exposure to violence can lead to a reduced emotional response to violence [6], reduced empathy for the victims of violence [4] and increased violent behavior [7]. When witnessing human suffering, the decision to intervene is determined firstly by perceiving there to be an urgent problem that needs to be addressed, followed by feeling personal responsibility to act [8]. Desensitization can interfere with this decision-making process by making incidents less likely to be noticed, by reducing the perceived seriousness of the suffering and by reducing feelings of personal responsibility [6]. It has previously been noted that agricultural communities may become desensitized to animal suffering as they are exposed to it on a regular basis [5]. However, this hypothesis has never been empirically studied despite potentially having important implications for animal welfare and farm efficiency.

Pig aggression is common in commercial farming as pigs fight to establish dominance relationships following regrouping [9]. In the UK and Ireland, growing pigs are typically regrouped at least once per production cycle, but this can reach as many as four times [10,11], whilst sows are returned to group housing during each gestation [12]. Regrouping occurs to optimize the use of space and to maintain homogeneity in groups (e.g., similar body weight or same gestational phase). Therefore, most farmers regroup animals regularly, and the exact frequency depends on the management of pig batches and farm size. As a result, intensive pig farmers will frequently witness animal suffering due to aggression during their working lives. Aggression between pigs often results in stress for the animals, which can compromise their growth performance [13–15], reproductive success [16–18] and immune competence [19,20], whilst injuries can impact upon carcass quality [21,22]. However, a recent survey of 167 UK pig farmers revealed that the majority of farmers did not perceive aggression between unfamiliar pigs to be a problem that needs to be addressed [10]. Furthermore, only a minority of farmers attempt to control aggression when regrouping, despite the existence of several effective aggression mitigation strategies [23,24]. These strategies require farmers to make specific changes to animal management or nutrition. For example, allowing litters to mix prior to weaning, housing pigs in large social groups, and enhancing levels of tryptophan in the feed can all reduce the occurrence or intensity of aggression at regrouping [23,25]. It is, therefore, possible that farmers underestimate the impact that aggression has on the welfare and productivity of their animals. The current study aims to investigate: (1) whether farmers underestimate the physical impact of pig aggression and; (2) if this response is influenced by the amount of experience of working with pigs.

2. Materials and Methods

2.1. Overview

We asked 90 farmers for their perceptual and emotional response to six video clips of aggressive encounters between pigs, employing a paper-based survey. Control groups of non-farmers with experience of working with pigs (10 pig veterinarians) and without experience of working with pigs (26 agricultural students and 24 animal science students) completed an amended version of the survey. Farmers' scores were compared against the scores of the comparison groups and against objective measures of severity (relative change in number of skin lesions and blood lactate as a result of the interaction) in order to investigate whether farmers underestimate the physical impact of aggression on pig welfare, and how the amount of experience of working with pigs may influence perceptions.

2.2. Ethical Approval

All animal experimentation was approved by Scotland's Rural College's (SRUCs) Animal Ethics Committee and the U.K. Government Home Office, ensuring compliance with EC Directive 86/609/EEC for animal experiments. This study was conducted in accordance with the Declaration of Helsinki. This study received internal ethical approval from the Human Ethical Review Committee at the University of Edinburgh (Project identification code: HERC_88_17), and informed consent was obtained for all participants.

2.3. Selection of Video Clips

Video footage was obtained from a separate research project carried out in 2015 at Scotland's Rural College (SRUC) whereby 168 growing pigs were video recorded in dyadic encounters comprising aggressive interactions. For each pig, measures of skin lesions and blood lactate were taken pre- and post-encounter to indicate relative change as a result of aggression. Skin lesions as a result of receiving bites (i.e., bite marks) are a good indicator of the severity of aggression [21]. Skin lesion count is a validated proxy measure for aggression that is moderately heritable and has been applied in animal welfare assessments [21,26]. Blood lactate gives a measure of physical fatigue. Further details of the dyadic encounters, lesion recording and lactate measurements are provided in [27].

A stepwise selection process was adopted to identify six video clips to be shown to observers. First, pigs that displayed a negative relative change in blood lactate, and therefore displayed a reduction in blood lactate following the fight, were eliminated from the dataset ($n = 26$, see Table 1 for descriptive statistics of the remaining dataset). Second, based on the severity of skin lesions and blood lactate, we identified from the remaining dataset the encounters in which both pigs obtained high (upper quartile, UQ), medium (interquartile range, IQR) or low (lower quartile, LQ) severity measures. Video clip 1 displayed a medium severity mutual fight and was always seen first. This 'dummy' clip acted as a practice and a common start point. Moreover, by displaying a typical aggressive encounter, this clip sets the scene for the following experimental clips. Clips 2–4 displayed pigs of low, medium and high severity encounters immediately after the fight ended. Videos of pigs with lesions and lactate in the IQR were also selected showing behavior during the actual occurrence of a fight or during bullying (winner chases the loser) to account for the different types of aggression seen on farms (clips 5–6). This ensured that all observers viewed fights that had ended and interactions that were in progress. The severity and content of each 20 s video clip can be seen in Table 2. Participants were asked to focus on one specific pig. The focal pig obtained severity measures which were as similar as possible to those of the non-focal pig. For exact measures of lesion score and blood lactate for both the focal and non-focal pigs, see Table 3. For a detailed description of the stepwise selection process and criteria, see Appendix A.

Table 1. Descriptive statistics regarding measures of relative change in lesions (number of lesions per pig) and blood lactate (mmol/L) following the fight, compared to before the fight, for the dataset ($n = 142$).

Measure	Blood Lactate	Lesion Score
Mean	7.76	57.37
Min	0	0
Quartile 1	2.15	12.75
Quartile 2	42.00	5.85
Quartile 3	13.20	77.25
Max	21.20	354

Table 2. Severity and content of each 20 s video clip displaying an aggressive encounter between two pigs. (LQ = lower quartile; IQR = interquartile range; UQ = upper quartile).

Clip	Blood Lactate	Lesion Score	Behavior
1 ('Dummy')	IQR	IQR	During mutual fight
2 ('Fight outcome: Low')	LQ	LQ	After fight
3 ('Fight outcome: Medium')	IQR	IQR	After fight
4 ('Fight outcome: High')	UQ	UQ	After fight
5 ('During fight: Mutual')	IQR	IQR	During mutual fight
6 ('During fight: Bullying')	IQR	IQR	During bullying

Table 3. Exact measures of relative change in lesion score (number of lesions per pig) and blood lactate (mmol/L) following the fight, compared to before the fight, for the pigs in each video clip.

Clip	Focal Pig		Non-Focal Pig	
	Blood Lactate	Lesion Score	Blood Lactate	Lesion Score
1	3.4	20	5.1	31
2	0.6	2	0.3	0
3	4.3	55	8.8	30
4	20.8	82	16.8	94
5	9.4	56	5.8	24
6	5	45	12	0

The order effects across the observation sessions were controlled for by creating six clip orders, as outlined in Table 4. Footage was edited using Windows Movie Maker (version 2012) and each clip was selected to be 20 s long and such that the focal pig was clearly identifiable. The clips were selected towards the end of the aggressive encounter (clips 1, 5 and 6) or immediately after (clips 2, 3 4), such that the behavior performed was as closely matched in time to the measures of lesions and lactate as possible. Images were played back with sound during observer scoring sessions. Before the clip began, a 'freeze-frame' showed the focal pig circled alongside a message stating '*please focus on this pig*'. Furthermore, during the clip, every time the focal pig made a major change to its position an arrow appeared pointing towards it.

Table 4. Each of the six clip orders.

Block	Clip Order					
	A	B	C	D	E	F
1	1	1	1	1	1	1
2	3	4	2	3	4	2
2	4	2	3	4	2	3
2	2	3	4	2	3	4
3	5	5	5	6	6	6
3	6	6	6	5	5	5

2.4. Survey Design

Answer sheets contained two main sections. Section 1 entitled 'demographics' collected information on the farmer's age (year of birth), gender, role on the farm, farm size and years of experience working with pigs. Farmers were also asked: 'Do you ever intervene during aggressive encounters between pigs on your farm? (Please tick ALL statements that you agree with from: No, there is no point; No, I never see aggressive encounters on my farm; No, it is too dangerous; Yes, when profitability is likely to be affected, and; Yes, to reduce injuries/stress for the animals)'. Section 2, entitled 'videos', was completed alongside watching the assigned movie. Following each video clip, the movie was paused and farmers were asked to place a downward line through three separate

100 mm visual analogue scales (VAS) at a point they felt best represented: (i) how much of a negative emotional reaction they had, from no negative reaction to strongest possible negative reaction; (ii) how exhausting they believed the fight was for the focal pig, from not exhausting at all to the most exhausting possible; (iii) how severe they believed the fight was for the focal pig, from not severe at all to the most severe seen on farms. Additionally, participants were asked what factors they used to judge the severity of the fight ('Tick all relevant factors from: number/severity of skin lesions, vocalizations, panting, other sounds (e.g., banging), facial expression, stress and others'). For clips 1, 5 and 6 (during fights), farmers were also asked; (iv) if they saw this fight on their farm, how much they would want to prevent it continuing, from not at all to the most possible. Sections 1 and 2 were amended slightly for non-farmers by removing all farm-related questions and replacing them with those relevant to the control group. For example, questions regarding their role on the farm and farm size were excluded and participants were alternatively asked about their occupation. Participants were instructed not to talk to each other in order to avoid possible effects of their discussions on their answers. For farmer and non-farmer response sheets, see Supplementary Materials.

2.5. Recruitment

Participants were recruited between February 2017 and November 2017. Ninety pig farmers were recruited whilst participating in six discussion group events held in the UK and Ireland, organized by Scotland's Rural College ($n = 26$), Teagasc ($n = 29$) and the Agricultural and Horticultural Development Board (AHDB) Pork ($n = 35$). Ten specialized pig veterinarians participated at the same discussion groups. Veterinarians provide an interesting comparison group due to their comparable years of experience working in the pig industry as the farmers. Farmers and veterinarians were unaware that they would be asked to participate in a study on pig aggression prior to attending the discussion groups. Sixty-one students participated in twelve groups following lectures at SRUC; 35 were students of Agriculture, and 26 studied Animal Science. The student populations provide interesting comparison groups due to their knowledge of farming and livestock, but lack of experience working directly with pigs. Students were unaware that they would be asked to participate in a study on pig aggression prior to attending the lectures. The order of presentation of video clips in Table 4 was replicated twice for students since there were 12 groups compared to the 6 groups of farmers and veterinarians. All responses were collected through 'face to face' recruitment; this does not allow response rate calculations. Furthermore, we had no control over the composition of the groups with respect to occupation, so it was not possible to balance each clip order to have the same number of people from each occupation (Table 5).

Table 5. Total number of participants who watched each of the six clip orders.

Clip Order	Total N			
	Farmers	Pig Veterinarians	Agricultural Students	Animal Science Students
A	26	4	6	0
B	9	0	1	8
C	7	1	6	2
D	20	0	4	6
E	20	3	1	2
F	8	2	8	6
Total	90	10	26	24

2.6. Demographics of the Final Sample

Nine agricultural students and two animal science students were excluded from the analysis due to reporting prior experience of working with pigs. In total, 150 participants with the following demographics were included:

(1) Pig farmers (*n* = 90) were mostly male (93.3%; female: 6.7%) and were on average 41.5 years old (s.d. = 14.13, range = 17–81 years) with 19.53 years of experience working with pigs (s.d. = 14.74, range = 0.5–65 years). There was a strong, positive correlation between years of experience and age (r = 0.871, *p* < 0.0001). Therefore, only age was included in the statistical analysis but it was considered informative of both age and experience effects. Farmers were mainly farm workers (41.1%), owners (32.2%) and managers (17.8%). The remaining farmers were contract farmers (6.7%) and retired (2.2%). A total of 28.9% were based in Scotland, 38.9% were based in England and 32.2% in Ireland. Additionally, 86.7% of farmers reported currently keeping sows, whilst 67.8% kept weaners (i.e., recently weaned piglets), 55.6% kept growers and 62.6% kept finishers. Therefore, most farmers kept pigs at more than one stage of production. The mean number of pigs kept at each stage of production can be found in Table 6.

(2) Specialized pig veterinarians (*n* = 10) were mostly female (70%; male: 30%) and were on average 40.6 years old (s.d. = 14.4, range = 24–64 years) with 15.3 years of experience working with pigs (s.d. = 17.6, range = 0.75–40 years). Of this group, 40% were based in Scotland and 60% were based in England.

(3) Agricultural students (*n* = 26; 46.2% male, 53.8% female; mean age = 21.4 years, s.d. = 1.65, range = 20–28) were in their 3rd (*n* = 23) and 4th (*n* = 3) years of study.

(4) Animal science students (*n* = 24; 16.7% male, 83.3% female; mean age = 22.2 years, s.d. = 2.97, range = 20–35) were in their 3rd (*n* = 15) and 4th (*n* = 8) years of study. All students were based in Scotland.

Table 6. Mean number of pigs kept at each stage of production at any one time (in brackets are the number of farmers that kept pigs at the specified stage of production), range and standard deviation (s.d.).

	Mean (Number)	Range	s.d.
Weaners	1929 (61)	150–10,000	1829.12
Growers	2850 (50)	10–30,000	5787.05
Finishers	3835 (56)	100–38,000	7109.85
Sows	1100 (78)	40–13,500	2127.45

2.7. Statistical Analysis

Statistical analyses were conducted in Statistical Package for the Social Sciences (SPSS, version 25, International Business Machines Corp., Armonk, NY, USA). Normality of the data was assessed by inspection of the residuals and data were transformed wherever necessary. Residual maximal likelihood (REML) models were run to investigate the factors that influence: (1) Emotional response; (2) Judgment of fight severity; (3) Judgment of exhaustion; and (4) Motivation to intervene if the interaction had occurred on their own farm. The fixed effects in the first three models were gender, age, occupation and the video clip. Occupation was not included as a fixed effect in the fourth model as only farmers were asked the question 'If you saw this fight on your farm, how much would you want to prevent it continuing?'. The clip order was included in each model as a random effect. The main effects were removed if *p* > 0.1 and the model was re-run until the simplest model was achieved. Post hoc analyses were conducted using least significant difference (LSD) tests with a Bonferroni correction made for multiple comparisons. Six Chi Square tests were carried out to ascertain the effects of occupation on the cues used to judge fight severity. The dependent variables were the use of: (1) lesions; (2) vocalizations; (3) panting; (4) other sounds (e.g., banging); (5) facial expression and; (6) stress when judging fight severity. The results were considered statistically significant where *p* < 0.05.

3. Results

A total of 78.9% of farmers indicated that they do intervene when they see pigs fighting on their own farm; 13.3% indicated that they did so when profitability was likely to be affected and 76.7% did so to avoid injuries/stress for the animals. Furthermore, 15.6% of farmers reported that they did not intervene during aggressive encounters on their farm; 7.8% believed there was no point, 2.2% never see aggressive encounters on their farm and 5.6% believed it is too dangerous.

The clip order had no effect on emotional response, judgment of severity, judgment of exhaustion or motivation to intervene. The results of all the main effects are described below.

3.1. Main Effects of Occupation

There were significant main effects of occupation on emotional response and judgment of exhaustion across video clips (Table 7). Pairwise comparisons revealed that farmers and animal science students expressed greater emotional response scores when compared to agricultural students ($p < 0.05$; Figure 1a). Farmers judged fight exhaustion to be higher than agricultural students ($p < 0.01$; Figure 1b). Occupations did not differ in their judgments of severity ($p > 0.05$) (Figure 1c). Participants employed a range of cues when judging severity (see Figure 2). There was no effect of occupation on use of skin lesions, vocalizations, panting, facial expressions or stress when judging the severity of aggressive encounters ($p > 0.05$). However, there was an effect of occupation on use of 'other sounds (e.g., banging)' ($p < 0.01$), with animal science students using this cue significantly more than farmers and agricultural students ($p < 0.05$).

Table 7. The results of four residual maximal likelihood (REML) models investigating the factors that influence: (1) emotional response; (2) judgment of exhaustion; (3) judgment of severity and; (4) motivation to intervene. The main effects were removed from the model if $p > 0.1$ unless involved in a significant interaction.

Main Effect	F (df)	p
Emotional response		
Gender	7.0 (1)	0.009
Occupation	4.5 (3)	0.004
Video clip	136.4 (4)	0.001
Judgment of exhaustion		
Gender	8.4 (1)	0.004
Occupation	4.8 (3)	0.002
Video clip	131.6 (4)	0.001
Judgment of severity		
Gender	6.8 (1)	0.010
Age	8.1 (1)	0.005
Video clip	153.6 (4)	0.001
Farmer motivation to intervene		
Gender	4.9 (1)	0.029
Age	4.9 (1)	0.030
Video clip	8.7 (1)	0.004

3.2. Main Effects of Video Clip

There were significant main effects of video clip on emotional response, exhaustion score, severity score and farmer motivation to intervene across occupations (Table 7). The mean emotional response, exhaustion score and severity score for the low severity outcome clip (lower quartile lactate and lesions) showing pigs after a fight (clip 2) were significantly lower than for the medium and high severity outcome clips (clips 3 and 4), as well as for the mutual fight and the bullying clips (clips 5 and 6; $p < 0.01$). Emotional response and severity scores for the medium and high severity outcome clips showing pigs after a fight (clips 3 and 4) did not differ ($p > 0.05$) but the exhaustion scores did ($p < 0.001$), whereby pigs with a higher lactate level and more lesions were regarded as being more exhausted. Emotional response, exhaustion score and severity score for both of the 'during fight' clips (clips 5 and 6) were greater than the scores for the 'post-fight' clips, and responses to the bullying clip

(clip 6) were significantly greater than to the mutual fight clip (clip 5) ($p < 0.001$; Figure 3a–c). Farmer motivation to intervene was significantly greater for the bullying clip than for the mutual fighting clip (Bullying: mean = 79.4, SE = 2.4; Mutual fight: mean = 72.0, SE = 2.5; $p < 0.01$).

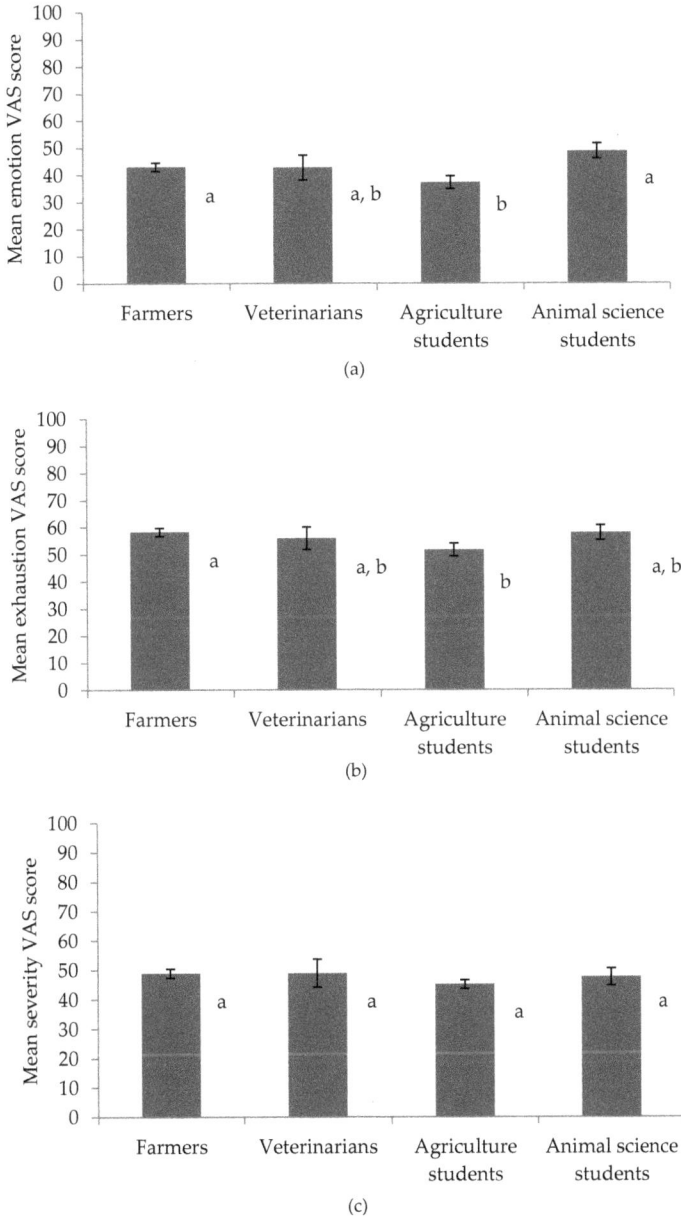

Figure 1. Mean visual analogue scale (VAS) scores for (**a**) emotion, (**b**) exhaustion, and (**c**) severity according to occupation; whereby occupations with different letters express a significant difference in mean response.

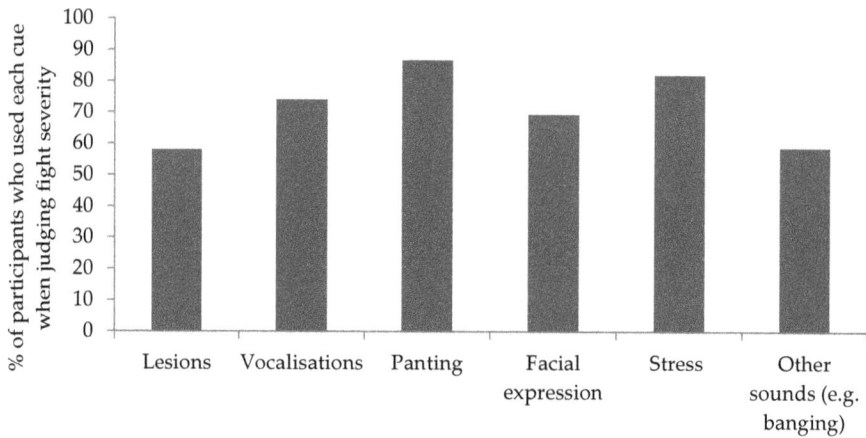

Figure 2. Percentage of participants who used each cue when judging fight severity.

3.3. Demographic Effects

Across occupations, women expressed significantly greater VAS scores compared to men for emotional response (Females: mean = 46.1, SE = 2.0; Males: mean = 41.6, SE = 1.4; $p < 0.01$), judgment of exhaustion (Females: mean = 59.2, SE = 1.9; Males: mean = 55.9, SE = 1.4; $p < 0.01$), and judgment of severity (Females: mean = 50.0, SE = 2.0; Males: mean = 47.2, SE = 1.4; $p < 0.05$). Female farmers also expressed greater motivation to intervene than male farmers (Females: mean = 88.8, SE = 3.5; Males: mean = 74.8, SE = 1.8; $p < 0.05$). There was a significant effect of age on farmer motivation to intervene ($p < 0.05$) with older farmers expressing greater motivation to intervene, although the significant positive correlation was weak (r = 0.148, $p < 0.05$). There was a significant effect of age on judgment of severity ($p < 0.01$) most likely linked to a cohort of young participants (agriculture students) who gave lower scores.

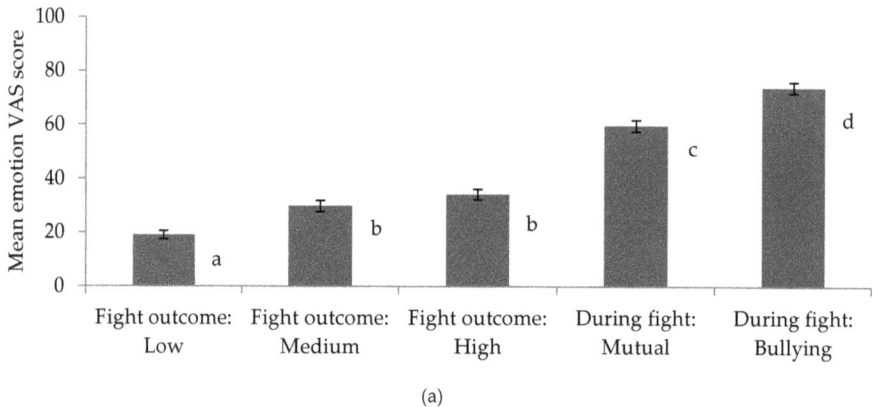

(a)

Figure 3. *Cont.*

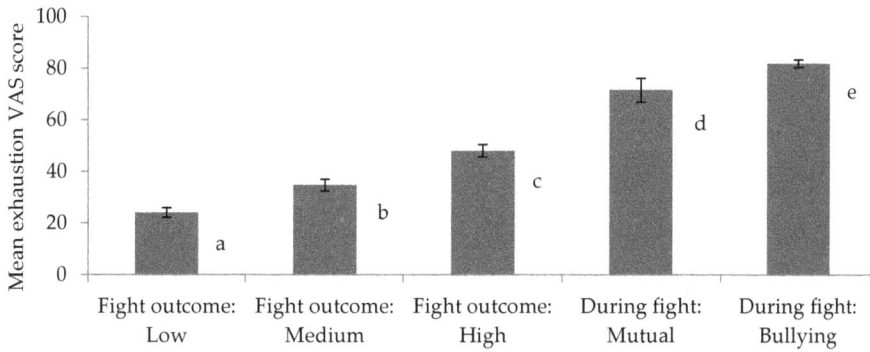

(b)

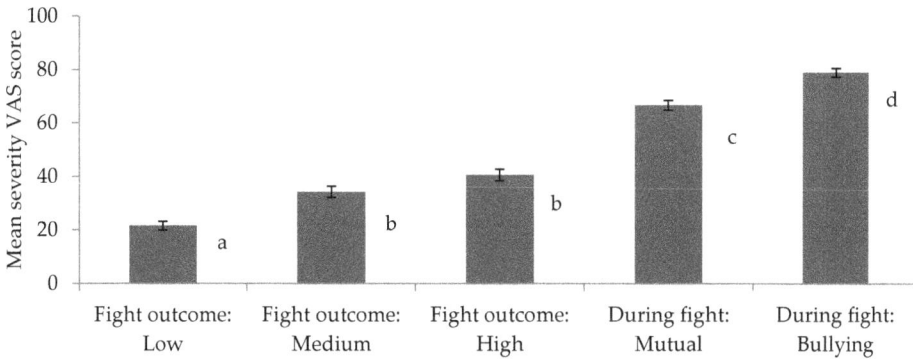

(c)

Figure 3. Mean visual analogue scale (VAS) score for (**a**) emotion, (**b**) exhaustion, and (**c**) severity according to video clip; whereby video clips with different letters express a significant difference in mean response.

4. Discussion

It is known that frequent exposure to suffering can disrupt human emotional, cognitive and behavioral responses to signs of distress [4,5], and the current study provides the first investigation into whether or not routine exposure to animal suffering disrupts farmer responses, using pig aggression as a case study. Our survey amongst 90 farmers and 60 control participants showed that farmers are not desensitized and in fact are motivated to change the situation when noticed. All participants assessed aggression as severe when having seen the fight, but underestimated the impact of aggression when the animals were viewed immediately after the fight had ended (assessed through objective animal-based measures).

4.1. Perceptions of Aggression

All comparison groups judged aggressive behavior in-action to be highly severe and exhausting for the animals, and experienced a negative emotional response to these interactions. Responses were particularly high for bullying aggression in comparison to mutual aggression, despite both encounters resulting in medium severity measures of blood lactate and skin lesions. Farmers reported that, if they saw these interactions on their farm, they would be highly motivated to prevent them continuing. Indeed, the majority of farmers reported that they do intervene when they see aggressive interactions on their farm, and their primary motivation for doing so was to reduce injuries and stress for the animals. Nevertheless, all comparison groups underestimated the impact of aggression as indicated by skin lesions and blood lactate when they did not see the fight occurring. Specifically, the perceived seriousness of injuries and exhaustion as a result of aggression, as well as judgments of their own emotional response, were lower when observing the animals immediately after a fight had ended, even when the outcomes were severe as indicated by objective measures. Furthermore, judgments of severity and emotional response did not differ between the medium and high severity post-fight outcome clips, suggesting that the participants perceived little difference between these outcomes.

Farmers are often engaged in a wide range of tasks performed in different buildings which makes it difficult for them to witness post-mixing fights as frequently as they actually occur. Farmers are expected to witness the injuries from such interactions during their regular animal inspections. However, results suggest that farmers are unlikely to fully recognize the severity of these outcomes and this may contribute to the limited uptake of recommendations from aggression research in commercial practice.

Farmers were not desensitized to aggression as a result of frequent exposure to the behavior; farmer perceptions of fight severity were comparable to those of participants with, and without, experience of working with pigs, and all participants employed a range of pig-based cues (lesions, facial expression, panting, stress and vocalizations) to a similar extent when making these judgments. Farmers and agriculture students did use 'other sounds (e.g., banging)' less than applied animal science students, which suggests that they relied on a more limited set of cues when making their judgments. However, it is unclear why this occurred. Farmers judged exhaustion to be significantly higher than agricultural students. Furthermore, farmers and applied animal science students experienced a greater negative emotional response to aggression when compared to agricultural students. Stakeholders differ in their knowledge, interests, values and norms regarding livestock. This can influence their perceptions of animal welfare [28–30] and may have contributed to the lower responses detected for agricultural students.

There were important differences between our comparison groups in their age and gender, and results confirmed that it was crucial to control for these differences in statistical analysis. Women on average gave higher scores than men for emotional response, judgment of severity and judgment of exhaustion, and female farmers were more motivated to intervene during fights than male farmers. This is consistent with evidence that, on average, females show more positive behaviors and attitudes toward animals; for example, by expressing greater empathy for animals [31,32], more opposition to animal use, and greater involvement with animal protection activities [33,34]. Furthermore, older farmers expressed greater motivation to intervene than younger farmers. Therefore, as age and years of experience were highly related, farmer experience may actually enhance their responses to fights. Age also influenced participant judgments of severity, which was determined by a subgroup of young participants who gave lower scores (agriculture students).

4.2. Animal Welfare Implications

The results indicate two important targets for implementing a change in practice. Firstly, farmers must be made aware of how to accurately determine the physical impact of aggression when they have not witnessed the fighting behavior. One useful tool for farmers to achieve this is scoring or estimating the number of visible lesions on affected animals. Counting lesions, or the simplified skin lesion score method, is an established and accurate measure of aggressive behavior which is regularly employed in research [21,35]. Secondly, researchers should calculate the economic and welfare impact of aggression as indicated by the lesions. If farmers observe the true frequency and intensity of fighting behavior on their own farm, and understand its impact on farm productivity, their motivation to control the issue is likely to increase. This advice regarding the recognition of aggression as a problem should be translated effectively to farmers and other stakeholders within the industry. Veterinarians are particularly important as they are the most valued source of information to farmers and highly influential in determining their animal welfare decisions [11,36].

4.3. Evaluation of Novel Methodology

The current study employed a novel methodology whereby perceptions of aggression exhibited in video clips were compared to objective, physiological measures of the severity of the welfare threat. There are many other animal welfare issues that have been resistant to change despite extensive research [3,37,38]. This novel methodology may represent a useful tool to assist in establishing stakeholder perceptions of these issues, in order to tailor successful interventions. There are a number of evaluative points regarding this technique, which are important to highlight here. First, ecological validity is limited by the use of video clips, which are removed from the real farm setting. However, the use of video footage with corresponding data allowed careful control over the experimental stimulus, which would not be possible in a real farm setting. Furthermore, by using pre-existing footage and data, we were able to avoid the use of animals for the purpose of the study (supporting the 3Rs of animal research: [39]), and maximize the utilization and impact of existing data. Second, although the self-reported measure of emotional response allowed for quick and easy data collection, this could be influenced by experimenter effects whereby participants might have responded in the way that they thought was being sought rather than how they really felt [40]. Future research could build upon the findings of this study by employing physiological measures of emotional response such as participant heart rate and galvanic skin response, which are less open to bias [41]. Third, the methodology allowed efficient data collection at pre-existing farmer discussion groups, as the procedure could be completed quickly with all group members participating simultaneously. Fourth, the method has quantified how well subjective scoring by observers compares to objective measures of the outcome of aggression (skin lesions and blood lactate). This makes the assumption that the objective measures are a closer approximation to the true experiences of the animal than the subjective scores but it is acknowledged that qualitative scores based on animal demeanor can also reflect welfare [42]. Finally, we did not include a non-professional group (e.g., consumers) as this was not within the aims of the current study. This study focused on how farmer exposure to pig aggression may have influenced their perceptions of aggression relative to others with experience of pigs (veterinarians) or with knowledge of agriculture but little experience of the pig industry (students). However, subsequent research examining consumer perceptions would be valuable.

5. Conclusions

Farmers were not desensitized to pig aggression. Farmers experienced a negative emotional response to seeing fights between pigs. They judged fights to be severe and exhausting for the animals and they were motivated to prevent them continuing. However, farmers and other observer groups underestimated the physical impact of aggression when they did not see the fight occurring and this may contribute to the limited uptake of methods to reduce aggression in commercial practice. Farmers are unlikely to see fights as frequently as they actually occur, and this likely limits their perception of aggression as a problem on their farm and their motivation to control aggression. In order to bridge the gap between research and practice, researchers must provide farmers with evidence of the economic and welfare impact of aggression as indicated by lesions, and farmers must be encouraged to estimate the impact of fights on their farm by counting lesions on the affected animals.

Supplementary Materials: Data for this project and the surveys are available at: https://osf.io/rh2sz/?view_only=d8b4c6f08e1448d487a4d204cf43077a.

Author Contributions: R.S.E.P. designed the experiment, conducted data collection, statistical analysis of the data, and writing of the original draft manuscript. S.P.T. provided supervision at all stages of the project, assisted with data collection, and reviewed and commented on the manuscript. I.C. provided supervision at all stages of the project and reviewed and commented on the manuscript. L.A.B. conducted data collection and reviewed and commented on the manuscript. F.A. provided supervision, and reviewed and commented on the manuscript.

Funding: This research was funded by Scotland's Rural College which receives funding from the Scottish Government.

Acknowledgments: The authors are grateful to the farmers who participated in this study. We thank the staff at Teagasc, Scotland's Rural College and AHDB Pork for help in the recruitment of farmers.

Conflicts of Interest: The authors declare no conflict of interest.

Appendix A

Stepwise selection of video clips

The final six video clips were selected using the following stepwise selection process:

(1) Video clips were collected based on data obtained from research carried out by an SRUC project in 2015. The descriptive statistics for the full dataset can be found in Table A1.

Table A1. Descriptive statistics regarding measures of relative change in skin lesions (number of lesions per pig) and blood lactate (mmol/L) following the fight, compared to before the fight, for the entire dataset ($n = 168$).

Measure	Lesion Score	Blood Lactate
Mean	49.89	6.43
Min	0	−2.80
Quartile 1	6.00	0.53
Quartile 2	30.00	4.20
Quartile 3	74.75	10.80
Max	354	21.20

(2) Pigs that displayed a negative relative change in blood lactate, and therefore displayed a reduction in blood lactate following the fight, were eliminated from analysis. In doing this, 26 pigs were eliminated. Descriptive statistics for the remaining dataset can be found in Table 1 (see Methods).

(3) Criteria for the video clips were set based on quartiles. Video criteria can be seen in Table A2.

Table A2. Description of criteria used to identify video clips (LQ = lower quartile; IQR = interquartile range; UQ = upper quartile).

Clip	Focal Pig		Non-Focal Pig	
	Lesion Score	Blood Lactate	Lesion Score	Blood Lactate
1	IQR	IQR	IQR	IQR
2	LQ	LQ	LQ	LQ
3	IQR	IQR	IQR	IQR
4	UQ	UQ	UQ	UQ
5	IQR	IQR	IQR	IQR
6	IQR	IQR	No criteria set	

(4) Contests that met the criteria for each clip were identified using the 'select cases' function in SPSS. The number of video clips to meet the criteria for each video clip can be seen in Table A3.

Table A3. Number of contests to meet the criteria for each video clip.

Clip	Number of Videos Identified
1	9
2	2
3	9
4	5
5	9
6	34

(5) All potential video clips were then watched to identify the six clips that best matched the selection criteria described in Table A4.

(6) Video clips were selected and Table 3 provides the exact objective measures for each of the pigs observed in the six video clips (see Methods).

Table A4. Description of video clip selection criteria and content (LS = lesion score).

Clip	Behaviour	Selection Criteria
1	*During fight:*	Both pigs engaged in mutual fighting behavior for a minimum period of 20 s. Both pigs remained in view of the camera for the duration of the 20 s. Both pigs obtained medium severity LS and lactate measures following the fight.
2	*After fight:*	During the 60 s immediately *after the fight ended*, both pigs were in view for a minimum period of 20 s. Both pigs displayed low severity LS and lactate measures following the fight.
3	*After fight:*	During the 60 s immediately *after* the fight, both pigs were in view for a minimum period of 20 s. Both pigs displayed medium severity LS and lactate measures following the fight.
4	*After fight:*	During the 60 s immediately *after the fight ended*, both pigs were in view for a minimum period of 20 s. Both pigs displayed high severity LS and lactate measures following the fight.
5	*During fight:*	During the 60 s immediately *before the fight ended*, both pigs engaged in mutual fighting behavior for a minimum period of 20 s. Both pigs remained in view of the camera for the duration of the 20 s. Both pigs obtained medium severity LS and lactate measures following the fight.
6	*During fight:*	During the 60 s immediately *before the fight ended*, one pig displayed bullying behavior whilst the other attempted to retreat for a minimum period of 20 s. Both pigs remained in view of the camera for the duration of the 20 s. The focal pig (recipient of bullying) obtained medium severity LS and lactate measures following the fight. No criteria were set for the non-focal (bullying pig) with regards to LS and lactate measures. This was because the two very different behaviors cannot be expected to result in the same measures.

References

1. Grandin, T. Transferring results of behavioral research to industry to improve animal welfare on the farm, ranch and the slaughter plant. *Appl. Anim. Behav. Sci.* **2003**, *81*, 215–228. [CrossRef]

2. Dwyer, C.M.; Conington, J.; Corbiere, F.; Holmoy, I.H.; Muri, K.; Nowak, R.; Rooke, J.; Vipond, J.; Gautier, J.M. Invited review: Improving neonatal survival in small ruminants: Science into practice. *Animal* **2016**, *10*, 449–459. [CrossRef]

3. Millman, S.T.; Duncan, I.J.H.; Stauffacher, M.; Stookey, J.A. The impact of applied ethologists and the International Society for Applied Ethology in improving animal welfare. *Appl. Anim. Behav. Sci.* **2004**, *86*, 299–311. [CrossRef]

4. Brockmyer, J.F. Playing Violent Video Games and Desensitization to Violence. *Child Adolesc. Psychiatr. Clin. N. Am.* **2015**, *24*, 65–77. [CrossRef] [PubMed]

5.   Fox, M.W. Empathy, Humaneness and Animal Welfare. In *Advances in Animal Welfare Science 1984*; Fox, M.W., Mickley, L.D., Eds.; Springer: Dordrecht, The Netherlands, 1985; pp. 61–73. [CrossRef]

6.   Carnagey, N.L.; Anderson, C.A.; Bushman, B.J. The effect of video game violence on physiological desensitization to real-life violence. *J. Exp. Soc. Psychol.* **2007**, *43*, 489–496. [CrossRef]

7.   Huesmann, L.R.; Moise-Titus, J.; Podolski, C.L.; Eron, L.D. Longitudinal relations between children's exposure to TV violence and their aggressive and violent behavior in young adulthood: 1977–1992. *Dev. Psychol.* **2003**, *39*, 201–221. [CrossRef] [PubMed]

8.   Latané, B.; Darley, J.M. *The Unresponsive Bystander: Why Doesn't He Help?* Appleton-Century Crofts: New York, NY, USA, 1970.

9.   McGlone, J.J. A quantitative ethogram of aggressive and submissive behaviours in recently regrouped pigs. *J. Anim. Sci.* **1985**, *61*, 559–565. [CrossRef]

10.   Camerlink, I.; Turner, S.P. Farmers' perceptions of aggression between growing pigs. *Appl. Anim. Behav. Sci.* **2017**, *192C*, 42–47. [CrossRef]

11.   Peden, R.S.E.; Akaichi, F.; Camerlink, I.; Boyle, L.A.; Turner, S.P. Factors influencing farmer willingness to reduce aggression between pigs. *Animals* **2019**, *9*, 6. [CrossRef]

12.   Greenwood, E.C.; Plush, K.J.; van Wettere, W.; Hughes, P.E. Hierarchy formation in newly mixed, group housed sows and management strategies aimed at reducing its impact. *Appl. Anim. Behav. Sci.* **2014**, *160*, 1–11. [CrossRef]

13.   Sherritt, G.W.; Graves, H.B.; Gobble, J.L.; Hazlett, V.E. Effects of Mixing Pigs during the Growing-Finishing Period. *J. Anim. Sci.* **1974**, *39*, 834–837. [CrossRef]

14.   Stookey, J.M.; Gonyou, H.W. The effects of regrouping on behavioural and production parameters in finishing swine. *J. Anim. Sci.* **1994**, *72*, 2804–2811. [CrossRef] [PubMed]

15.   Coutellier, L.; Arnould, C.; Boissy, A.; Orgeur, P.; Prunier, A.; Veissier, I.; Meunier-Salaun, M.C. Pig's responses to repeated social regrouping and relocation during the growing-finishing period. *Appl. Anim. Behav. Sci.* **2007**, *105*, 102–114. [CrossRef]

16.   Turner, A.I.; Hemsworth, P.H.; Tilbrook, A.J. Susceptibility of reproduction in female pigs to impairment by stress and the role of the hypothalamo-pituitary-adrenal axis. *Reprod. Fertil. Dev.* **2002**, *14*, 377–391. [CrossRef] [PubMed]

17.   Turner, A.I.; Hemsworth, P.H.; Tilbrook, A.J. Susceptibility of reproduction in female pigs to impairment by stress or elevation of cortisol. *Domest. Anim. Endocrinol.* **2005**, *29*, 398–410. [CrossRef]

18.   Einarsson, S.; Brandt, Y.; Lundeheim, N.; Madej, A. Stress and its influence on reproduction in pigs: A review. *Acta Vet. Scand.* **2008**, *50*. [CrossRef]

19.   de Groot, J.; Ruis, M.A.W.; Scholten, J.W.; Koolhaas, J.M.; Boersma, W.J.A. Long-term effects of social stress on antiviral immunity in pigs. *Physiol. Behav.* **2001**, *73*, 145–158. [CrossRef]

20.   Tuchscherer, M.; Puppe, B.; Tuchscherer, A.; Kanitz, E. Effects of social status after mixing on immune, metabolic, and endocrine responses in pigs. *Physiol. Behav.* **1998**, *64*, 353–360. [CrossRef]

21.   Turner, S.P.; Farnworth, M.J.; White, I.M.S.; Brotherstone, S.; Mendl, M.; Knap, P.; Penny, P.; Lawrence, A.B. The accumulation of skin lesions and their use as a predictor of individual aggressiveness in pigs. *Appl. Anim. Behav. Sci.* **2006**, *96*, 245–259. [CrossRef]

22.   Terlouw, E.M.C.; Arnould, C.; Auperin, B.; Berri, C.; Le Bihan-Duval, E.; Deiss, V.; Lefevre, F.; Lensink, B.J.; Mounier, L. Pre-slaughter conditions, animal stress and welfare: Current status and possible future research. *Animal* **2008**, *2*, 1501–1517. [CrossRef]

23.   Peden, R.S.E.; Turner, A.I.; Boyle, L.A.; Camerlink, I. The translation of animal welfare research into practice: The case of mixing aggression between pigs. *Appl. Anim. Behav. Sci.* **2018**, *204*, 1–9. [CrossRef]

24.   Ison, S.H.; Bates, R.O.; Ernst, C.W.; Steibe, J.P.; Siegford, J.M. Housing, ease of handling and minimising inter-pig aggression at mixing for nursery to finishing pigs as reported in a survey of North American pork producers. *Appl. Anim. Behav. Sci.* **2018**, *205*, 159–166. [CrossRef]

25.   Marchant-Forde, J.N.; Marchant-Forde, R.M. Minimizing inter-pig aggression during mixing. *Pig News Inf.* **2005**, *26*, 63–71.

26.   Turner, S.P.; Camerlink, I.; Baxter, E.M.; D'Eath, R.B.; Desire, S.; Roehe, R. Breeding for pig welfare: Opportunities and challenges. In *Advances in Pig Welfare*; Špinka, M., Ed.; Woodhead Publishing: Duxford, UK, 2018; pp. 399–414. [CrossRef]

27. Camerlink, I.; Turner, S.P.; Farish, M.; Arnott, G. Aggressiveness as a component of fighting ability in pigs using a game-theoretical framework. *Anim. Behav.* **2015**, *108*, 183–191. [CrossRef]
28. Te Velde, H.; Aarts, N.; Van Woerkum, C. Dealing with ambivalence: Farmers' and consumers' perceptions of animal welfare in livestock breeding. *J. Agric. Environ. Ethics* **2002**, *15*, 203–219. [CrossRef]
29. Vanhonacker, F.; Verbeke, W.; Van Poucke, E.; Tuyttens, F.A.M. Do citizens and farmers interpret the concept of farm animal welfare differently? *Livest. Sci.* **2008**, *116*, 126–136. [CrossRef]
30. Duijvesteijn, N.; Benard, M.; Reimert, I.; Camerlink, I. Same Pig, Different Conclusions: Stakeholders Differ in Qualitative Behaviour Assessment. *J. Agric. Environ. Ethics* **2014**, *27*, 1019–1047. [CrossRef]
31. Colombo, E.S.; Crippa, F.; Calderari, T.; Prato-Previde, E. Empathy toward animals and people: The role of gender and length of service in a sample of Italian veterinarians. *J. Vet. Behav. Clin. Appl. Res.* **2017**, *17*, 32–37. [CrossRef]
32. Hills, A.M. The Motivational Bases of Attitudes toward Animals. *Soc. Anim.* **1993**, *1*, 111–128. [CrossRef]
33. Heleski, C.R.; Mertig, A.G.; Zanella, A.J. Stakeholder attitudes toward farm animal welfare. *Anthrozoos* **2006**, *19*, 290–307. [CrossRef]
34. Herzog, H.A. Gender differences in human-animal interactions: A review. *Anthrozoos* **2007**, *20*, 7–21. [CrossRef]
35. Turner, S.P.; Roehe, R.; Mekkawy, W.; Farnworth, M.J.; Knap, P.W.; Lawrence, A.B. Bayesian analysis of genetic associations of skin lesions and behavioural traits to identify genetic components of individual aggressiveness in pigs. *Behav. Genet.* **2008**, *38*, 67–75. [CrossRef] [PubMed]
36. Alarcon, P.; Wieland, B.; Mateus, A.L.P.; Dewberry, C. Pig farmers' perceptions, attitudes, influences and management of information in the decision-making process for disease control. *Prev. Vet. Med.* **2014**, *116*, 223–242. [CrossRef] [PubMed]
37. Rushen, J. Changing concepts of farm animal welfare: Bridging the gap between applied and basic research. *Appl. Anim. Behav. Sci.* **2003**, *81*, 199–214. [CrossRef]
38. Dawkins, M.S. D.G.M. Wood-Gush Memorial lecture: Why has there not been more progress in animal welfare research? *Appl. Anim. Behav. Sci.* **2007**, *53*, 59–73. [CrossRef]
39. Russell, W.M.S.; Burch, R.L. *The Principles of Humane Experimental Technique*; Methuen: London, UK, 1959.
40. Rosenthal, R. *Experimenter Effects in Behavioral Research*; John Wiley: New York, NY, USA, 1976.
41. Lee, C.; Yoo, S.K.; Park, Y.; Kim, N.; Jeong, K.; Lee, B. Using neural network to recognize human emotions from heart rate variability and skin resistance. In Proceedings of the 2005 27th Annual International Conference of the IEEE Engineering in Medicine and Biology Society, Shanghai, China, 17–18 January 2006; Volumes 1–7, pp. 5523–5525. [CrossRef]
42. Wemelsfelder, F.; Hunter, A.E.; Paul, E.S.; Lawrence, A.B. Assessing pig body language: Agreement and consistency between pig farmers, veterinarians, and animal activists. *J. Anim. Sci.* **2012**, *90*, 3652–3665. [CrossRef]

Article

Text Mining Analysis to Evaluate Stakeholders' Perception Regarding Welfare of Equines, Small Ruminants, and Turkeys

Emanuela Dalla Costa [1,*], Vito Tranquillo [2], Francesca Dai [1], Michela Minero [1], Monica Battini [1], Silvana Mattiello [1], Sara Barbieri [1], Valentina Ferrante [3], Lorenzo Ferrari [3], Adroaldo Zanella [4] and Elisabetta Canali [5]

[1] Dipartimento di Medicina Veterinaria, Università degli Studi di Milano, 20133 Milano, Italy; francesca.dai@unimi.it (F.D.); michela.minero@unimi.it (M.M.); monica.battini@unimi.it (M.B.); silvana.mattiello@unimi.it (S.M.); sara.barbieri@unimi.it (S.B.)
[2] Istituto Zooprofilattico Sperimentale della Lombardia e dell'Emilia Romagna-Sezione diagnostica di Bergamo, 24125 Bergamo, Italy; vito.tranquillo@izsler.it
[3] Dipartimento di Scienze e Politiche Ambientali, Università degli Studi di Milano, 20133 Milano, Italy; valentina.ferrante@unimi.it (V.F.); lorenzo.ferrari@unimi.it (L.F.)
[4] Department of Preventive Veterinary Medicine and Animal Health (VPS), University of São Paulo, São Paulo, SP 05508-270, Brazil; adroaldo.zanella@usp.br
[5] Dipartimento di Scienze Agrarie e Ambientali - Produzione, Università degli Studi di Milano, Territorio, Agroenergia, 20133 Milano, Italy; elisabetta.canali@unimi.it
* Correspondence: emanuela.dallacosta@unimi.it

Received: 25 March 2019; Accepted: 4 May 2019; Published: 8 May 2019

Simple Summary: Consumers are currently more sensitive regarding the way animals are kept and handled and, in general, have increased their awareness towards animal welfare. There is a close link between welfare of animals and stakeholders' perception of their needs. This study aimed at investigating stakeholders' perception of the welfare of equines, small ruminants, and turkeys using a text mining approach to analyze their answers to open-ended questions. A total of 270 surveys were collected from respondents of 32 different countries. Independently from the species, respondents considered that an animal needs appropriate nutrition to be fit, healthy, and productive. The "environment" was considered particularly important for turkey stakeholders, whereas housing factors were relevant for goat stakeholders. Horse stakeholders also considered "exercise", "pasture" and "proper training" important. Although the sample was too small to analyze variation in welfare perception among stakeholders of different species, text mining analysis seems to be a promising method to investigate stakeholders' perception of animal welfare, as it emphasizes their real perception, without the constraints deriving by close-ended questions.

Abstract: Welfare of animals significantly depends on how stakeholders perceive their needs and behave in a way to favor production systems that promote better welfare outcomes. This study aimed at investigating stakeholders' perception of the welfare of equines, small ruminants, and turkeys using text mining analysis. A survey composed by open-ended questions referring to different aspects of animal welfare was carried out. Text mining analysis was performed. A total of 270 surveys were filled out (horses = 122, sheep = 81, goats = 36, turkeys = 18, donkeys = 13). The respondents (41% veterinarians) came from 32 different countries. To describe welfare requirements, the words "feeding" and "water" were the most frequently used in all the species, meaning that respondents considered the welfare principle "good feeding" as the most relevant. The word "environment" was considered particularly important for turkeys, as well as the word "dry", never mentioned for other species. Horses stakeholders also considered "exercise" and "proper training" important. Goat stakeholders' concerns are often expressed by the word "space", probably because goats are often intensively managed in industrialized countries. Although the sample was too small to be representative, text mining analysis seems to be a promising method to investigate stakeholders' perception of animal welfare, as it emphasizes their real perception, without the constraints deriving by close-ended questions.

Keywords: animal welfare; stakeholder perception; text mining; horse; donkey; goat; sheep; turkey

---

1. Introduction

Citizen concern regarding animal welfare has been increasing in many parts of the world [1]; in general, consumers are more sensitive regarding the way farm animals are kept and handled [2–4]. Harper and colleagues reported that "animal welfare is used by consumers as an indicator of important product attributes, such as safety and the impact on health" [3]. However, the welfare of animals significantly depends on how stakeholders perceive their needs and behave with them. Indeed, stakeholders' perception and animal welfare are closely linked because if stakeholders do not perceive that their animals have certain needs, it is less likely that they will protect them. Different stakeholders, such as farmers, veterinarians, and animal owners, may have a different perception regarding what an animal needs to be fit, healthy, and productive. Therefore, identifying different stakeholders' perception is necessary to lead to improvement of animal welfare at farm level.

Stakeholders' engagement was fostered during the AWIN (Animal Welfare Indicators) research project, an EU-funded research project aimed to improve animal welfare through the development of practical on-farm animal welfare assessment protocols [5]. During the AWIN project (2011–2015), stakeholders' input was proactively sought in several participatory activities with the aim to increase the acceptability of the project outcomes, but also to identify potential barriers to the practical application of the protocols [6,7]. As a first step of the AWIN project, stakeholders' opinion on animal welfare was gained through an open-ended online survey [6,8,9].

Surveys have often been used to assess perceptions on welfare issues in different animal species, such as small ruminants [10,11], cattle [2,10] and horses [12–19]. The use of online surveys in animal welfare science presents several advantages: They are quick and easy to create, their creation and distribution is cheap (they can be shared via email, websites, and social networks), and they allow researchers to reach a wide number of respondents all around the world.

The response rate and a possible bias in the sample of respondents (e.g., people without internet access cannot answer) are recognized issues when using online surveys to collect stakeholders' opinion [20]. Other possible problems are: incomplete questionnaires resulting in incomplete data collection [21], and surveys with answers provided (closed-ended surveys) could not reflect all the possible answers of the respondents [1]. Open-ended questions may overcome this last problem, as they allow respondents to use their own words to express their feelings, attitudes and understanding of the subject and to include more information and details [1]. Therefore, surveys based on open-ended questions permit researchers to better understand the respondents' true feelings on a topic and/or an issue. One of the possible difficulties of open-ended questionnaires is transforming the answers in statistical data, for further analysis. In order to solve this issue, new statistical techniques were developed, such as opinion mining [22,23] and text mining [24,25].

Text mining analysis is an approach to text analysis that has been frequently used in different scientific fields, from economics to history and from social science to biology [26–30]. The term text mining refers to a "process of distillation of useful information from a text" [26]. It is a set of quantitative methods that use the words present in a text as "units" of analysis. It applies to different types of texts (e.g., books, tweets, mails, open-ended surveys). The text is first "tokenized", i.e., reduced to a sequence of simple terms deprived of those words (stop words) that serve the sensible and comprehensible definition of a period (e.g., article, adverb, number, punctuation). Stop words are commonly found in documents; they add little value to understanding the meaning of a sentence.

For instance, in the sentence: "we have to ensure movement freedom, it's compulsory" the words: "we, have, to, it's" are considered stop words. This document called Corpus is then transformed into a term document matrix (TDM), a matrix that shows for each single term how many times it appears in a single document. From this matrix, all the types of textual analysis are obtained, including: Word frequency (how many times the same word appears in the same textual document), word association (it refers to the term pairings), text clustering (the application of cluster analysis to textual documents), sentiment analysis (systematically identify, extract, quantify, and study affective states and subjective information), and many more [25,26,31].

This work is a first attempt to approach the issue of stakeholders' perception of animal welfare using text mining analysis. The study aimed at investigating, through open-ended questions, stakeholders' perception of what equine, small ruminants, and turkeys need to be fit, healthy, and productive, and which signs would be observed using a text mining approach to analyze the answers.

2. Materials and Methods

2.1. Web Survey

For each species (horses, donkeys, sheep, goats, and turkeys), a survey composed of 14 open-ended questions (max. 150 characters for each answer) referring to different aspects of animal welfare was carried out. The target population included in the present study refers in particular to farmers, veterinarians, and animal owners in order to collect information regarding animal welfare of different species (horses, donkeys, sheep, goats, and turkeys). To take the survey, respondents were required to be over the age of 18. The survey, translated in five languages (English, French, Italian, Portuguese, and Spanish), was published on a web platform (http://www.questionari.unimi.it/awin/), and it was freely accessible for 15 months (December 2012–March 2014). The survey link was anonymous, as this option was reported to be the easiest way to disseminate the survey and gather answers. The web link was shared via email, social networks, and hosted in websites of several academic and international organizations (such as FAO, International Society for Equitation Science, Italian Equestrian Federation, Istituto Zooprofilattico Sperimentale dell'Abruzzo e del Molise "Giuseppe Caporale") in order to reach different stakeholders.

Questions were related to four main topics: (1) needs (what do animals need to be fit, healthy, and productive?); (2) behavior (how might an animal behave/react in response to the following situations: noise, isolation, presence of known/unknown animal/person?) (3) emotions (how might an animal feel in response to the following situations: noise, isolation, presence of known/unknown animal/person?); and (4) welfare indicators (looking at your neighbor's animals, which signs would you observe to assess: accommodation, feeding, health, manifestation of normal and abnormal behavior?). All the questions included in the survey are presented in Supplementary Table S1. In order to investigate stakeholders' perception of welfare, in this manuscript, we focused only on welfare and its measures as reported by stakeholders; for this reason, only the answers referring to the six questions related to topics 1 and 4 (Table 1) are presented. For topic 4, due to the number of questionnaires collected, only horses were considered.

Table 1. List of the questions selected from the questionnaire whose results are reported in the manuscript.

Question	Type of Question
In your opinion, what do sheep/goats/turkeys/donkeys/horses need to be fit, healthy and productive?	Open text (max 150 characters)
Looking at your neighbour's horse, which signs would you observe to assess [1]	
- The conditions of accommodation [1]	Open text (max 150 characters)
- Feeding conditions [1]	Open text (max 150 characters)
- Health conditions [1]	Open text (max 150 characters)
- The manifestation of normal behaviour [1]	Open text (max 150 characters)
- The manifestation of abnormal behaviour [1]	Open text (max 150 characters)

[1] Only results for horses are reported for this topic.

2.2. Text Mining Analysis

Data were imported into Microsoft Excel (Microsoft Corporation, Redmond, USA, 2010) and then a Text mining analysis was performed using R statistical software [32] by "tm" R package version 0.7-6 (https://CRAN.R-project.org/package=tm) [31]. A "document" was defined as every answer to single questions. A pre-process of answers was undertaken, to get the Corpus for each single question: English translation from other languages (Italian, Spanish, Portuguese, and French), exclusion of words according to a preselection list of so-called "stop-words", cleaning (e.g., numbers, punctuation). For the question "In your opinion, what do animals need to be fit, healthy, and productive?" the term document matrix was performed, then a frequency words analysis was carried out, meaning that the most used words were identified. For the questions "Looking at your neighbour's animal, which signs would you observe to assess accommodation, feeding, health, manifestation of normal and abnormal behaviour?", only answers for horses were considered in the present paper as, for the other species, the number of respondents was too limited to allow meaningful analysis. The TDM initially was performed, then a frequency words analysis, followed by the term association analysis was carried out. Words with similar meaning (i.e., feed, food and feeding, etc.) were merged together. In the text mining analysis, terms association is like a correlation; it means that there is a co-presence of terms in the same document. The analysis of association therefore measures the co-presence between a term of interest defined by us and all the other words. Association study refers to the term pairings (when the term x appears, the other term y is associated with it), and it is not directly related to the frequency of terms. Unlike statistical correlation, text mining association ranges between 0 and 1: Score 1 means that two words always appear together in the documents, while a score approaching 0 means the terms seldom appear in the same document. The more often two terms appear in the same document, the stronger their association is. For each question, the association study was performed between the most frequent word and all the other terms. Associations with a value greater than or equal to 0.20 were considered relevant.

3. Results

A total of 270 surveys were properly filled out (122 for horses, 81 for sheep, 36 for goats, 18 for turkeys, and 13 for donkeys). The respondents came from 32 different countries spread in the five continents (Table 2), but the majority of them were from European countries (70%), among whom 41% were Italian.

Table 2. Percentage (%) of respondents for each geographic area.

Area	Respondents (N = 270)
Europe	70%
North America	12%
South and Central America	4%
Oceania	9%
Asia	3%
Africa	1%

In general, stakeholders were mainly veterinarians (41%), but some differences in stakeholders' role were observed among species. For example, for horses, most of the respondents were private owners and trainers (34% and 13%, respectively), whereas for goats, they were mainly farmers (47%); for turkeys, contract farmers (11%) were also included (Table 3).

Table 3. Stakeholder roles (%) for each species included in the survey (N = 270).

	Sheep (N = 81)	Goats (N = 36)	Horses (N = 122)	Donkeys (N = 13)	Turkeys (N = 18)	Overall (N = 270)
Veterinarian	56%	33%	35%	23%	50%	41%
Farmer	23%	47%	7%	62%	28%	21%
Contract farmer	-	-	-	-	11%	1%
Technician	21%	20%	11%	15%	11%	16%
Owner	-	-	34%	-	-	15%
Trainer	-	-	13%	-	-	6%

Gender of respondents was balanced in all species (Table 4), except for horses (85% women). Most of the respondents (77%) were aged between 31 and 60 years, while 17% and 6% of respondents were under 30 and over 60, respectively.

Table 4. Gender of respondents (%) for each species included in the survey (N = 270).

	Sheep (N = 81)	Goats (N = 36)	Horses (N = 122)	Donkeys (N = 13)	Turkeys (N = 18)	Overall (N = 270)
Male	60%	58%	15%	31%	56%	38%
Female	40%	42%	85%	69%	44%	62%

3.1. In Your Opinion, What do Animals Need to be Fit, Healthy, and Productive?

The results of the word frequency analysis on the answers of the stakeholders of each species to the question "In your opinion, what do animals need to be fit, healthy and productive?" are reported in Figure 1. Words are presented according to the four welfare principles identified by Welfare Quality® [33].

To describe welfare requirements, the words "feeding" and "water" were the most frequently used in all the species, meaning that respondents considered the welfare principle "good feeding" as the most relevant to guarantee fitness, health, and production. Horse stakeholders used the word "forage" to highlight the need of fiber in the horse diet. As for the principle "good housing", "clean", "shelter", "environment", "space", and "bedding/litter" appeared to be of primary importance. The word "environment" was considered particularly important for turkeys, as well as the word "dry", which is never mentioned by stakeholders of other species. Horse stakeholders also considered "exercise" and the presence of "pasture" important. The principle "good health" was represented by words like "care" and "health". Interestingly, for donkeys, the word "deworming" was used. The welfare principle "appropriate behaviour" was addressed by equine and turkey stakeholders using the words "company/social"; for horses, "proper training" was also mentioned. Interestingly, small ruminants' stakeholders did not use any word referring to this welfare principle. Stakeholders of different species, except horses, frequently used the word "management", which is a general term, not clearly linked with a specific welfare principle.

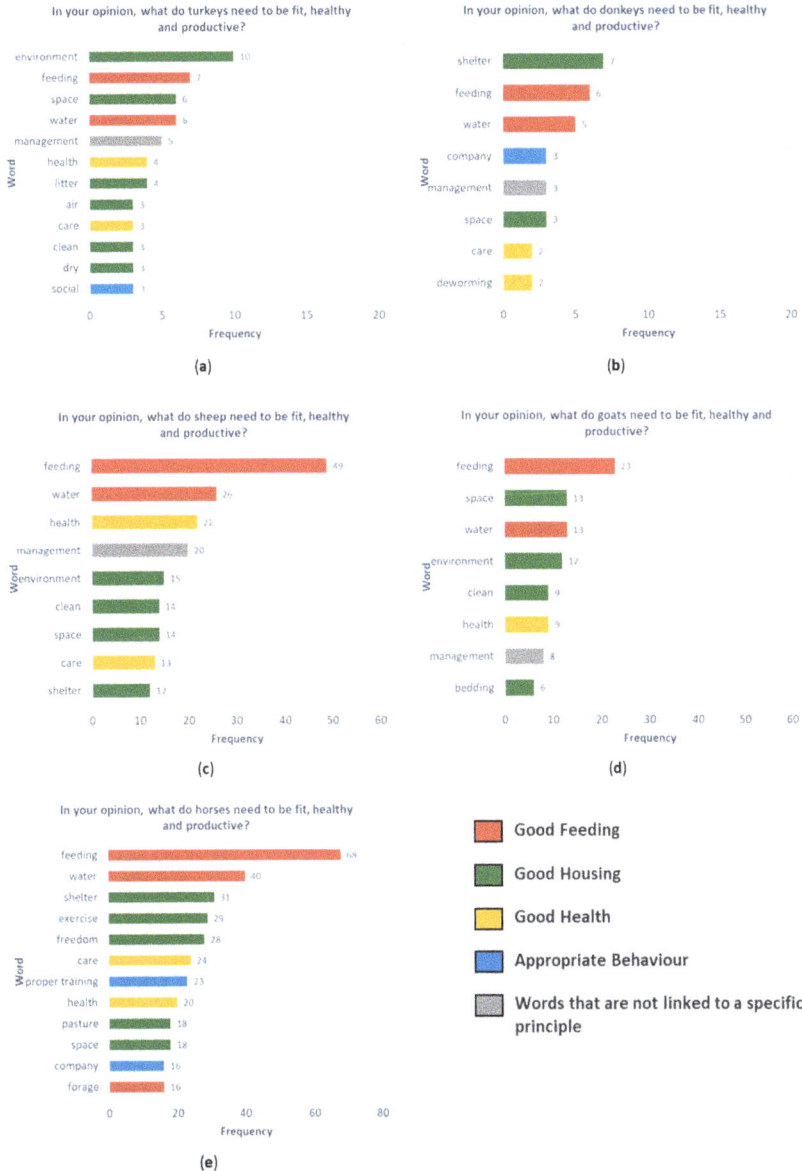

Figure 1. The graphs report the frequency of words used to answer the question "In your opinion, what do turkeys (**a**), donkeys (**b**), sheep (**c**), goats (**d**), and horses (**e**) need to be fit, healthy, and productive?" To highlight the association between each word and welfare principles [34], different colors are used: Red for "good feeding", green for "good housing", yellow for "good health", blue for "appropriate behaviour"; grey is used for the words that are not linked to a specific principle.

3.2. Looking at Your Neighbor's Horse, Which Signs would You Observe to Assess Accommodation, Feeding, Health, Manifestation of Normal and Abnormal Behaviour?

The results of the word frequency analysis of these five questions for horses are reported in Figure 2. An example of answer to the question "looking at your neighbor's horse, which signs would you observe to assess the condition of accommodation?" was: "Presence and cleanliness of bedding, repair and maintenance of buildings, pastures and equipment". The most frequent word used by horse stakeholders to evaluate the conditions of accommodation (Figure 2a) was "clean", followed by "bedding", "box", "shelter", and "water". The text mining association showed that the term "clean" was associated with words referring to resource-based measures such as "floors" (0.33), "manger" (0.33), "drinkers" (0.32), "walls" (0.32), and "water" (0.31).

An example of an answer to the question "Looking at your neighbour's horse, which signs would you observe to assess the feeding conditions?" was: "A lot of roughage. Free access to clean hay that is not dusty or mouldy. Or access to grazing". The words "roughage", "water", "quality", and "clean" were the most used by horse stakeholders to evaluate the feeding conditions (Figure 2b). The term "roughage" was associated with the words: "access" (0.52), "available" (0.32), and "daily" (0.32). Additionally, for this question, most of the words used by stakeholders referred to resource-based measures; three words (body score, fat, and coat) referred to animal-based measures.

"Body score and coat shine/condition. Horses posture, movement (lameness)" is an example of answer to the question "Looking at your neighbour's horse, which signs would you observe to assess health conditions?" The assessment of health condition (Figure 2c) was described by horse stakeholders with words such as "coat", "body score", "eyes", and "hooves". The term "coat" was associated with the words: "vet" (0.29), "clean" (0.26), and "worming" (0.26).

An example of an answer to the question "Looking at your neighbour's horse, which signs would you observe to assess the manifestation of normal behaviour?" was: "Horses are relaxed but still interested in their surroundings, they are eating hay or grazing, socializing together and allo-grooming too". The most frequent word used (Figure 2d) was "group", which can be considered as a management-based measure (even though it is strictly related to social behavior, which is an animal-based measure), followed by "grazing", "calm", "relaxed", and "playing". Most of the words used by the stakeholders referred to animal-based measures. The term "group" was associated with the words: "happy" (0.38) and "curious" (0.30).

Finally, an example of answer to the question "Looking at your neighbour's horse, which signs would you observe to assess the manifestation of abnormal behaviour?" was: "Biting on doors, walls, posts. Kicking at walls. Constant or frequent vocalizing. Weaving, windsucking, pacing up and down, holding head high, eyes wide open, withdrawn, uninterested in surroundings, not interested in food, laying ears flat back when people approach, sudden or explosive movements". Abnormal behavior (Figure 2e) was described using words referring to animal-based measures such as "weaving", "aggression", and "stereotypies". The term "weaving" was associated with the word "tic" (0.65), "eat" (0.31), and biting (0.27).

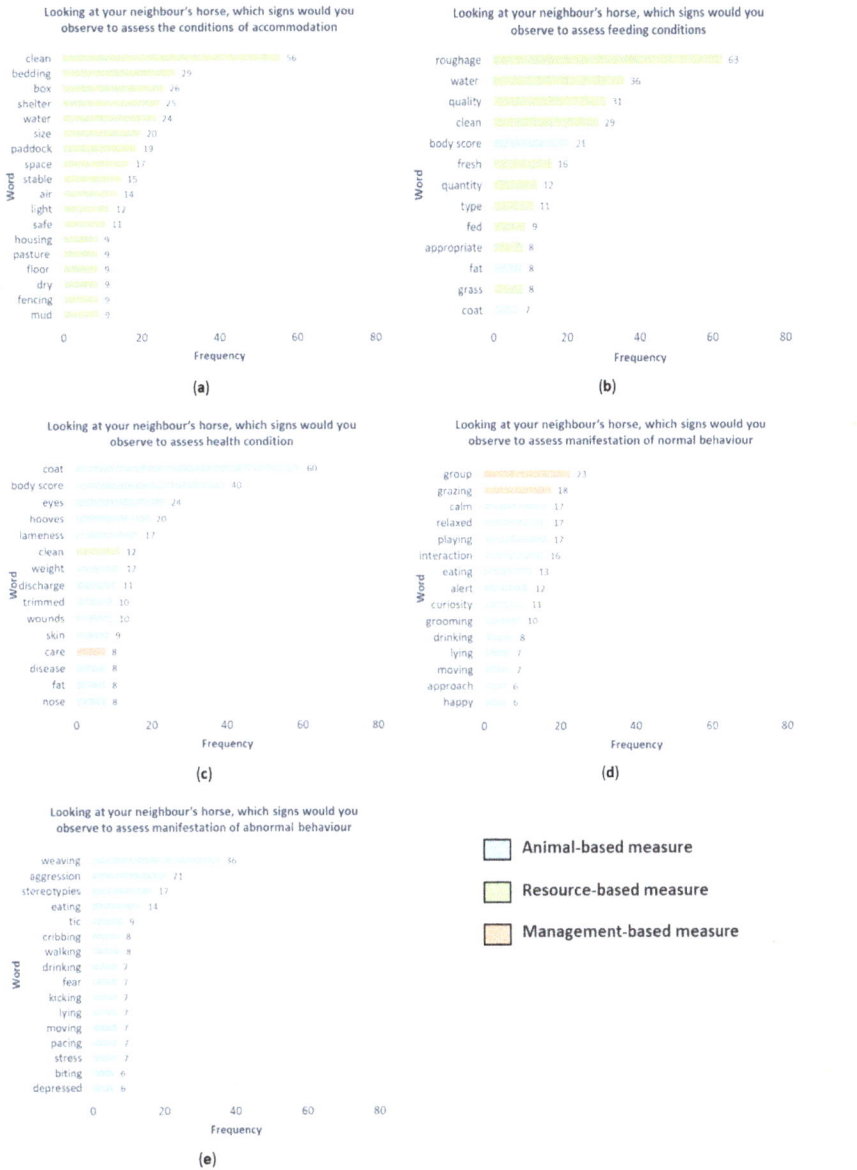

Figure 2. The graphs report the frequency of words for each of the five questions: "Looking at your neighbour's horse, which signs would you observe to assess the conditions of accommodation (**a**), feeding conditions (**b**), health condition (**c**) manifestation of normal behaviour, (**d**) and manifestation of abnormal behaviour (**e**)?". Different colors highlight the type of measure (animal-, resource- or management-based).

4. Discussion

The aim of the present study was to investigate stakeholders' perception of welfare of equids, small ruminants, and turkeys. Most surveys were filled out by horse stakeholders that seemed to present a higher interest in welfare compared to other species. Considering the geographic area and role of respondents, it could be hypothesized that survey sampling could have been affected by the country of origin and scientific background of the AWIN partners. Answers collected through an open-ended survey were analyzed using a text mining approach. To the authors' knowledge, this is the first time that text mining analysis has been applied to investigate which words were the most used by stakeholders of different species to describe animal welfare. The results suggest that respondents' perception regarding what an animal needs to be fit, healthy, and productive is primarily linked with the principle "good feeding" [33]. In fact, words like "feeding" and "water" were the most used by respondents. It is well known that the importance of appropriate nutrition is paramount to safeguarding welfare of farm animals [35], as it is a basic need to be healthy and productive. Feeding was also considered one of the most important attributes indicated in an Australian survey for sheep and goats by farmers and veterinarians [36], who represent the majority of our respondents for these two species. Welfare indicators related to nutrition were also considered among the most important welfare indicators in a survey on stakeholders' perception of the welfare of extensively managed sheep [37]; in the same study, water provision was also included among the most important factors affecting welfare. For small ruminants, water was one of the most frequently used words for both species: For sheep, this is probably due to the fact that they are often managed in extensive systems, also in Mediterranean countries, where scarcity of water may actually be a limiting factor for animal welfare. For this reason, new indicators were developed and included in the AWIN welfare assessment protocol for goats [38] and in the AWIN welfare assessment protocol for sheep [39]. Unfortunately, although stakeholder differences in gender and role, as well as many other factors, may significantly affect the perception of animal welfare [33], the limited sample size of this survey did not allow an investigation in this sense, and the following results are pooled for all the questionnaires.

For horses, together with "feeding" and "water", stakeholders recognized that providing "forage" is also necessary. The recognition of the importance of foraging is shown also by the use of other words such as "pasture", "grazing", "roughage", and "grass". Horses naturally spend the majority of their time grazing [40,41], and it is reported that foraging food with high fiber content is not only a physical need (80% of the time, the gastric tract should be filled), but also a psychological need for them [42]. In term of signs to evaluate feeding conditions, horse stakeholders referred mostly to resource-based indicators, emphasizing how the quality of feeding ("quality", "clean", "fresh", and "appropriate") is important. Nutrition is a critical component of horse health [43], and it is reported that good quality of feeding could be a reasonable target for decreasing the prevalence of pulmonary diseases in horses [44].

Words referring to the principle "good housing", such as "shelter", "space", "clean", "air", and "bedding", were frequently used by stakeholders of different species. This principle seems to be paramount for turkeys' stakeholders: "environment" was the word most used. The fact that the environment is a key factor in maintaining poultry welfare is well known: Jones and colleagues demonstrated that a good ventilation, to control temperature and humidity, is even more important than density in determining poultry welfare [45]. Turkeys may show large behavioral adjustments as a response to inadequate environmental conditions, with consequences including gait deterioration, higher frequency of injuries, increased aggression levels or feather pecking occurrences [46].

Factors referring to "good housing" seemed more important for goats than for sheep stakeholders. In fact, words referring to this principle in goats are ranked higher than in sheep (Figure 1c,d), and they are cited for a total of 40 times in 36 questionnaires, whereas in sheep, they are mentioned only 55 times in 81 questionnaires. This is probably because in most industrialized countries, goats are often intensively managed, and they are kept indoors with only occasional access to pasture [47], and therefore, stakeholders' concerns are often expressed by the word "space". To evaluate the condition of the accommodation, horse stakeholders referred only to resource-based measures. In particular, they referred to "clean floors and walls", "clean manger", and "clean drinkers and water". A lack of animal-based measures referring to the principle "good housing" of equines was already recognized [48]; for this reason, most of the indicators included in the AWIN welfare assessment protocol for horses [49] and in the AWIN welfare assessment protocol for donkeys [50] were resource-based measures such as "bedding" and "box dimensions".

For the principle "good health", stakeholders used the same words ("health" and "care") across species. For sheep, and even less for goats, these terms were not of primary importance. This is in contrast with the results of a survey carried out in Australia [36], where a health issue, parasite control, reached the highest importance value among stakeholders. However, it has to be noticed that the issue "parasite control" was suggested to the stakeholders from a fixed list, whereas in our case, there was no fixed list, and stakeholders were totally free to generate their own terms and to express their personal concerns. Donkey stakeholders considered also "deworming" as an important issue to address. In fact, it is well known that gastrointestinal parasites are common in equines, and they affect both their health and welfare [51–53]. Hence, the word "deworming" was used in association with the term "coat" by horse stakeholders to describe health conditions; this reflects that the presence of gastrointestinal parasites could affect the overall health of the subject.

Interestingly, no word was generated by small ruminant stakeholders referring to appropriate behavior. This principle was mentioned only by stakeholders of turkeys, donkeys, and horses. Particular emphasis was given to social behavior, with the words "company" and "social" used by equine and turkey stakeholders, respectively. Offering the possibility to horses to interact with conspecifics (e.g., "group" and "interaction") was highlighted by respondents when asked to evaluate normal horse behavior. Despite the fact that stakeholders recognized that horses are highly social species, and thus, contact with conspecifics plays an important role in their welfare [54–57], single boxes, which restrict horses' possibility of socialization, are still a standard housing system [58]. Another important aspect for horse stakeholders was "proper training". Horses are kept for very different purposes, ranging from sport and leisure to use in animal-assisted therapies; for these reasons, they are exposed to training. Choosing appropriate training methods together with proper equipment is fundamental to guaranteeing not only good performances but also welfare of sport horses. In fact, several studies suggest that the use of inappropriate equipment such as bit, reins or nosebands may cause pain and stress [59–61] affecting horse response to training, reducing performances and causing the manifestation of potentially dangerous behaviors, and impairing horse welfare.

The major constraint of this work, in common with other online questionnaires [20], is the limited sample: Although the survey was widely publicized, the number of respondents was too small to allow meaningful analysis of differences in welfare perceptions among different species' stakeholders and among individual stakeholders' characteristics within each species. In particular, for some species, such as turkeys and donkeys, few answers were collected. One possible explanation could be related to the advertising used to spread out the questionnaire. For example, of the 41 million donkeys present in the world, it is estimated that 96% are owned in developing countries [62], where internet connection is not so widely spread as in Europe. As for turkeys, another possible explanation is that the poultry supply chain consists of large numbers of animals concentrated in very few stakeholders. Furthermore, turkey stakeholders do not seem so willing to share their ideas regarding poultry welfare: This can be the reason most of the answers related to turkeys came from Italy, where AWIN researchers have been working to build a good relationship with the stakeholders and to focus their attention toward animal welfare. Finally, the lack of interest of the poultry sector in issues related to animal welfare could be linked to the fact that turkeys are seen much more in terms of productivity with respect to other species. However, increasing numbers are demanding animal welfare assurances for the products the poultry industry produces [63]. The industry must address these concerns or risk alienating clients and customers.

5. Conclusions

Stakeholders' involvement is fundamental for any action intended to improve animal welfare; this work portrays the stakeholders' perception, highlighting the need of proper dissemination of scientific knowledge. The words used by stakeholders were related to indicators included in the AWIN welfare assessment protocols. In stakeholders' opinion, small ruminants need good feeding, but also a good environment, such as clean bedding, shelter, and enough space to be fit, healthy, and productive. For turkey stakeholders, the environment (dry and clean litter, clean air, space) is paramount to guaranteeing good welfare. Finally, stakeholders consider a horse fit, healthy, and productive if feeding is appropriate (including the possibility to eat roughage), at the same time allowing the possibility to interact with conspecifics and spend time at pasture. A limitation of this study was that the sample was too small to be representative of different stakeholders' perception worldwide. However, text mining analysis seems to be a promising method to investigate stakeholders' perception of animal welfare, as it emphasizes their real perception, without the constraints derived by close-ended questions. With a larger sample, it would be possible to undertake a more detailed analysis of welfare perception variations among species and to explore underlying reasons for these differences.

Supplementary Materials: The following are available online at http://www.mdpi.com/2076-2615/9/5/225/s1, Table S1: List of the questions reported in the questionnaire.

Author Contributions: Conceptualization, M.M., S.M., S.B., V.F., A.Z., and E.C.; methodology, V.T., M.M., S.M., S.B., V.F., A.Z., and E.C.; formal analysis, V.T.; data curation, E.D.C., F.D., M.B., and L.F.; writing—original draft preparation, E.D.C., F.D., and M.B.; writing—review and editing, E.D.C., V.T., F.D., M.M., M.B., S.M., S.B., V.F., L.F., A.Z., and E.C.; supervision, M.M., S.M., S.B., V.F., A.Z., and E.C.; project administration, S.M., A.Z., and E.C.; funding acquisition, A.Z.

Funding: This research was funded by the Animal Welfare Indicators (AWIN) Project, FP7-KBBE-2010-4, Grant n. 266213.

Acknowledgments: The authors wish to thank all the organizations which hosted and promoted the web survey, and Daniela Battaglia for her great contribution to the initial idea of the survey. Authors are grateful to CTU (the university eLearning and audio-visual production Center of Milan) for their precious technical support.

Conflicts of Interest: The authors declare no conflict of interest.

Ethical Statement: Ethical approval was not deemed necessary as: Only adults (>18 years) were involved; all respondents took part at the questionnaire on a voluntary basis, and participants could withdraw at any time; information regarding the purpose, intent, motivation, sponsoring organizations, potential use of data, and methods of data collection were provided to participants prior entering the questionnaire; all data were collected anonymously, and it was not possible to identify the respondents in the raw research data.

References

1. Buddle, E.A.; Bray, H.J.; Ankeny, R.A. "I Feel Sorry for Them": Australian Meat Consumers' Perceptions about Sheep and Beef Cattle Transportation. *Animals* **2018**, *8*, 171. [CrossRef]
2. Spooner, J.M.; Schuppli, C.A.; Fraser, D. Attitudes of Canadian citizens toward farm animal welfare: A qualitative study. *Livest. Sci.* **2014**, *163*, 150–158. [CrossRef]
3. Harper, G.C.; Makatouni, A. Consumer perception of organic food production and farm animal welfare. *Br. Food J.* **2002**, *104*, 287–299. [CrossRef]
4. Gracia, A. The determinants of the intention to purchase animal welfare-friendly meat products in Spain. *Anim. Welf.* **2013**, *22*, 255–265. [CrossRef]
5. Zanella, A. AWIN—Animal Health and Welfare—FP7 Project. *Impact* **2016**, *2016*, 15–17. [CrossRef]
6. Dalla Costa, E.; Dai, F.; Lebelt, D.; Scholz, P.; Barbieri, S.; Canali, E.; Zanella, A.J.; Minero, M. Welfare assessment of horses: The AWIN approach. *Anim. Welf.* **2016**, *25*, 481–488. [CrossRef]
7. Battini, M.; Stilwell, G.; Vieira, A.; Barbieri, S.; Canali, E.; Mattiello, S. On-Farm Welfare Assessment Protocol for Adult Dairy Goats in Intensive Production Systems. *Animals* **2015**, *5*, 934–950. [CrossRef]
8. Battini, M.; Barbieri, S.; Canali, E.; Dai, F.; Dalla Costa, E.; Ferrante, V.; Ferrari, L.; Mattiello, S.; Minero, M. Outcomes of a web-survey for collecting stakeholders' opinion on welfare requirements for sheep, goats, turkeys, donkeys, and horses. *Ital. J. Anim. Sci.* **2017**, *16*, 53.
9. Dai, F.; Tranquillo, M.; Dalla Costa, E.; Barbieri, S.; Canali, E.; Minero, M. Outcomes of a web-survey to collect stakeholders' opinion on welfare requirements for horses. In *Book of Abstract of the 69th Annual Meeting of the European Federation of Animal Science (EAAP)*; Wageningen Academic Publishers: Wageningen, The Netherlands, 2018; p. 505.

10. Doughty, A.K.; Coleman, G.J.; Hinch, G.N.; Doyle, R.E. Stakeholder perceptions of welfare issues and indicators for extensively managed sheep in Australia. *Animals* **2017**, *7*, 28. [CrossRef] [PubMed]
11. Heleski, C.R.; Mertig, A.G.; Zanella, A.J. Stakeholder attitudes toward farm animal welfare. *Anthrozoos* **2006**, *19*, 290–307. [CrossRef]
12. Li, X.; Zito, S.; Sinclair, M.; Phillips, C.J.C. Perception of animal welfare issues during Chinese transport and slaughter of livestock by a sample of stakeholders in the industry. *PLoS ONE* **2018**, *13*, e0197028. [CrossRef] [PubMed]
13. Collins, J.; Hanlon, A.; More, S.J.; Wall, P.G.; Duggan, V. Policy Delphi with vignette methodology as a tool to evaluate the perception of equine welfare. *Vet. J.* **2009**, *181*, 63–69. [CrossRef] [PubMed]
14. Collins, J.A.; Hanlon, A.; More, S.J.; Wall, P.G.; Kennedy, J.; Duggan, V. Evaluation of current equine welfare issues in Ireland: Causes, desirability, feasibility and means of raising standards. *Equine Vet. J.* **2010**, *42*, 105–113. [CrossRef] [PubMed]
15. Padalino, B.; Henshall, C.; Raidal, S.L.; Knight, P.; Celi, P.; Jeffcott, L.; Muscatello, G. Investigations Into Equine Transport-Related Problem Behaviors: Survey Results. *J. Equine Vet. Sci.* **2016**, *48*, 166–173. [CrossRef]
16. Padalino, B.; Rogers, C.; Guiver, D.; Bridges, J.; Riley, C.; Padalino, B.; Rogers, C.W.; Guiver, D.; Bridges, J.P.; Riley, C.B. Risk Factors for Transport-Related Problem Behaviors in Horses: A New Zealand Survey. *Animals* **2018**, *8*, 134. [CrossRef] [PubMed]
17. Padalino, B.; Raidal, S.L.; Hall, E.; Knight, P.; Celi, P.; Jeffcott, L.; Muscatello, G. Risk factors in equine transport-related health problems: A survey of the Australian equine industry. *Equine Vet. J.* **2016**, *49*, 507–511. [CrossRef]
18. Padalino, B.; Raidal, S.L.; Hall, E.; Knight, P.; Celi, P.; Jeffcott, L.; Muscatello, G. A Survey on Transport Management Practices Associated with Injuries and Health Problems in Horses. *PLoS ONE* **2016**, *11*, e0162371. [CrossRef]
19. Lee, J.; Houpt, K.; Doherty, O. A survey of trailering problems in horses. *J. Equine Vet. Sci.* **2011**, *21*, 235–238. [CrossRef]
20. Christley, R.M. Questionnaire survey response rates in equine research. *Equine Vet. J.* **2016**, *48*, 138–139. [CrossRef]
21. Robert, M.; Hu, W.; Nielsen, M.K.; Stowe, C.J. Attitudes towards implementation of surveillance-based parasite control on Kentucky Thoroughbred farms—Current strategies, awareness and willingness-to-pay. *Equine Vet. J.* **2015**, *47*, 694–700. [CrossRef]
22. Liu, B.; Zhang, L. A survey of opinion mining and sentiment analysis. In *Mining Text Data*; Springer: Boston, MA, USA, 2012; Volume 9781461432, pp. 415–463.
23. Pang, B.; Lee, L. Opinion mining and sentiment analysis. *Comput. Linguist.* **2009**, *35*, 311–312. [CrossRef]
24. Li, H.; Yamanishi, K. Mining from open answers in questionnaire data. In Proceedings of the Seventh ACM SIGKDD International Conference on Knowledge Discovery and Data Mining—KDD '01, San Francisco, CA, USA, 26–29 August 2001; pp. 443–449.
25. Feldman, R.; Sanger, J. The text mining handbook: Advanced approaches in analyzing unstructured data. *Choice Rev. Online* **2013**, *44*, 5644–5684.
26. Kwartler, T. *Text Mining in Practice with R*; John Wiley and Sons Ltd.: Hoboken, NJ, USA, 2017.
27. Kao, A.; Poteet, S.R. *Natural Language Processing and Text Visualization*; Springer: London, UK, 2007.
28. Berry, M.W. *Survey of Text Mining: Clustering, Classification, and Retrieval*, 2nd ed.; Springer: London, UK, 2007.
29. Aggarwal, C.C.; Zhai, C. Mining Text Data. In *Mining Text Data*; Springer: New York, NY, USA, 2012; pp. 43–76.
30. Lazard, A.J.; Scheinfeld, E.; Bernhardt, J.M.; Wilcox, G.B.; Suran, M. Detecting themes of public concern: A text mining analysis of the Centers for Disease Control and Prevention's Ebola live Twitter chat. *Am. J. Infect. Control* **2015**, *43*, 1109–1111. [CrossRef]
31. Feinerer, I.; Hornik, K.; Meyer, D. Text Mining Infrastructure in R. *J. Stat. Softw.* **2015**, *25*, 1–54.
32. R Development Core Team. *R: A Language and Environment for Statistical Computing*; R Foundation for Statistical Computing: Vienna, Austria, 2008.
33. Botreau, R.; Veissier, I.; Butterworth, A.; Bracke, M.; Keeling, L. Definition of criteria for overall assessment of animal welfare. *Anim. Welf.* **2007**, *16*, 225–228.
34. Welfare Quality®. *Welfare Quality® Assessment Protocol for Cattle*; Welfare Quality Consortium: Lelystad, The Netherlands, 2009.

35. Davidson, N.; Harris, P. Nutrition and Welfare. In *The Welfare of Horses*; Springer: Dordrecht, The Netherlands, 2007; pp. 45–76.

36. Phillips, C.J.C.; Wojciechowska, J.; Meng, J.; Cross, N. Perceptions of the importance of different welfare issues in livestock production. *Animal* **2009**, *3*, 1152–1166. [CrossRef] [PubMed]

37. Serpell, J.A. Factors Influencing Human Attitudes to Animals and Their Welfare. *Anim. Welf.* **2004**, *13*, 145–151.

38. AWIN. *AWIN Welfare Assessment Protocol for Goats*; AWIN: Berlin, Germany, 2015.

39. AWIN. *AWIN Welfare Assessment Protocol for Sheep*; AWIN: Berlin, Germany, 2015.

40. Benhajali, H.; Richard-Yris, M.; Ezzaouia, M.; Charfi, F.; Hausberger, M. Foraging opportunity: A crucial criterion for horse welfare? *Animal* **2009**, *3*, 1308. [CrossRef] [PubMed]

41. Löckener, S.; Reese, S.; Erhard, M.; Wöhr, A.C. Pasturing in herds after housing in horseboxes induces a positive cognitive bias in horses. *J. Vet. Behav. Clin. Appl. Res.* **2016**, *11*, 50–55. [CrossRef]

42. Jensen, P.; Toates, F.M. Who needs "behavioural needs"? Motivational aspects of the needs of animals. *Appl. Anim. Behav. Sci.* **1993**, *37*, 161–181. [CrossRef]

43. Hoffman, C.J.; Costa, L.R.; Freeman, L.M. Survey of Feeding Practices, Supplement Use, and Knowledge of Equine Nutrition among a Subpopulation of Horse Owners in New England. *J. Equine Vet. Sci.* **2009**, *29*, 719–726. [CrossRef]

44. Séguin, V.; Garon, D.; Lemauviel-Lavenant, S.; Lanier, C.; Bouchart, V.; Gallard, Y.; Blanchet, B.; Diquélou, S.; Personeni, E.; Ourry, A. How to improve the hygienic quality of forages for horse feeding. *J. Sci. Food Agric.* **2012**, *92*, 975–986. [CrossRef] [PubMed]

45. Jones, T.A.; Donnelly, C.A.; Stamp Dawkins, M. Environmental and management factors affecting the welfare of chickens on commercial farms in the United Kingdom and Denmark stocked at five densities. *Poult. Sci.* **2005**, *84*, 1155–1165. [CrossRef] [PubMed]

46. Marchewka, J.; Watanabe, T.T.N.; Ferrante, V.; Estevez, I. Review of the social and environmental factors affecting the behavior and welfare of turkeys (Meleagris gallopavo). *Poult. Sci.* **2013**, *92*, 1467–1473. [CrossRef] [PubMed]

47. Battini, M.; Vieira, A.; Barbieri, S.; Ajuda, I.; Stilwell, G.; Mattiello, S. Animal-based indicators for on-farm welfare assessment for dairy goats. *J. Dairy Sci.* **2014**, *97*, 6625–6648. [CrossRef]

48. Dalla Costa, E.; Murray, L.A.M.; Dai, F.; Canali, E.; Minero, M. Equine on-farm welfare assessment: A review of animal-based indicators. *Anim. Welf.* **2014**, *23*, 323–341. [CrossRef]

49. AWIN. *AWIN Welfare Assessment Protocol for Horses*; AWIN: Berlin, Germany, 2015.

50. AWIN. *AWIN Welfare Assessment Protocol for Donkeys*; AWIN: Berlin, Germany, 2015.

51. Getachew, M.; Trawford, A.; Feseha, G.; Reid, S.W.J. Gastrointestinal parasites of working donkeys of Ethiopia. *Trop. Anim. Health Prod.* **2009**, *42*, 27–33. [CrossRef]

52. Ayele, G.; Feseha, G.; Bojia, E.; Joe, A. Prevalence of gastro-intestinal parasites of donkeys in Dugda Bora District, Ethiopia. *Livest. Res. Rural Dev.* **2006**, *18*, 14–21.

53. Mezgebu, T.; Tafess, K.; Tamiru, F. Prevalence of Gastrointestinal Parasites of Horses and Donkeys in and around Gondar Town, Ethiopia. *Open J. Vet. Med.* **2013**, *03*, 267–272. [CrossRef]

54. Van Dierendonck, M.; Goodwin, D. Social contact in horses: Implications for human-horse interactions. In *The Human-Animal Relationship (Animals in Philosophy and Science)*; de Jonge, F.H., Ed.; Ruud van den Bos: Assen, The Netherlands, 2005; pp. 65–82.

55. McGreevy, P.D.; Masters, A.M. Risk factors for separation-related distress and feed-related aggression in dogs: Additional findings from a survey of Australian dog owners. *Appl. Anim. Behav. Sci.* **2008**, *109*, 320–328. [CrossRef]

56. Mcdonnell, S.M. *A Practical Field Guide to Horse Behavior. The Equid Ethogram*; Eclipse Press: Lexington, KY, USA, 2003.

57. Knubben, J.M.; Furst, A.; Gygax, L.; Stauffacher, M. Bite and kick injuries in horses: Prevalence, risk factors and prevention. *Equine Vet. J.* **2008**, *40*, 219–223. [CrossRef] [PubMed]

58. Dalla Costa, E.; Dai, F.; Lebelt, D.; Scholz, P.; Barbieri, S.; Canali, E.; Minero, M. Initial outcomes of a harmonized approach to collect welfare data in sport and leisure horses. *Animal* **2017**, *11*, 254–260. [CrossRef] [PubMed]

59. Christensen, J.W.; Zharkikh, T.L.; Antoine, A.; Malmkvist, J. Rein tension acceptance in young horses in a voluntary test situation. *Equine Vet. J.* **2011**, *43*, 223–228. [CrossRef] [PubMed]

60. Randle, H.; Wright, H. Rider perception of the severity of different types of bits and the bitless bridle using rein tensionometry. *J. Vet. Behav.* **2013**, *8*, e18. [CrossRef]

61. McGreevy, P.; Warren-Smith, A.; Guisard, Y. The effect of double bridles and jaw-clamping crank nosebands on temperature of eyes and facial skin of horses. *J. Vet. Behav. Clin. Appl. Res.* **2012**, *7*, 142–148. [CrossRef]

62. FAOSTAT. Food and Agriculture Organization of the United Nations. 2016. Available online: http://faostat.fao.org (accessed on 1 January 2019).

63. Tabler, G. Farm Animal Welfare Issues Affect Poultry Producers. Available online: https://thepoultrysite.com/articles/farm-animal-welfare-issues-affect-poultry-producers (accessed on 4 May 2019).

animals

MDPI

Article

Citizens' and Farmers' Framing of 'Positive Animal Welfare' and the Implications for Framing Positive Welfare in Communication

Belinda Vigors

Scotland's Rural College (SRUC), West Mains Road, Edinburgh EH9 3RG, UK; belinda.vigors@sruc.ac.uk

Received: 4 March 2019; Accepted: 1 April 2019; Published: 4 April 2019

Simple Summary: The words used to communicate farm animal welfare to non-specialists may be more important than knowledge of welfare itself. Framing research finds that human perception is influenced, not by *what* is said, but by *how* something is said. By increasing the emphasis placed on animals having positive experiences, positive animal welfare changes the framing of farm animal welfare. Yet, we do not know how such framing of animal welfare may influence the perceptions of key animal welfare stakeholders. In response, this study uses qualitative interviews to explore how citizens and farmers frame positive animal welfare and what this means for the effective communication of this concept. This study finds that 'positive' evokes associations with 'negatives' amongst citizens. This leads them to frame positive animal welfare as animals having 'positive experiences' or being 'free from negative experiences'. Farmers rely more on their existing frames of animal welfare and integrate positive welfare into this. As such, most farmers frame positive welfare as 'good husbandry', a smaller number frame it as 'proactive welfare improvement', and a small number frame it as an 'animal's point of view'. The implications of such internal frames for effectively transferring positive welfare from science to society are further discussed.

Abstract: Human perception can depend on how an individual frames information in thought and how information is framed in communication. For example, framing something positively, instead of negatively, can change an individual's response. This is of relevance to 'positive animal welfare', which places greater emphasis on farm animals being provided with opportunities for positive experiences. However, little is known about how this framing of animal welfare may influence the perception of key animal welfare stakeholders. Through a qualitative interview study with farmers and citizens, undertaken in Scotland, UK, this paper explores what positive animal welfare evokes to these key welfare stakeholders and highlights the implications of such internal frames for effectively communicating positive welfare in society. Results indicate that citizens make sense of positive welfare by contrasting positive and negative aspects of welfare, and thus frame it as animals having 'positive experiences' or being 'free from negative experiences'. Farmers draw from their existing frames of animal welfare to frame positive welfare as 'good husbandry', 'proactive welfare improvement' or the 'animal's point of view'. Implications of such internal frames (e.g., the triggering of 'negative welfare' associations by the word 'positive') for the effective communication of positive welfare are also presented.

Keywords: farmer perception; citizen perception; qualitative research; free elicitation narrative interviews

1. Introduction

Getting people to act upon, and according to, research evidence is not easy. You do not have to look far within the farm animal welfare literature for examples of key stakeholders (e.g., citizens and farmers) failing to apply evidence-based recommendations in practice (e.g., [1,2]). However, it is often not *what* is said but *how* something is said that determines human perception and action. Decades of framing research has revealed that specific words or phrases can evoke a clear set of associations or 'ways of viewing the world' in someone's mind, and this influences how they understand, evaluate, and thus respond to information [3,4]. Consequently, the way in which information is presented (e.g., the words or phrases used in communication and the salience of positives and negatives) can influence and change how an individual interprets and evaluates something and the choices they make in response [5]. However, this knowledge has not yet received extensive consideration within animal welfare science [6]. This is a noteworthy issue as (i) the internal frames in thought an individual holds can influence how they interpret and respond to research evidence and information and (ii) the way in which research evidence is described and framed in communication may determine the actions animal welfare stakeholders take [5,7], thus affecting the lives of farm animals.

Frames in thought are the internal "cognitive schemas of interpretation, mental filters or 'mind-sets'" (p.3, [7]) an individual draws from to make sense of and assimilate information. The way in which an individual internally frames something can affect their behaviour by influencing how they evaluate information and determining what they consider most relevant for such evaluation [5]. For instance, an individual framing farm animal welfare in terms of an animal's affective state will place considerable emphasis on how an animal feels [8], thus influencing how they evaluate the welfare of an animal and how they interpret associated information. However, frames in communication (i.e., the properties of interpersonal communication) can influence what internal frames an individual draws from (or indeed, provide new frames) by making sub-sets of information more or less salient [3,9]. For example, Mellor [10] highlights how the Five Freedoms framework was not formulated to be an absolute standard or represent ideal states, but wider society has interpreted and assimilated the freedoms as being fully achievable. This is likely because 'freedom' evokes connotations of 'absence from', hence, the salience of 'freedom' in the communication of the framework caused individuals to interpret that negative experiences should be eliminated [10]. Therefore, framing in communication has potentially influenced individual interpretation and response to one of the most conspicuous frameworks in animal welfare.

One of the key research areas in the framing literature is whether it is better to accentuate negatives or accentuate positives in communication [11]. This is because a plethora of studies, in numerous contexts, find a *framing effect*, where the alternative framing of objectively equivalent information in positive and negative terms produces divergent responses within individuals [11–15]. This effect of framing in communication is particularly relevant to the newly emergent concept of positive animal welfare. Positive welfare can be broadly understood as a movement to shift animal welfare's more traditional focus on avoiding the negative aspects of an animal's life (e.g., pain) towards a focus on enhancing the positive aspects of an animal's life (e.g., pleasure) [16]. In doing so, positive welfare is arguably changing the 'framing' of animal welfare, both in terms of the internal mental frames it intends to evoke (e.g., positive affect rather than negative affect) and in terms of the words, phrases and language used to frame animal welfare in communication (e.g., the opportunity for animals to have a *meaningful* and *pleasurable* life [17]). Although there are differences within the literature on what positive welfare entails (e.g., enhancing positive affect and promoting pleasurable behaviours such as play [18,19]), the shared motivation of this literature is to make clear that animal welfare needs to be updated to include greater consideration of the positive aspects of an animal's life (e.g., [16,20,21]). Indeed, in most cases, creating a convincing case for positive welfare has relied on distinguishing it from animal welfare's more traditional emphasis on negatives [10,17].

By emphasising and increasing the salience of positive rather than negative aspects of welfare, positive welfare arguably seeks to re-frame the attributes associated with animal welfare. However, we have little insight into how key stakeholders in society may respond to such attribute framing. This is a pertinent and contemporary issue for the furtherance of positive welfare research, in particular, as indicators and frameworks for assessing positive welfare at the animal level are developing [17,21].

Before positive welfare can be effectively taken forward and promoted in society, there is a timely need to better understand how critical stakeholders, such as farmers and citizens, perceive and interpret positive animal welfare. Moreover, we should be concerned with how their current meaning frames—influenced by their values, beliefs and social interactions—impact their perception of positive welfare, as this may influence the way in which it is adopted and/or applied, and indeed, resisted. Furthermore, understanding how key stakeholders understand and frame positive welfare may prove beneficial for the effective communication and transfer of positive welfare research from science to society (e.g., farmers and citizens). In light of the knowledge of framing and its effect on human behaviour, it is arguably not evidence of positive welfare that may motivate key animal welfare stakeholders to engage in positive welfare behaviours, but *how* that research is communicated. As Blackmore et al. (p.14, [22]) explained, "the presentation of facts or data is rarely sufficient in motivating our concern and action". Knowing what internal meaning frames such stakeholders draw from to make sense of positive welfare is one key step in ensuring it is communicated appropriately.

In response, this paper seeks to explore and uncover how farmers and citizens frame positive welfare (i.e., frames in thought), and the implications of this for the effective framing of positive welfare in communication. As little is known about how key social actors frame positive welfare and there is a limited consensus within the literature on what should be included within a positive welfare framework [10,16,17,21,23], it is too early to use experimental manipulation of frames to test the impact of positive welfare on behaviour. Moreover, it is likely that positive welfare may be construed and understood in different ways by different people as they draw from personal meaning frames, experiences, values and norms to make sense of it. As such, there is a need to recognise this and employ methods which draw out, exemplify and accept such diverse subjective interpretations of positive welfare in society. Therefore, a qualitative approach is most applicable, as it enables exploration into *how* individuals interpret and make sense of phenomena and, is particularly well suited to the study of topics which are 'ill-defined' or 'poorly understood' [24]. For this reason, the free association narrative interview method [25] is applied to access farmers' and citizens' associations with positive animal welfare and how they frame positive welfare based on these internal points of reference. Uncovering these internal frames, or 'frames in thought', is essential for understanding how best to frame positive welfare in communication.

In sum, this paper explores both *how* positive welfare is framed in thought by key animal welfare stakeholders and suggests how *ought* positive welfare be framed in communication by animal welfare science.

2. Methodology

This article draws from qualitative interview data collected as part of a research study, undertaken in Scotland, UK, to explore experts' (farmers) and lay-persons' (citizens) perspectives on positive animal welfare. The study explored a number of factors including how livestock farmers and citizens freely interpret and 'frame' positive animal welfare, and the factors underlying and influencing this framing and the extent to which aspects of positive welfare are evident in their current perceptions of animal welfare. This present article primarily focuses on the outputs relating to how positive welfare is framed by farmers and citizens and the implications of this for the effective communication of positive welfare.

2.1. Method

The interview protocol was designed according to the 'free association narrative interview method' described by Hollway and Jefferson [25]. This is an open approach to interviewing, which combines the story-telling role of narratives with the psychoanalytical principle of free association:

> "Which assumes that unconscious connections will be revealed through the link that people make if they are free to structure their own narratives". [25]

Unlike most interview-based methods, which assume that participants can 'tell it like it is', it posits that interviewees are unlikely to understand the question in the same way as the researcher [25]. As such, a critical aspect of interview design was ensuring participants had space to recount their understanding of the target phenomenon, free from any *a priori* research assumptions affecting data collection (e.g., by asking narrow or closed questions). Although there is no singular way to conduct narrative interviews, the generally recommended approach is to first pose a primary narrative inducing question, where the researcher listens passively (e.g., does not offer any prompts) whilst actively taking notes on key points generated by the participant [26]. Once it is clear the participant has completed the recounting of this main narrative, the researcher may then use these notes to ask further questions and probe particular points [26]. In this regard, a particular strength of this interview method is that participants are not restricted to the researcher's agenda or the questions the researcher asks. Rather, it encourages participants to tell their story and to take ownership of how that story is told [25].

In order to explore the associations the term 'positive animal welfare' brought to participants' minds and how this influenced their framing and interpretation of it, the interview design aimed to encourage participants to 'think-out-loud' about positive welfare. For this reason, a singular free elicitation question: "Positive animal welfare—can you tell me about what comes to mind when you hear that?" was posed. No prior explanation of positive welfare was provided, and no additional prompts were given after this main narration question was asked. This left the participant free to set the agenda of discussion and construct their own narrative around positive animal welfare. By doing so, the researcher was provided with a lens into what factors the participants associated positive welfare with, what they drew from to make sense of it and their overarching frame of positive welfare. Once this main narration was completed, further details on points raised by participants could be asked by the researcher (e.g., can you give an example of what you mean by 'natural'). Through constructing and recounting the narratives of what positive welfare 'brings to mind', participants revealed the experiences and associations most relevant to their framing of positive welfare and the importance of the subjective meanings they attach to it [26]. This approach produced rich and diverse discussions across participants and provided insights into the meaning frames they drew from to make sense of and assimilate positive welfare.

2.2. Sample

Farmers and citizens were the chosen case studies as they represent two key stakeholder groups in animal welfare, with farmers directly impacting animal welfare through their behaviour and management practices and citizens indirectly affecting welfare through their preferences [27,28]. The study used a purposive heterogeneous sampling approach to select an information-rich and diverse range of research participants [29,30]. As positive welfare is a relatively new concept, this approach enabled as much insight as possible to be gained from a variety of different farming systems and perspectives and a variety of different people in society.

Farmers were recruited through various means, including advertisement on social media, referral by farm advisers and other industry stakeholders, attendance at farmer events (e.g., monitor farm meetings) and by directly contacting individual farmers via email. Face-to-face interviews were completed with 13 dairy farmers, 10 beef and sheep farmers, 2 free range egg producers, 2 mixed farmers (i.e., pig, poultry, beef and sheep) and 1 large pig farmer (total n = 28). These interviews took place in the homes of individual farmers in Scotland between March and September 2018 and lasted, on average, 55 minutes. The majority were males (86%), while a small number were females (11%) (one participant preferred not to indicate their gender). In reflection of the heterogeneous sampling approach, a variety of farming systems were evident within each sector (i.e., zero-grazed or grazed in dairy, indoor-wintered or outdoor-wintered in the beef and sheep sector, and organic or non-organic in the poultry sector). Further demographic details, specific to each participating farmer, can be found in Table A1 in Appendix A.

Citizens were initially recruited through advertisement on social media outlets and in public spaces (e.g., libraries, local shops and community centres) in large urban, small towns and accessible rural areas in Scotland [31]. However, this resulted in a very poor response rate with only five responses and three individuals volunteering to participate. As such, a further 12 citizen participants were recruited using the market research company 'Testing-Time', resulting in 15 citizen interviews in total. Citizen interviews took place in Scotland between April and September 2018. Three interviews were completed face-to-face, while the remaining took place over Skype at the preference of participants. The average interview duration was 50 minutes. In total, 60% of participants (n = 9) were females and 40% (n = 6) were males. The majority of the sample lived in large urban areas (n = 11) with the remaining evenly split between small towns (n = 4) and accessible rural areas (n = 4). Reflecting heterogeneous sampling, citizen participants differed in their dietary preferences with six stating they consumed meat, six indicating they have reduced meat consumption in the past year, while two participants were vegetarian and one vegan. Participants also differed in their experiences of farming—four participants came from a farming background (i.e., worked on farms in the past or had family members who farmed); seven noted they had visited farms (i.e., on educational trips); and four stated they had no experience of livestock farming at all. The specific demographic details of each citizen can be found in Table A2 in Appendix A.

The research was approved, before commencement, by the SRUC Social Science Research Ethics Committee. All interviews were audio-recorded with participant permission, except in the case of one farmer interview (who preferred not to be recorded) where detailed notes were taken instead. As with any qualitative research, the aim was not to determine outputs that could be generalised to a population but to reveal meaning and understanding, in the context, of how different individuals framed positive animal welfare. As such, once a saturation point had been reached (i.e., no additional data, from which to develop properties of a category, emerged [32]), no further participants were sought.

2.3. Data Analysis

Each recorded interview was transcribed in full and entered into the qualitative data analysis package MaxQDA to assist analysis. To distance the exploration of farmers' and citizens' framing of positive welfare from any *a priori* definitions of positive welfare, the constant comparison method was used to analyse the data [32]. This entailed reading each interview and categorising or 'coding' sections of text according to the nature of the points raised. For example, codes emerging from farmers' narratives in response to 'what positive welfare brings to mind' included healthy animals, improving welfare, stress reduction and happy animals (amongst others). Once this was completed, coded sections were compared between interviews and related codes, and comments were grouped to form overarching key themes. This ensured the generation of insights that were grounded in the data [33]. These key, overarching themes then formed the basis for uncovering how farmers and citizens framed positive welfare, as outlined in the subsequent sections.

3. Findings

This section describes how citizens and farmers interpret and make sense of positive animal welfare and the different ways in which citizens and farmers frame positive welfare. The first sub-section illustrates the sense-making process evident in participants' free association narratives before the subsequent two sections detail how citizens and farmers frame positive welfare and the themes characterising each frame. Throughout, participants' narratives are directly drawn from and presented to ensure a transparent link between their views and the researcher's interpretation of these views. The data is analysed in more detail in the subsequent discussion section.

3.1. Participants' Sense-Making of Positive Animal Welfare

In seeking to understand how farmers and citizens made sense of and gave meaning to positive welfare, it is important to first note that neither group had heard of the term 'positive animal welfare' before. As one citizen explained:

"When you said positive animal welfare, I thought that's a strange term; I've never heard that before". (Citizen 2)

Similarly, farmers often reflected that the interview had been:

"The first time I've ever heard that term used". (Dairy 6)

As such, in seeking to give meaning to and respond to a term previously unknown to them, many participants focused on the word 'positive', as this represented what made this term strange, relative to their existing constructs of animal welfare.

Amongst citizens, it was evident that the term 'positive' triggered associations with 'negative'; participating citizens rarely made sense of positive animal welfare *without* using a comparative opposite. For example:

*"So the welfare should be **not** to exploit them, **not** to obviously harm the animals"*. (Citizen 4)

*"Have a good quality of life, for me, that means sort of as natural a life as possible, you know, young **not** being taken away from parents, having access to the outdoors and the fresh air and all that sort of thing, and **not** being fed anything artificial"*. (Citizen 6)

*"I think it is the animal's experiences, positive experiences rather than like **negative**"*. (Citizen 7)

*"Just a nice kind of hill for them, like fenced around, so **nothing** can get to them that is going to **hurt** them, a few sheep in there; **not** crowded at all"*. (Citizen 9)

*"It's the exact sort of **opposite** of that imagery; I've got the horror image in my head right now and then the other image is the exact **opposite**"*. (Citizen 3)

As highlighted by the comments in bold (emphasis added), when de-constructing 'positive animal welfare', citizens made sense of it by juxtaposing positive aspects (e.g., 'a natural life') with negative aspects (e.g., no harm). In this way, many citizens deduced that positive welfare was something which did not involve such negatives. Indeed, for some citizens, this comparative association between a 'positive' and a 'negative' of welfare was explicit:

*"Positive animal welfare? Well, what would be **negative animal welfare**? That would be the question"*. (Citizen 2)

Overall, citizens' free association narratives revealed that the term 'positive' engendered potentially automatic associations with 'negative'. As will be shown in the subsequent section, this influenced how they framed 'positive animal welfare' overall.

Farmers demonstrated a much more nuanced approach in their sense-making of positive welfare. Some focused directly on and engaged in semantic analysis of the word 'positive' as a means to understand the whole term 'positive animal welfare':

"Positive means to add. So therefore, you are adding above the baseline". (Dairy 10)

"I suppose positive: higher than you need to be, crossing a bar". (Beef & Sheep 10)

However, in most cases, farmers drew from their personal constructs of animal welfare to make sense of 'positive animal welfare' as a whole. In other words, unlike the prior examples which deduced positive welfare by first thinking what 'positive' means *per se*, the majority of farmers' sense-making centred on the deduction of 'what are the positive aspects or outputs of my current animal welfare practices?' For some farmers, although not quite as overt as citizens' juxtaposition of 'positive' versus 'negative', positive welfare was surmised to be a combination of preventing negatives and enabling positives:

> *"Positive welfare . . . . It's trying to prevent anything bad from happening to them, you're trying to maintain comfort, you're trying to maintain health and well-being".* (Dairy 6)

Nevertheless, the majority of farmers focused on their 'ideals' of welfare to give meaning to positive welfare:

> *"Keeping them well-fed, bedded, watered and their health looked after . . . . So, they're cared for and looked after to the best of our abilities; if they're ill, or there's a problem, that you look after them".* (Beef & Sheep 7)

Interestingly, pig and poultry producers' sense-making of positive welfare centred around the five freedoms:

> *"You know, if you want to be scientific about it you've got the five freedoms, which, yeah, I wouldn't disagree with. You know, if you go through those five freedoms that is basically what you are trying to give the animal".* (Organic Free Range Egg 1)

> *"You've got something called the six freedoms I think it is. And it is to do with the freedom from fear, freedom from hunger, freedom from thirst, freedom from stress, freedom to exhibit natural behaviours and there is one more that I can't think of. And that is basically it. As long as pigs have all of those, then to me that is good enough . . . . Humans would be the same, pigs are the same, sheep would be the same. As long as they can do all of those things, they should be relatively happy, I guess would be the term".* (Pig 1)

Overall, it was evident that citizens and farmers focus on their associations with the word 'positive' to make sense of positive animal welfare. For citizens this conjured up associations with 'negative', triggering an overt comparison of welfare opposites (e.g., negatives versus positives, or the 'dos and don'ts') to make sense of and distinguish positive welfare. Farmers, on the other hand, drew from their semantic associations of the word positive (e.g., 'to add', 'raise the bar') and their existing constructions of the negative welfare aspects they sought to reduce and the positive aspects they sought to provide. Moreover, they did not usually separate negative and positive but rather saw both as mutually inclusive aspects of overall good welfare.

3.2. Citizens' Framing of Positive Animal Welfare

Citizens clearly framed positive welfare in one of two ways: (1) positive welfare as 'positive experiences' for farm animals, or (2) positive welfare as being 'free from negative experiences'. Of the four participants who stated they had a farming background, three of them demonstrated a 'free from negatives' frame, while one framed positive welfare as 'positive experiences'. Amongst the seven participants who expressed they had experienced livestock farms (e.g., through educational visits), five framed positive welfare as 'positive experiences' and two as 'free from negatives'. The four participants who stated they had no experience of livestock farming were evenly split between both frames, with two demonstrating the 'positive experiences' frame and two demonstrating the 'free from negative experiences' frame. The vegan participant framed welfare as 'positive experiences', while one vegetarian demonstrated the 'positive experiences' frame and the other the 'free from negatives' frame. Similarly, the six participants who stated they had been actively reducing their meat consumption were evenly split between the two frames.

3.2.1. Positive Experiences

The 'positive experiences' frame was made evident by the emphasis individual citizens placed on farm animals having opportunities for positive experiences. To present this mental representation of positive welfare, participants focused on and made salient several factors which they considered provided positive experiences to farm animals. These included (i) a natural–outdoor environment:

"Just that animals are being cared for in the best way for them, for me that conjures up ideas of the most natural . . . having access to the outdoors and fresh air and all that sort of thing". (Citizen 6)

which was further seen as a means to ensure animals had desired (ii) space:

"I think it is more animals with free space in large fields"; (Citizen 8)

which was further inextricably linked as a way in which animals could be provided with opportunities to exert (iii) autonomy over their environment:

"Just being outside in the fresh air and having the sun, I think these things are good for humans so I think that would extend to animals as well. Freedom to roam, trying to have as normal a life as possible, what a cow would do day to day, allowing them the space and freedom to do that". (Citizen 2)

"Positive, a positive experience for an animal . . . . Maybe like in the case of pigs, a little mud bath or something like that, something that they can choose to do, they are not being forced to do it. They don't need to do it to survive. They are not being forced to do it by a farmer". (Citizen 9)

In addition, the representation of a traditional or pastoral farm as being the best way to create a (iv) positive human–animal relationship between farmer and farm animal was further made salient by participants within this frame:

"The first thing that comes to mind is the happy farmer and happy animals . . . .I think it's the love between them and if they are showing love to the animals it is like both sides are happy . . . so I guess in a broader sense it is just a happier environment, more than just the animal. I feel like it is a happy environment on both ends, where animals are being looked after by people who care for the animals". (Citizen 4)

These factors were not exclusive but coalesced to create this frame, or mental representation, of positive animal welfare. Specifically, a natural–outdoor environment was framed as a prerequisite for providing animals with the desired space for them to exert autonomy within their environment, which was further supported and enabled by a positive or caring human–animal relationship.

"So I think positive animal welfare to me, is not just open fields and all that, which it is, that is usually the first image that comes to mind, but it is also one of these small little farms where they have a quality life, and it is a lot smaller and you know someone who knows every single one of the cows and has probably even named them, I don't know. So, you know, there is a bond between the farmer say, with the animals, and the animals with the farmer, for sure. And there is love, that probably does [matter], it does you know". (Citizen 1)

Thus, participants who framed positive animal welfare as animals having the opportunity for 'positive experiences' placed emphasis not just on the prevention of negatives but on the factors which they considered gave animals a positive life namely (i) access to a natural–outdoor environment, (ii) space, (iii) autonomy and (iv) a positive human–animal relationship.

3.2.2. Free from Negative Experiences

The 'free from negatives' frame differed from the 'positive experiences' frame in that little to no emphasis was placed on the provision of positive experiences. Rather, the emphasis here was that not experiencing any negatives would provide an animal with a positive life. In other words, positive welfare was framed as the absence or elimination of negative experiences.

In this regard, citizens made salient the importance of ensuring (i) no harm:

"Treat them how you want to be treated . . . .so the welfare should be not to exploit them, not to obviously harm animals. . . . So before it was obviously killed and what not, you did the best for it, you didn't kind of treat it bad or negatively, you didn't give it a bad life" (Citizen 5)

"What I think it means is that animals are treated in the way they should be treated and not tortured in a farm let's say. So the picture in my head is animal's free out in nature or the other scenario would be the animals that are treated, they may be not in nature, but at least they are treated properly and they are not being tortured"; (Citizen 14)

(ii) eliminating negative affect:

"Because I think not letting them feel fear. I'm sure around the world some animals have been really abused. So like having that fear that they could have . . . . That is definitely not a positive experience"; (Citizen 7)

and (iii) preventing health issues:

"I guess, to me, more kind of like their health, so they are not kept in a way that is going to make them ill or that is going to be detrimental to them". (Citizen 11)

Thus, participants demonstrating this frame interpreted the prevention of harms, negative affect and health issues as being a positive thing, in and of itself. As such, positive animal welfare was framed as animals being free from negative experiences.

3.3. Farmer Framing of Positive Animal Welfare

Amongst farmers' responses to positive animal welfare, three frames were evident: positive welfare as (i) good husbandry; (ii) proactive welfare improvement; and (iii) the animal's point of view. 'Good husbandry' was the predominant frame with seven of the dairy farmers, six of the beef and sheep farmers, and all of the pig and poultry farmers demonstrating this frame. A smaller number—five dairy farmers and four beef and sheep farmers—framed positive welfare as 'proactive welfare improvement'. The remaining two mixed farmers and one dairy farmer framed positive welfare as considering the 'animal's point of view'.

Interestingly, across all participating farmers, a pre-existing frame of animal welfare as 'happy–healthy–productive' animals was persistent and transcended across all three frames of positive animal welfare. Specifically, each farmer framed the productivity and performance of their animals as the key 'objective' indicator that their animals were well cared for. However, productivity and performance were given meaning by the emphasis placed on 'happy–healthy' animals. Namely, 'productiveness' came from an interaction between 'happy' and 'healthy', whereby healthy animals were framed as happy animals and an unhappy animal was framed as being unlikely to be healthy. Farmers used and applied this pre-existent frame of animal welfare to affirm and demonstrate that the welfare of their animals was positive. In other words, a productive farm indicates well cared for animals (i.e., happy and healthy) and therefore, a productive farm shows that animals are being positively cared for and managed:

"So I suppose the weigh scales will tell you how happy they are . . . because if they are happy they are putting on weight, which is what I am trying to do".

(Beef and sheep 6)

"And animal welfare and health and all those things around about is our priority. Because we want to create a, we want to create the best environment for our livestock. Because if they are happy, then they are going to produce more milk, be healthier and make our life a lot easier".

(Dairy 11)

"Well why I consider my hens have a good life is, well it's very simply because we, whilst my hens work for me, everything that we do here is all about hen welfare, and about trying to get, and it's a terrible thing to say, but you know I told you earlier, profitability and output from animals is very directly linked to their health and welfare".

(Free Range Egg 2)

Overall, the farmers group demonstrated some differences between participants in terms of how they frame positive welfare and therefore interpret it, but each participating farmer drew from the same overarching animal welfare frame of 'happy–healthy–productive' animals. This, as is shown in the subsequent sections, impacts the way in which each positive welfare frame is constructed.

3.3.1. Good Husbandry

The 'Good Husbandry' frame was the most prevalent frame of positive welfare. This frame was made clear by the emphasis these individuals placed on factors they considered meaningful for providing the best possible care or, good husbandry, for their animals. This included a strong emphasis on (i) health:

"Yes, performance. I suppose it goes back to measure, manage, measure. I look at welfare as a sort of health issue, and I suppose if they're healthy, they perform better"; (Beef and Sheep 7)

which, as previously discussed, was closely tied with and considered a prerequisite to having (ii) happy animals:

"But we would hope that we are doing everything and more to, if you create the best environment that you can create, within reason, and can keep them as happy and healthy as you can"

(Dairy 11)

"Just treating your cow, treating them right, treating them well. As I say you want just happiness. It does sound silly, but happy cow happy herd really. Positive animal welfare, well it's you just want things to be good. They're never going to be perfect but you can aim for it". (Dairy 9)

As alluded to in the prior narrative, farmers also stressed their own role within the 'good husbandry' frame, where they emphasised the importance of (iii) doing the best they can, for their animals:

"I suppose it's offering the best for them. Like for me, it's having a shed that you'd be happy if anybody came in and took a photograph over the gate in the winter time and the same in your fields. So nothing that you wouldn't want somebody to go and see, whether it's slat mats; that they're well stocked and they're clean". (Beef and Sheep 3)

Ensuring (iv) resource needs are well met:

"They don't have any further needs, they have their food there, they have got plenty of food, they've got a nice bed, they've got water, and they are content"; (Dairy 11)

and reducing (v) stress:

"Lack of illness, just that lack of stress really is my take on it". (Beef and Sheep 5)

"So for me, welfare is actually quieter cattle, far easier to work with ... they're not stressed. To me, that's more how I'd say I treat welfare" (Beef and Sheep 4)

were also presented as key themes within the 'good husbandry' frame of positive welfare.

As illustrated by the above narratives, each theme is not discussed in isolation but are interconnected aspects of farmers' interpretation of positive animal welfare. Namely, a happy animal is a healthy animal, which is what farmers want to achieve by providing their animal with the 'best', which further requires them to minimise stress, so the animal can, in turn, be healthy and happy and the farmer, simultaneously, can know they are doing their best. Thus, there is a cyclical relationship between each of the above-mentioned factors which, when taken as a whole, forms the positive animal welfare as 'good husbandry' frame. In short, farmers used these factors to construct and convey their interpretation of:

"Positive animal welfare ... [as being] about looking after your animals in a good manner and good welfare". (Beef and Sheep 7)

Thus, it is evident how the 'happy–healthy–productive' frame transcends and influences farmers' interpretation of positive animal welfare here. However, as will be made evident by the subsequent sections, the 'good husbandry' frame is distinct in that farmers here did not consider positive welfare to be much different to or anything beyond their current practices.

3.3.2. Proactive Welfare Improvement

Those who framed positive welfare as 'proactive welfare improvement' saw it as something which involved going beyond the standard levels of animal welfare. However, this did not mean their overall frame of welfare was completely distinct from those discussed in the previous section. Possibly due to the ubiquitous nature of the 'happy–healthy–productive' frame, many farmers demonstrating the 'proactive welfare improvement' frame also emphasised 'good husbandry' themes such as minimising stress:

"Trying to make them live comfortably for as long as they can, trying to be stress-free, pain-free, injury-free", (Dairy 6)

ensuring resource needs are met:

"Just providing enough, providing all the things they need, the food and water, the other animals to keep them company and the disease, the treatment of disease whenever possible. Things like clipping feet and trimming feet when they need it, just the things that you need to do to keep them content", (Beef and Sheep 10)

and the importance of health and happiness:

"It means that the animals should be happy, contented and healthy". (Dairy 1)

However, unlike the previously discussed group, farmers here differed in that they interpreted positive animal welfare to mean going beyond the norm and doing something to actively improve or make changes to their current practices. In particular, they emphasised that:

"It is about being proactive" (Beef and Sheep 2)

where 'being proactive' meant doing things to improve welfare beyond basic standards or norms:

"I guess taking what we've got and improving on it. Accepting where we are and trying to get better", (Beef and Sheep 9)

"And then that is your farm, you have got this slightly higher, you are doing better than you have to, yeah high, I suppose positive: higher than you need to be, crossing a bar, that is a definite again". (Beef and Sheep 10)

"I don't think you can be successful without exhibiting some positive animal welfare. With that kind of thing, I am sure there are aspects that every farm can improve and most farmers will strive to improve them . . . . I think we are conscious of, well I am conscious of anyway, the next thing that you should be improving is the thing that you feel least comfortable showing to somebody". (Dairy 6)

Interestingly, farmers within this frame often anchored this interpretation of positive welfare in their own personal motivations, where they emphasised the personal value they place on actively and continuously seeking to improve or do things differently:

"If you think you have already done all you can do, and you can't get any better, then you will never get any better. But if you have some kind of goal ... There is always a goal and an achievement that you should look to and if that means, if that is positive animal welfare then that is positive thinking. You have to think that you could improve somewhere, something, otherwise what are you working towards? You will never sustain yourself if you reach a level and then decide that you've reached. Like if we thought we'd bred the best cow we were ever going to breed, then we might as well give up now because we are never going to breed another one. That is the way I see it .... [Overall] ... .It sounds to me like it's something that is going to be a benefit to us. So if it's positive animal welfare it is positive for the animal but it is also positive for our system as well. That is essentially what it means". (Dairy 8)

In sum, farmers who demonstrated a 'proactive welfare improvement' frame felt positive welfare was about more than just the appropriate provision of resource needs or the maintenance of health (i.e., good husbandry); they saw positive welfare as actively thinking about how to improve welfare or go beyond basic welfare standards. In many instances, this was closely related to their personal, perhaps intrinsic, motivations and self-image as a proactive individual.

3.3.3. Animal's Point of View

A small number of farmers, just three in total, framed positive welfare as considering the 'animal's point of view' when making decisions. Again, it was not that these individuals did not emphasise the importance of 'good husbandry' or 'happy–healthy–productive' animals, it was that they made salient additional factors which set them apart from those discussed in the previous two sections. Specifically, they continuously returned to and emphasised how, when making decisions, they tried to think about it from the animal's point of view or take into consideration an animal's preferences:

"Positive animal welfare. I think it is putting, the way I see that is you would look at the cow's perspective on life rather than yours. What would the cow want out of life rather than what you would want the cow to have? What would the cow want to do?". (Dairy 5)

"And if animals, I mean you can't tell me that when you have lambs at about 4–6 weeks, and they start jumping up on the silage bales, they are not playing. You know, they are not jumping up on the silage bales to get food. Or to have sex, which is what all the psychologists seem to think motivates all of us .... You know, they are playing, they run around and they form a gang ... so, farms need to be places where they can do that". (Mixed 1)

"Positive animal welfare .... It's giving all animals, no matter what their species, a positive life. Positive environment. Just actively constantly thinking what's right for this animal in this system. Within this production system, what's the positive?". (Mixed 2)

Interestingly, when discussing 'positive animal welfare' and their association of it with an 'animal's point of view', they often juxtaposed it against some of the more basic welfare provisions (e.g., resource needs) to emphasise how, even when animals are appropriately provided for, they still felt it was important to consider their preferences and point of view. For example, dairy farmer 5 continued from their above narrative with:

"What would the cow want to do? I feel as if we underestimate that. We think we know, and we think we can, because it is an animal, we herd it and we tell it what to do and all the rest of it. Might be right, but I think we should still look at what the cow wants. I know that my cows, and again ... we've got mats and everything else, scraper systems, and washed down every day, everything

is washed every day and clean and the cows are clean. But the cow is still, and you know they've·
got lovely high quality, again prize-winning silage and a TMR mix that has all been designed by
nutritionists and all the rest, and analysed and sampled and footered about with, and the length of the
straw that we chop in to it, it's got to the right length of it, it's all there. But they are still right keen to
go out of the house, so you've got to let the cow go out. You've got to let it go out . . . because that cow
will reward you by you giving it it's choice".

Thus, farmers within this frame emphasise that even with the best quality of care, or the best
provision of resource needs, or the best management systems, the animal won't have a positive life if
their preferences and point of view are not taken into consideration. This is what differentiates this
group from the previously discussed frames.

3.4. Farmers' Response to Positive Animal Welfare

Farmers, in addition, volunteered their opinion on positive animal welfare. Here, the former
finding regarding citizens' contrasting of positive and negative is particularly noteworthy considering
the concerns many farmers expressed about the term 'positive animal welfare':

"I don't know, the positive bit doesn't seem right. The word positive doesn't seem right. The positive,
the first thing you think of when someone says positive is negative. I would say. I would say by
emphasising that, you would make people think; well there is positive but that means there must be
negative. What word would you put instead? I don't know, I can't think of another word, but to me,
that, whenever you say positive you invite the comparison towards negative".(Beef and Sheep 10)

Consequently, many farmers were perturbed, and understandably defensive, about what 'positive
animal welfare', due to its associations with negative, could engender in wider society:

"Why would they want to do that? Is it that they think that we are not having a positive welfare
outlook?". (Beef and Sheep 1)

"When I hear the term positive welfare, ahh, I suppose I start to be fearful of that term; positive animal
welfare, because I think it sort of nearly, we get painted, farmers get pained in an unfair light most of
the time". (Beef and Sheep 6)

"My only concern would be . . . would it be perceived as, well why are you promoting that? Why is
that not happening already? Kind of a thing". (Dairy 3)

However, in addition to concerns that positive welfare evokes the idea that negative welfare
exists, some farmers welcomed positive welfare when they gave meaning to it as something which
would help promote a positive image of farming in society:

"I think positive animal welfare, I think that's what we need to promote to a wider public; we are
going beyond the, beyond what we do, beyond what the public perceive us to be doing"; (Dairy 12)

or something which would differentiate them from others:

"It would be nice to think that we could get recognised more for the extra effort and that. And we're
not saying that other people don't go that extra mile . . . . But how do we get recognised between
trying to do the best we can do, and also maybe somebody further down the road that is just getting by
and probably could do a lot more to make that a nice place, a nicer environment"; (Dairy 11)

or as something that would enable a desired re-framing of welfare within the farming community:

"Relief—because for too long the word welfare is something that as farmers we have shied away from.
We say 'Oh, there's none of that here, I know we've got it but'. Welfare is the chronically lame sheep,
it's a dead lamb that has never been seen and has never been lifted. It is something that you don't want

to see. It is something that you just don't want to know about and it's the little guy with a clipboard at the market that is kind of peering over the pen and gives you a look. And that is what welfare has meant. Whereas that is not what it is at all, that is cruelty. Welfare is about the things that actually make us profitable as a business. It's about the things that help us stay in business . . . . I think it is a bit of a stumbling block for the industry . . . we need to get our heads over that. And I think it is a shame that welfare has become about something that you want to hide because somebody might pick on you for having a lame sheep. And we all have lame sheep". (Beef and Sheep 9)

Thus, farmers demonstrated diverse opinions about positive welfare, with some worried by how it may be perceived by wider society and others interested by its potential to communicate a positive message, differentiate their products, or promote a change in mindset.

4. Discussion

This paper set out to explore how positive animal welfare was framed by farmers and citizens, and the implications for the effective framing of positive welfare in communication. Findings reveal that both groups relied heavily on their interpretation of the word 'positive' and the associations it brought to mind to make sense of 'positive animal welfare' as a whole. Thus, a framing effect is arguably evident; the inclusion of the word 'positive' shaped and influenced how these key stakeholder groups made sense of positive welfare and the attributes of welfare they emphasised and gave importance to.

The reliance of citizens on the word 'positive' for sense-making may be an outcome of the limited knowledge and first-hand experience they have of farm animal welfare [34]. However, the automatic associations with 'negative welfare', 'positive animal welfare' evoked may be a natural process of linguistic inference. Within the study of linguistics, polarity is considered a key feature of adjectives, meaning descriptive words have a comparative opposite [35]. A positive adjective signals to a person that a negative form does exist but in this instance, it is not negative [35]. In the context of positive animal welfare, this manifested in citizens' apparent need to think through what is *not* positive (i.e., what is negative) to make sense of positive animal welfare overall. Regardless of the underlying reason, the positive–negative associations brought to mind by the term 'positive animal welfare' was central to how citizens constructed their framing of positive welfare.

Beyond the positive–negative comparisons made by citizens, citizen experience of farming may account for their framing of positive welfare. The 'free from negatives' frame was mainly demonstrated by participants with a farming background. Similarly, Nijland et al. [6] found that individuals from rural areas are more likely to frame farming in 'realistic' terms with a focus on the extrinsic value of animals (e.g., production). Thus, those with a farming background, due to their socialisation within the welfare norms and values of farming communities, may prioritise the elimination of negative factors affecting welfare (e.g., health issues), as many farmers do [36,37]. Interestingly, nearly all participants who had visited farms (i.e., on educational trips) demonstrated the 'positive experiences' frame. Ventura et al. [38] noted that farm visits can reduce citizens' concerns about the provision of basic resource needs, such as access to food and water, but increase their concerns about animal-based provisions, such as pasture access and cow–calf separation. Thus, visiting and experiencing farms may have led such participants to place greater emphasis on the provision of positive experiences for animals, rather than the prevention of negative experiences. In addition, this finding may further support arguments within the literature that reducing the knowledge distance between farmer and consumer (e.g., through farm visits or educational product information) helps improve citizens' understanding of farming practices and animal welfare [39,40]. More generally, the majority 'positive experiences' frame is consistent with findings that citizens tend to focus more on the positive aspects of an animal's life [41].

To some extent, citizens' interpretation and framing of positive welfare could be argued as being reflective of the scientific literature on positive animal welfare. For instance, participants within the 'positive experiences' frame freely elicited some of the positive welfare opportunities noted within the literature, such as animal autonomy [42] and positive human–animal relations [17,42]. However, positive welfare so framed brought to citizens' minds images of traditional, small-farms with animals living a 'natural' life outdoors. This is consistent with previous research on citizens' attitude to animal welfare (e.g., [36,37,43]) and indicates that their understanding of what positive welfare entails derives from personal values that small-scale extensive farming is the best for animals' quality of life [44]. Interestingly, participating citizens, particularly those in the 'free from negatives' frame, emphasise an *elimination* of negative experiences. This is evidenced in their repeated use of overt negatives, such as 'no' and 'not'. However, despite its emphasis on the positives in an animal's life, the positive welfare literature does not suggest that the prevention of negative aspects of welfare is no longer important, nor does it suggest these should or can be eliminated [10]. Rather it calls for a minimisation of the negatives and an enhancement of the positives, whereby promotion of the latter may incidentally produce the former [17]. Thus, the addition of 'positive' to 'animal welfare' may lead to similar issues associated with the Five Freedoms framework [10]; the language and words used lead people to deduce that negative aspects can and should be eliminated. As such, the phrasing of positive welfare appeared to influence citizens' expectations of what it would entail (e.g., elimination of negatives and promotion of positives).

Farmers' framing of positive animal welfare is 'fuzzier' than that of citizens; there were overlaps and similarities across the three farmer frames. This is likely because of the pre-existing norms and values farmers associate with animal welfare, as evidenced in the 'happy–healthy–productive' frame transcending between individual farmers. Indeed, the 'good husbandry' frame has similar attributes to the 'happy–healthy–productive' frame and the existing literature (e.g., [37,45]). For example, Skarstad et al. [46] similarly found that farmers emphasise the adequate provision of resource needs (i.e., food and water), keeping animals healthy and 'generally taking good care of the animals' (p. 80). However, farmers in the 'good husbandry' group discussed these themes directly in the context of positive animal welfare, indicating they assimilate positive welfare as being comparable to what they already strive to do (e.g., provide animals with good care). As such, the majority of farmers, when posed with the concept of positive welfare, integrated it into their existing frames of welfare and used it to reinforce their ideals as good farmers who looked after their animals 'positively'.

The superordinate nature of the 'happy–healthy–productive' frame and the relatively large number of farmers demonstrating the 'good husbandry' frame indicates the importance of pre-existing social norms and values in this context; these were the internal meaning frames farmers drew from to make sense of positive animal welfare. However, despite the strength of these pre-existing frames, it is notable that three different frames of positive welfare emerged. This may be a result of differences in personal values and motivational drivers between farmers. Indeed, farmers in the 'proactive welfare improvement' frame often discussed their own personal motivations in this context, where they emphasised the value they placed on seeking ways to continuously improve their current practices. Indeed, Hansson et al. [47] found that the emphasis farmers placed on different economic use or non-use values of animal welfare was strongly related to their personal values (as opposed to personality traits). Notably, they found that animal-centred farmers gave the most importance to non-use values of animal welfare (i.e., economic value not derived from the direct use of the animal). Thus, the 'animal's point of view' frame of positive welfare was potentially shaped by the personal predilection of these farmers to give priority to animals' wants and desires, regardless of the direct economic value of such acts. Different motivational orientations may thus account for why farmers demonstrated different frames of positive welfare. For instance, when investigating the motivational drivers of European farmers, Baur et al. [48] found that most farmers were more conservative than the general population and tended to be less open to change. This study's finding that most farmers have a 'good husbandry' frame may be a reflection of such conservative motivational orientations, where positive animal welfare was conservatively integrated into existing welfare norms. Nevertheless, the remaining farmers presented very different values and motivations. Further research in this area would thus be of benefit, to explore how such individual motivations and values may affect the way farmers assimilate and respond to positive animal welfare.

The key point is that frames—both in thought and in communication—provide an individual with a reference point for what is important [49]. Amongst citizens, it was evident that the communication of 'positive' and its automatic association with 'negative' was central to their understanding of positive welfare. In addition, their internal frames of reference about animal welfare influenced what they perceived positive welfare entailed—a small, traditional farm with animals experiencing a 'natural' outdoor environment. Within farmers' narratives, it was evident that their pre-existing frames of animal welfare (e.g., ensuring resource needs are met) referenced what was important for welfare and thus formed the reference point from which positive welfare was judged. Consequently, positive welfare was assessed from the aspects of welfare farmers considered positive or good (e.g., maintaining health and well-being). Overall, farmers and citizens demonstrate two interpretation approaches whereby they looked to their linguistic and semantic associations with 'positive' and their pre-existing frames of animal welfare. This reflects the wider human decision-making literature, which notes that individuals often rely on what is most salient in the decision context (i.e., 'positive') and their previous experiences (i.e., existing animal welfare norms) to interpret ambiguous information (e.g., positive animal welfare) [50,51]. This has implications for the assimilation of positive welfare amongst individuals as such internal frames act as "mentally stored clusters of ideas that guide individuals' processing of information" ([3], p.53). Consequently, as evidenced in this study's findings, farmers and citizens did not approach positive welfare from a neutral state but assimilated it in a way which was consistent with their understanding of 'positive' and their own attitudes to farm animal welfare [52].

A particularly noteworthy consequence of the communication of 'positive welfare' is citizens' association with 'negative welfare'. This is all the more pertinent given the concerns many farmers raised about positive welfare producing such comparisons within society. As will be discussed in the following section, the knowledge of such internal frames reveals implications for the effective framing of positive welfare in communication.

5. Implications for the Framing of Positive Welfare in Communication

The internal frames individuals hold matter for how they interpret information and thus, how they act. As such, if animal welfare research is to be effectively transferred to key stakeholders, it is essential to consider what story or frame is engaged or reinforced by the message frames used to communicate it [53]. As suggested by this study's findings, the term 'positive animal welfare' currently reinforces the comparison of negative and positive facets of animal welfare amongst citizens, and somewhat reinforces farmers' perception that science and wider society is critical of their current welfare practices. As Entman (p.52, [3]) explains, "to frame is to select some aspects of a perceived reality and make them more salient in a communicating text, in such a way to promote a particular problem definition, causal interpretation, moral evaluation, and/or treatment recommendation for the item described". Therefore, effectively framing positive welfare in communication requires a consideration of what aspects to make salient and how they will be received within the internal frames held by the target audience. The findings of this study highlight several implications in this regard.

First and foremost, effectively framing positive welfare in communication requires careful consideration of the salience of 'positive', and its indirect emphasis of 'negative', on individual perception. Citizens and farmers have limited opportunity or indeed inclination, to learn that positive animal welfare derives its name from human positive psychology. Consequently, its meaning is ambiguous, and direct reference to 'positive' engenders potentially detrimental framing effects; citizens make comparisons with negatives and farmers often defensively affirm their efficacy as competent animal caretakers. As it is not what is communicated but how something is communicated that matters, inclusion of the word 'positive' appears to disproportionately sway an individual's thoughts to what 'positive' means to them in the context of animal welfare (e.g., extensive farming for citizens and good husbandry to farmers). Consequently, the essence of what positive welfare is trying to achieve (e.g., a balance of positive and negative experiences [20]) is lost in their cognition and assimilation of the term. As such, there is arguably a need to revolutionise the language used to communicate animal welfare. Rather than words that lead to comparative associations, effective message framing would make salient the overarching end goal of positive animal welfare (e.g., 'animal well-being' or 'flourishing farm animals') and as discussed below, consider the internal frames and motivations of the target audience.

For frames in communication to be effective, they should be "easily accessible and resonate with the existing beliefs of the audience" (p.3, [54]). However, this study found that 'positive animal welfare' is, unsurprisingly, unfamiliar to citizens and farmers, making it an ineffective communication frame. As such, it may be better to frame positive welfare in line with the target audience's existing associations with farm animal welfare. For instance, citizens often associate good welfare with positive affective states [44], natural behaviours [55] and positive human–animal relationships [43]. Positive welfare supports and seeks to promote each of these factors [17]. Effective communication of positive welfare to citizens should thus make salient its association with these factors, so that these beliefs are communicated as applicable to their evaluation of positive welfare [9]. Framing positive welfare in line with farmers existing animal welfare beliefs may be even more critical, as farmers often resist external knowledge cultures [56,57] and make decisions in accordance with existing social and cultural capital (i.e., 'the rules of the game') [58]. As evident in both this study and previous research, farmers almost ubiquitously frame animal welfare in terms of health and productivity [45,46]. Thus, framing positive welfare in terms of its potential to enhance animal health and productivity [59] may prove effective, because it resonates with farmers' existing beliefs. Additionally, as suggested by participating farmers, framing positive welfare's potential to differentiate farmers on the market may be an effective communication strategy as it evokes farmers' desire to convey their social and cultural capital as 'good farmers' to the external social world [58,60]. Using the existing belief frames of a target audience to frame a communicating message works, because it evokes the available information stored in the memory of the individual. This provides them with an appropriate point of reference from which to evaluate information, and thus supports more effective assimilation and understanding of the concept [9].

A further implication for effective communication framing is the underlying motivational orientation of the target audience. Research has found that an individual's motivation to do something increases when that activity fits with their motivational orientation [61]. Central to this is the concept of regulatory focus, which posits that an individual can pursue the achievement of a goal with either a promotion-focused orientation or a prevention-focused orientation [62]. A promotion-focused individual "is concerned with advancement and accomplishment, with the presence and absence of positive outcomes" (p.171, [63]). Conversely, a prevention-focused individual is concerned with the "absence and presence of negative outcomes, with protection, safety and responsibilities" (p.171, [63]). For instance, farmers within the 'proactive welfare improvement' frame appear to hold a promotion-focus (e.g., want to advance and improve), while farmers in the 'good husbandry' frame appear to hold a prevention-focus (e.g., want to prevent negative outcomes or maintain current standards). Critically, leveraging the particular regulatory focus people are sensitive to can increase motivation [61]. Indeed, research finds that messages framed positively are more influential under a promotion-focus, while messages framed negatively are more influential under a prevention-focus [64]. This may account for why 'good husbandry' farmers did not frame positive welfare differently to pre-existing frames of welfare; its positive emphasis may not have matched with their prevention orientation. Conversely, 'proactive welfare improvement' farmers assimilated positive welfare as being something above and beyond existing welfare provisions, potentially as its positive emphasis matched their promotion orientation. This is supported by wider research findings that message effectiveness is enhanced when the message frame is compatible with the regulatory focus of the individual (e.g., prevention or promotion) [63–65]. Therefore, particular subsets and attributes of positive welfare (and indeed animal welfare more generally) may be best communicated in either prevention or promotion frames. For instance, health concerns may be best framed with a prevention focus, while enhancing positive affect in animals may be best framed with a promotion focus.

As Chong and Druckman (p.109, [9]) argue "frames in communication matter — that is, they affect the attitudes and behaviours of their audiences". Thus, how positive welfare is communicated may determine how key animal welfare stakeholders understand and act in response. Yet, as noted in this study, 'positive animal welfare' represents an unknown and ambiguous concept to many. Consequently, the phrase may not effectively communicate what positive welfare is about or what it seeks to achieve. Moreover, the salience of the 'positive' appears to reinforce the undesired frame of negative welfare. To effectively frame positive welfare in communication, it is arguably better to reflect what positive welfare aims to achieve (e.g., a good quality of life, animal well-being, etc.) rather than where it originated (i.e., positive animal welfare reflecting the notion of positive psychology). Moreover, the effective communication and transfer of positive animal welfare from science to society may benefit from tailoring its message to the existing frames and motivations of the target audience. In this regard, a key question for welfare scientists is—what message should positive welfare convey and what behaviours should it motivate? This may require a deeper reflection on the vision for the future of positive welfare and the role of different societal actors within this.

6. Conclusions

Framing is an important theoretical framework for understanding why individuals interpret and perceive research evidence as they do and how best to communicate such research to different actors in the society. Positive animal welfare brings a change in frame to the traditional study of animal welfare, where greater emphasis is placed on the positive experiences in an animal's life. However, little is known about how citizens and farmers may understand this concept or how such re-framing of welfare may impact their behavioural responses. This study was but an inaugural step into the exploration and examination of positive animal welfare in society and its effects on the perceptions, attitudes and behaviours of key animal welfare stakeholders. As no previous research, to the author's knowledge, has asked or explored what 'positive animal welfare' means to citizens and farmers, it makes a fundamental contribution to the further development and expansion of positive animal welfare. Such explorations of the framing of positive welfare are timely, as positive welfare is a relatively recent concept that has not yet proliferated into wider society. This provides a great window of opportunity to ensure that positive welfare is developed in sympathy with the existing norms, values and meaning frames of key welfare stakeholders and is communicated in a way that resonates with and motivates, rather than alienates, such individuals. Understanding the internal frames individuals hold is essential for shaping the effective communication of positive welfare, and in turn, influencing the human behaviours which affect farm animal welfare.

Funding: This research was supported by funding from the Scottish Government's Rural and Environment Science and Analytical Services Division (RESAS).

Acknowledgments: The researcher gratefully acknowledges the input of each of the research participants and their generosity in sharing their time and thoughts.

Conflicts of Interest: The author declares no conflict of interest.

Appendix A

Table A1. Farmers' demographic information.

Sector Dairy	Gender	Age	Farm Size (Ha)	Number of Animals	System
1	Male	30–40	130	100–200	Pasture
2	Male	50–60	137	200–300	Pasture
3	Male	18–30	62	100–200	Pasture
4	Male	30–40	343	700–800	Zero-grazed
5	Male	50–60	283	100–200	Pasture

Table A1. *Cont.*

Sector Dairy	Gender	Age	Farm Size (Ha)	Number of Animals	System
6	Male	30–40	160	300–400	Pasture and robotic milking
7	Male	40–50	344	300–400	Pasture and zero-grazed, non-robotic and robotic milking
8	Male	30–40	100	100–200	Zero-grazed
9	Female	18–30	307	300–400	Zero-grazed
10	Male	40–50	776	1000–1500	Outdoor 365 days/ year
11	Female	30–40	687	400–500	Pasture and zero-grazed
12	Male	40–50	176	100–200	Organic and robotic milking
13	Male	40–50	283	800–1000	Zero-grazed
Beef and Sheep					
1	Male	60–70	178	600–700	Indoor-wintered
2	Male	40–50	438	200–300	Indoor-wintered
3	Male	30–40	95	200–300	Outdoor-wintered
4	Male	50–60	230	400–500	Outdoor-wintered
5	Female	30–40	4	<100	Indoor-wintered
6	Male	50–60	100	400–500	Outdoor-wintered
7	Male	40–50	1011	200–300	Indoor-wintered
8	Male	60–70	60	400–500	Outdoor-wintered
9	Male	40–50	500	1000–1500	Outdoor-wintered & Indoor-wintered
10	Prefer not to say	40–50	750	1000–1500	Indoor-wintered
Poultry (egg)					
1	Male	30–40	141	10000–15000	Free range and organic
2	Male	50–60	95	120000–130000	Free range
Mixed					
1	Male	40–50	54	200–300	Free range (pig and poultry), organic (all species), indoor-wintered (beef), outdoor-wintered (sheep)
2	Male	30–40	230	1000–1500	Free range (poultry), straw-housed (pig), outdoor-wintered (sheep)
Pig					
1	Male	30–40	555	2000–3000	Housed (slats and straw)

Table A2. Citizens' demographic information.

Citizen	Gender	Age	Farming Experience	Urban Rural	Dietary Preferences
1	Female	40–50	None	Large urban	Reduced meat consumption
2	Female	30–40	None	Large urban	Reduced meat consumption
3	Female	18–30	Farming background	Rural	Vegetarian
4	Male	18–30	Visited farms	Large urban	Consume meat
5	Male	18–30	Farming background	Large urban	Reduced meat consumption
6	Female	50–60	Farming background	Rural	Consume meat
7	Female	30–40	Visited farms	Small town	Consume meat
8	Female	18–30	Visited farms	Small town	Vegetarian
9	Male	18–30	Visited farms	Large urban	Consume meat
10	Male	50–60	Farming background	Large urban	Consume meat
11	Female	18–30	None	Large urban	Reduced meat consumption
12	Male	18–30	Visited farms	Large urban	Vegan
13	Male	40–50	Visited farms	Large urban	Consume meat
14	Female	18–30	None	Large urban	Reduced meat consumption
15	Female	30–40	Visited farms	Large urban	Reduced meat consumption

References

1. Peden, R.S.E.; Akaichi, F.; Camerlink, I.; Boyle, L.A.; Turner, S.P. Factors influencing farmer willingness to reduce aggression between pigs. *Animals* **2019**, *9*, 6. [CrossRef]

2.   Bracke, M.B.M.; de Lauwere, C.C.; Wind, S.M.M.; Zonerland, J.J. Attitudes of dutch pig farmers towards tail biting and tail docking. *J. Agric. Environ. Ethics* **2013**, *26*, 847–868. [CrossRef]
3.   Entman, R.M. Framing: Toward clarification of a fractured paradigm. *J. Commun.* **1993**, *43*, 51–58. [CrossRef]
4.   Druckman, J.N. The implications of framing effects for citizen competence. *Polit. Behav.* **2001**, *23*, 225–256. [CrossRef]
5.   Chong, D.; Druckman, J. A theory of framing and opinion formation in competitive elite environments. *J. Commun.* **2007**, *57*, 99–118. [CrossRef]
6.   Nijland, H.J.; Aarts, N.; van Woerkum, C.M.J. Exploring the framing of animal farming and meat consumption: On the diversity of topics used and qualitative patterns in selected demographic contexts. *Animals* **2018**, *8*, 17. [CrossRef] [PubMed]
7.   Kahneman, D.; Tversky, A. Prospect Theory: An analysis of decision under risk. *Econometrica* **1979**, *47*, 263–291. [CrossRef]
8.   Weary, D.; Robbins, J. Understanding the multiple conceptions of animal welfare. *Anim. Welf.* **2019**, *28*, 33–40.
9.   Chong, D.; Druckman, J.N. Framing theory. *Annu. Rev. Polit. Sci.* **2007**, *10*, 103–126. [CrossRef]
10.   Mellor, D.J. Updating animal welfare thinking: moving beyond the "Five Freedoms" towards "A life worth living". *Animals* **2016**, *6*, 21. [CrossRef]
11.   Levin, I.P.; Schneider, S.L.; Gaeth, G.J. All frames are not created equal: a typology and critical analysis of framing effects. *Organ. Behav. Hum. Decis. Process.* **1998**, *76*, 149–188. [CrossRef]
12.   Donovan, R.J.; Jalleh, G. Positive versus negative framing of a hypothetical infant immunization: The influence of involvement. *Health Educ. Behav.* **2000**, *27*, 82–95. [CrossRef] [PubMed]
13.   Gamliel, E.; Peer, E. Positive versus negative framing affects justice judgments. *Soc. Justice Res.* **2006**, *19*, 307–322. [CrossRef]
14.   Kirchler, E.; Maciejovsky, B.; Weber, M. Framing effects, selective information, and market behavior: An experimental analysis. *J. Behav. Finance* **2005**, *6*, 90–100. [CrossRef]
15.   Lee, H.-C. Positive or negative? The influence of message framing, regulatory focus, and product type. *Int. J. Commun.* **2018**, *12*, 788–805.
16.   Lawrence, A.B.; Newberry, R.C.; Špinka, M. Positive welfare: What does it add to the debate over pig welfare? In *Advances in Pig Welfare*; Špinka, M., Ed.; Elsevier Science and Technology: Cambridge, UK, 2017.
17.   Yeates, J.W.; Main, D.C.J. Assessment of positive welfare: A review. *Vet. J.* **2008**, *175*, 293–300. [CrossRef]
18.   Boissy, A.; Manteuffel, G.; Jensen, M.B.; Moe, R.O.; Spruijt, B.; Keeling, L.J.; Winckler, C.; Forkman, B.; Dimitrov, I.; Langbein, J.; et al. Assessment of positive emotions in animals to improve their welfare. *Physiol. Behav.* **2007**, *92*, 375–397. [CrossRef] [PubMed]
19.   Napolitano, F.; Knierim, U.; Grass, F.; de Rosa, G. Positive indicators of cattle welfare and their applicability to on-farm protocols. *Ital. J. Anim. Sci.* **2009**, *8*, 355–365. [CrossRef]
20.   Green, T.C.; Mellor, D.J. Extending ideas about animal welfare assessment to include 'quality of life' and related concepts. *N. Z. Vet. J.* **2011**, *59*, 263–271. [CrossRef]
21.   Mellor, D.J. Positive animal welfare states and reference standards for welfare assessment. *N. Z. Vet. J.* **2015**, *63*, 17–23. [CrossRef]
22.   Blackmore, E.; Underhill, R.; McQuilkin, J.; Leach, R. *Common Cause for Nature: Values and Frames in Conservation*; Public Interest Research Centre: Machynlleth, Wales, UK, 2013; pp. 1–180.
23.   Mellor, D.J. Enhancing animal welfare by creating opportunities for positive affective engagement. *N. Z. Vet. J.* **2015**, *63*, 3–8. [CrossRef]
24.   Christley, R.M.; Perkins, E. Researching hard to reach areas of knowledge: Qualitative research in veterinary science. *Equine Vet. J.* **2010**, *42*, 285–286. [CrossRef]
25.   Hollway, W.; Jefferson, T. The free association narrative interview method. In *The SAGE Encyclopaedia of Qualitative Research Methods*; Given, L., Ed.; Sage: Sevenoaks, CA, USA, 2008; pp. 296–315.
26.   Jovchelovitch, S.; Bauer, M.W. Narrative interviewing. In *Qualitative Researching with Text, Image and Sound: A Practical Handbook*; SAGE: London, UK, 2000; pp. 57–74.
27.   Akaichi, F.; Revoredo-Giha, C. Consumers demand for products with animal welfare attributes: Evidence from homescan data for Scotland. *Br. Food J.* **2016**, *118*, 1682–1711. [CrossRef]
28.   Devitt, C.; Kelly, P.; Blake, M.; Hanlon, A.; More, S.J. An Investigation into the human element of on-farm animal welfare incidents in Ireland. *Sociol. Rural.* **2015**, *55*, 400–416. [CrossRef]

29. Robinson, O.C. Sampling in interview-based qualitative research: A Theoretical and Practical Guide. *Qual. Res. Psychol.* **2014**, *11*, 25–41. [CrossRef]

30. Palinkas, L.A.; Horwitz, S.M.; Green, C.A.; Wisdom, J.P.; Duan, N.; Hoagwood, K. Purposeful sampling for qualitative data collection and analysis in mixed method implementation research. *Adm. Policy Ment. Health* **2015**, *42*, 533–544. [CrossRef] [PubMed]

31. Scottish Government, S.A.H. Scottish Government Urban Rural Classification. Available online: http://www2.gov.scot/Topics/Statistics/About/Methodology/UrbanRuralClassification (accessed on 5 February 2019).

32. Glaser, B.G.; Strauss, A.L. *The Discovery of Grounded Theory: Strategies for Qualitative Research*; Aldine: Chicago, IL, USA, 1967.

33. Boeije, H. A purposeful approach to the constant comparative method in the analysis of qualitative interviews. *Qual. Quant.* **2002**, *26*, 391–409. [CrossRef]

34. Van Poucke, E.; Vanhonacker, F.; Nijs, G.; Braeckman, J.; Verbeke, W.; Tuyttens, F. Defining the concept of animal welfare: Integrating the opinion of citizens and other stakeholders. In *Proceedings of the 6th Congress of the European Society for Agricultural and Food Ethics*; Wageningen Academic Publishers: Wageningen, The Netherlands, 2006; pp. 555–559.

35. Cruse, D.A. *Lexical Semantics*; Cambridge University Press: Cambridge, UK, 1986.

36. Spooner, J.M.; Schuppli, C.A.; Fraser, D. Attitudes of Canadian citizens toward farm animal welfare: A qualitative study. *Livest. Sci.* **2014**, *163*, 150–158. [CrossRef]

37. Te Velde, H.; Aarts, N.; Van Woerkum, C. Dealing with ambivalence: farmers' and consumers' perceptions of animal welfare in livestock breeding. *J. Agric. Environ. Ethics* **2002**, *15*, 203–219. [CrossRef]

38. Ventura, B.A.; von Keyserlingk, M.A.G.; Wittman, H.; Weary, D.M. What difference does a visit make? changes in animal welfare perceptions after interested citizens tour a dairy farm. *PLoS ONE* **2016**, *11*, e0154733. [CrossRef]

39. Mochizuki, M.; Osada, M.; Ishioka, K.; Matsubara, T.; Momota, Y.; Yumoto, N.; Sako, T.; Kamiya, S.; Yoshimura, I. Is experience on a farm an effective approach to understanding animal products and the management of dairy farming? *Anim. Sci. J.* **2014**, *85*, 323–329. [CrossRef] [PubMed]

40. Musto, M.; Cardinale, D.; Lucia, P.; Faraone, D. Influence of different information presentation formats on consumer acceptability: The case of goat milk presented as obtained from different rearing systems. *J. Sens. Stud.* **2015**, *30*, 85–97. [CrossRef]

41. Miele, M.; Veissier, I.; Evans, A.; Botreau, R. Animal welfare: establishing a dialogue between science and society. *Anim. Welf.* **2011**, *20*, 103.

42. Edgar, J.L.; Mullan, S.M.; Pritchard, J.C.; McFarlane, U.J.C.; Main, D.C.J. Towards a 'Good Life' for farm animals: Development of a resource tier framework to achieve positive welfare for laying hens. *Animals* **2013**, *3*, 584–605. [CrossRef] [PubMed]

43. Ellis, K.A.; Billington, K.; McNeil, B.; McKeegan, D.E.F. Public opinion on UK milk marketing and dairy cow welfare. *Anim. Welf.* **2009**, *18*, 267–282.

44. Miele, M. *Report Concerning Consumer Perceptions and Attitudes Towards Farm Animal Welfare*; European Animal Welfare Platform: Brussels, Belgium, 2010; pp. 1–16.

45. Spooner, J.M.; Schuppli, C.A.; Fraser, D. Attitudes of Canadian pig producers toward animal welfare. *J. Agric. Environ. Ethics* **2014**, *27*, 569–589. [CrossRef]

46. Skarstad, G.A.; Terragni, L.; Torjusen, H. Animal welfare according to Norwegian consumers and producers: definitions and implications. *Int. J. Sociol. Food Agric.* **2007**, *15*, 74–90.

47. Hansson, H.; Lagerkvist, C.J.; Vesala, K.M. Impact of personal values and personality on motivational factors for farmers to work with farm animal welfare: a case of Swedish dairy farmers. *Anim. Welf.* **2018**, *27*, 133–145. [CrossRef]

48. Baur, I.; Dobricki, M.; Lips, M. The basic motivational drivers of northern and central European farmers. *J. Rural Stud.* **2016**, *46*, 93–101. [CrossRef]

49. Carter, M.J. The hermeneutics of frames and framing: An examination of the media's construction of reality. *SAGE Open* **2013**, *3*, 2158244013487915. [CrossRef]

50. Tversky, A.; Kahneman, D. The Framing of decisions and the psychology of choice. *Science* **1981**, *211*, 453–458. [CrossRef] [PubMed]

51. Higgins, E.T. Knowledge activation: Accessibility, applicability, and salience. In *Social Psychology: Handbook of basic principles*; Higgins, E.T., Kruglanski, A.W., Eds.; The Guildford Press: New York, NY, USA, 1996; pp. 133–168.

52. Hameleers, M.; Vliegenthart, R. Framing the participatory society: Measuring discrepancies between interpretation frames and media frames. *Int. J. Public Opin. Res.* **2018**, *30*, 257–281. [CrossRef]

53. Meade, D. *Framing Nature Toolkit*; Public Interest Research Centre: Machynlleth, Wales, UK, 2018.

54. Sullivan, M.; Longnecker, N. Choosing effective frames to communicate animal welfare issues. In *Proceedings of the ResearchGate; Science Communication Program*; The University of Western Australia: Crawley, Australia, 2010; pp. 1–6.

55. Lassen, J.; Sandøe, P.; Forkman, B. Happy pigs are dirty!—Conflicting perspectives on animal welfare. *Livest. Sci.* **2006**, *103*, 221–230. [CrossRef]

56. Tsouvalis, J.; Seymour, S.; Watkins, C. Exploring knowledge-cultures: Precision farming, yield mapping, and the expert–farmer interface. *Environ. Plan. Econ. Space* **2000**, *32*, 909–924. [CrossRef]

57. Anneberg, I.; Vaarst, M.; Sørensen, J.T. The experience of animal welfare inspections as perceived by Danish livestock farmers: A qualitative research approach. *Livest. Sci.* **2012**, *147*, 49–58. [CrossRef]

58. Shortall, O.; Sutherland, L.-A.; Ruston, A.; Kaler, J. True cowmen and commercial farmers: exploring vets' and dairy farmers' contrasting views of 'good farming' in relation to biosecurity. *Sociol. Rural.* **2018**, *58*, 583–603. [CrossRef]

59. Manteuffel, G.; Langbein, J.; Puppe, B. Increasing farm animal welfare by positively motivated instrumental behaviour. *Appl. Anim. Behav. Sci.* **2009**, *118*, 191–198. [CrossRef]

60. Christensen, T.; Denver, S.; Sandøe, P. How best to improve farm animal welfare? Four main approaches viewed from an economic perspective. *Anim. Welf.* **2019**, *28*, 95–106. [CrossRef]

61. Aaker, J.L.; Lee, A.Y. Understanding Regulatory Fit. *J. Mark. Res.* **2006**, *43*, 15–19. [CrossRef]

62. Higgins, E.T. Promotion and prevention: regulatory focus as a motivational principle. In *Advances in Experimental Social Psychology*; Zanna, M.P., Ed.; Elsevier: Amsterdam, The Netherlands, 1998; Volume 30, pp. 1–46, ISBN 978-0-12-015230-8.

63. Lin, H.-F.; Shen, F. Regulatory focus and attribute framing: Evidence of compatibility effects in advertising. *Int. J. Advert.* **2012**, *31*, 169–188. [CrossRef]

64. Lee, A.Y.; Aaker, J.L. Bringing the frame into focus: the influence of regulatory fit on processing fluency and persuasion. *J. Pers. Soc. Psychol.* **2004**, *86*, 205–218. [CrossRef] [PubMed]

65. Kim, Y.-J. The role of regulatory focus in message framing in antismoking advertisements for adolescents. *J. Advert.* **2006**, *35*, 143–151. [CrossRef]

Article

Veterinary Students' Perception and Understanding of Issues Surrounding the Slaughter of Animals According to the Rules of Halal: A Survey of Students from Four English Universities

Awal Fuseini *, Andrew Grist and Toby G. Knowles

School of Veterinary Science, University of Bristol, Langford, Bristol BS40 5DU, UK;
andy.grist@bristol.ac.uk (A.G.); toby.knowles@bristol.ac.uk (T.G.K.)
* Correspondence: awalfus@yahoo.com; Tel.: +44-074-7419-2392

Received: 18 April 2019; Accepted: 27 May 2019; Published: 30 May 2019

Simple Summary: Veterinarians play a vital role in safeguarding the welfare of animals and protecting public health during the slaughter of farm animals for human consumption. With the continued increase in the number of animals slaughtered for religious communities in the UK, this study evaluates the perception and level of understanding of religious slaughter issues by veterinary students at various stages of their studies. The results showed a significant effect of the year of study on respondents' understanding of the regulations governing Halal slaughter and other related issues. Whilst the prevalence of vegetarianism and veganism were higher than the general UK population, the majority of respondents were meat eaters who indicated that they prefer meat from humanely slaughtered (pre-stunned) animals, irrespective of whether slaughter was performed with an Islamic prayer (Halal) or not.

Abstract: The objective of this study was to evaluate the perception and level of understanding of religious slaughter issues, and the regulations governing the process, amongst veterinary students in England. A total of 459 veterinary students in different levels, or years of study (years 1–5), were surveyed. On whether there is a need for food animals to be stunned prior to slaughter, the majority of respondents 437 (95.2%) indicated that they would want all animals to be stunned before slaughter, including during religious slaughter, 17 (3.6%) either did not have an opinion or indicated 'other' as their preferred option and 5 (1.1%) indicated that religious slaughter should be exempt from stunning in order to comply with traditional religious values. The results showed a significant association between respondents' year of study and (i) their understanding of UK animal welfare (at slaughter) regulations, (ii) their recognition of stunning as a pain-abolishing procedure and (iii) the likelihood of them wittingly purchasing and consuming meat from animals that have been stunned prior to slaughter, and also classified as Halal.

Keywords: animal welfare; stunning; religious slaughter; veterinary students; Halal meat

---

1. Introduction

The slaughter of food animals, whether stunned or not, is an emotive issue that has long divided opinion. Those against the production of farm animals, their slaughter and the subsequent consumption of meat have often cited the effect of livestock agriculture on the environment [1–4], a decline in the population of wild animals as a consequence of cultivating animal feed [5] or simply put a case for animal rights [6] or religiosity [7]. However, the perceived importance of meat in the diet of man cannot be underestimated. Meat is seen by many as an important source of proteins, amino acids, vitamins and other essential nutrients required for the sustenance of life [8,9]. The slaughter of animals in many industrialised economies is a highly regulated procedure; these regulatory measures are put in place to protect the welfare of animals (and the health and safety of operatives) and to ensure that meat is fit for human consumption. Within the European Union, the protection of animals at the time of slaughter is regulated under Council Regulation, EC1099/2009 [10] which specifies acceptable pre-slaughter procedures and approved slaughter methods for food animals. To protect animal welfare, EC1099/2009 requires the stunning of all animals prior to death (itself caused by bleeding out), with the exception of animals slaughtered in accordance with religious rites, this being mainly for Shechita and Halal slaughter. Halal slaughter is practised by followers of the Islamic faith, and animals are required to be alive prior to bleeding and a prayer is said by the slaughterer at the time of neck-cutting on every animal. Shechita slaughter, on the other hand, is practised by followers of the Jewish faith, and again animals are required to be alive and a prayer is said; however, during Shechita slaughter, there is no requirement for the prayer to be recited on every animal. Whilst some Muslims accept stunning during Halal slaughter, the Jewish community unanimously reject all forms of stunning. Member states can apply a derogation to permit slaughter without stunning, and this derogation is in place in the English domestic regulation, the Welfare of Animals at the Time of Killing (WATOK 2015) Regulation [11]. Research using, for example, Electroencephalogram (EEG) recordings of the electrical activity of brain has demonstrated that this method of slaughter does compromise animal welfare [12–14]. Contrary to the findings of Gibson and colleagues [12], other researchers [15] subjectively observed the behaviour of some three thousand cattle slaughtered without stunning (in line with the Shechita rules) and concluded that, 'in their opinion', the animals did not exhibit overt behaviours that were consistent with pain. The UK and other member states have used the derogation to permit slaughter without stunning under strict conditions. In Great Britain, the domestic regulation (WATOK) [11,16,17] requires animals to be individually and mechanically restrained during slaughter without stunning, and that ruminants must not be moved after the neck-cut until they lose sensibility. As a regulation, sheep must stand still for at least 20 s whilst cattle must not be moved for at least 30 s following the neck-cut. The essence of the standstill time is to ensure that animals lose sensibility due to blood loss before they are moved in order to avoid any additional pain or distress associated with the process. Further, abattoirs slaughtering animals without stunning must also have a backup stunner (a requirement also of abattoirs using pre-slaughter stunning) to be used in the event of delayed loss of consciousness after the neck-cut.

Animal rights and welfare groups continue to publicise the negative aspects of slaughter (both stun and nonstun) with a view to highlighting welfare compromises during slaughter. Over the last decade or so, UK-based animal welfare charity, Animal Aid, has released several covert recordings [18] taken in abattoirs that have highlighted animal suffering and have used this as an argument for veganism [19]. Harper and Makatouni [20] suggested that consumers are becoming well informed about the welfare aspects of livestock agriculture and are opting for welfare-friendly products. Some consumers are well informed about the role official veterinarians (OVs) play in safeguarding animal health and welfare. Wall [21] reported that the role of OVs is paramount in the implementation of the 'one health' initiative which is a collaborative, multidisciplinary approach to ensuring the optimal attainment of good health and welfare of human and nonhuman animals globally. OVs receive specialist veterinary training to be licenced to work in abattoirs to safeguard animal welfare and human health. However, Spinka [22] noted that there are gaps in the level of knowledge of OVs across the EU due to differences in the modules taught and the depth of subjects covered by different EU universities.

As potential future enforcers of religious slaughter laws in the UK (and other parts of the world), veterinary students at four English universities were recruited to participate in this study. The aim was to evaluate their perception and understanding of the regulations governing religious slaughter as it stands in the UK. The paper further examines the difference in the level of understanding of these issues amongst different year groups. To the best of our knowledge, there are no prior published data on veterinary students' perception and understanding of religious slaughter.

2. Materials and Methods

2.1. Data Collection and Sampling Methods

A total of 459 veterinary students from four universities in England participated in the study; University of Bristol ($n = 344$), University of Nottingham ($n = 57$), University of Liverpool ($n = 45$) and Royal Veterinary College ($n = 13$). Prior to the survey, all students were provided information on the aims and objectives of the study and all respondents gave informed consent to participate in the study. Respondents' data were anonymised. Data were collected using 'SurveyMonkey' online software by sending a weblink to students to allow them to participate at a time convenient to themselves. One of the possible limitations of this study is that respondents were not asked about their religion. The University of Bristol's Ethical Review Board granted ethical approval for the study (ID75001).

2.2. Data Analysis

Responses to questions are reported as percentages with actual number of respondents contributing following in brackets. Exact chi square tests were used to test for associations between categorical variables; where there were ordered categorical variables, a test for trend was carried out using an exact Gamma statistic (IBM SPSS Statistics Version 25, IBM, Inc., New York, NY, USA).

3. Results

One respondent was dropped from the analysis because he/she did not answer a majority of the questions. Elsewhere, where there were occasional missed questions; those respondents are not included in the count or calculation of percentage. Omissions are treated as missing at random (there were 12 questions requiring a response, and there were 8 individual (respondents) omissions in total across 7 of these questions). The mean age of respondents was 22 with a range of 18–39. Respondents were recruited from four universities in England offering veterinary degrees, and the majority of respondents indicated their programmes of study as veterinary science 98.9% (454), with the remaining 1.1% (5) selecting 'other' (Veterinary Medicine and Surgery, Veterinary Medicine, Bioveterinary science and Veterinary Medicine with Intercalation) as their programmes of study. The levels or years of study of respondents were 22.9% (105) in the fourth year, 22.7% (104) in the third year, 21.2% (97) in the second year, 18.3% (84) in the fifth year and 14.9% (68) in the first year. Respondents' homes of origin were from cities and towns across the UK with the larger proportions of respondents indicating they came from London 6.1% (28), Bristol 6.1% (28), Nottingham 2.8% (13) and Manchester 2.8% (13) areas.

Respondents were asked whether they were meat eaters and if so, their level of meat consumption. One respondent did not answer this question. The majority of respondents indicated that they were meat eaters, 81.3% (372), whilst 19.2% (88) indicated that they did not eat meat and 0.9% (4) chose only dietary exclusions by choice. Of the 372 meat eaters, 62.5% (286) indicated that they ate meat regularly, whilst 18.8% (86) said they ate meat occasionally. The 88 respondents who indicated that they did not eat meat were 15.3% (70) vegetarians and 3.9% (18) vegans. Respondents were presented with the statement: 'At slaughter, the death of an animal takes place because the major blood vessels are severed, and critical blood loss occurs. This process is thought to be painful' and were then asked if they agreed with a series of statements. Two respondents did not answer these questions. The majority of respondents, 90.4% (413) indicated that 'pre-slaughter stunning abolishes the pain associated with the neck-cut during slaughter', whilst 9.6% (44) selected the option 'pre-slaughter stunning cannot abolish the pain associated with the neck-cut during slaughter'. There was a significant association between respondents' year of study and their response to whether stunning is capable of abolishing the pain associated with the neck-cut (chi sq. $= 33.0$, df $= 4$, $p < 0.001$). The proportion of those agreeing that pre-slaughter stunning abolishes the pain associated with the neck-cut during slaughter were 79.4% (54), 80% (77), 95.2% (99), 98.1% (103) and 95.2% (80) of years 1, 2, 3, 4 and 5, respectively (proportions per year). The perception and understanding of respondents with regard to halal slaughter in the UK was evaluated (one respondent did not answer this question). The majority of respondents,

36.9% (169) selected the following option: 'the majority of animals are not stunned, but some Muslims will accept meat from stunned animals as being Halal', 29% (133) selected 'all animals are required to be slaughtered without stunning', 23.1% (106) selected 'the majority of animals are stunned as most Muslims accept meat from stunned animals as Halal', whilst 10.9% (50) respondents indicated that they were not sure about the situation of Halal slaughter in the UK. The results indicated a significant association between the year of study and respondents' understanding regarding the situation with Halal slaughter in the UK; in later years of study, students understanding tended to improve (chi sq. = 84.2, df = 12, $p < 0.001$). Respondents' awareness about the permissibility of slaughter without stunning for religious slaughter was evaluated (see Table 1). One respondent did not answer the question, 90.4% (414) indicated 'Yes', whilst 9.6% (44) indicated 'No'. There was a significant association between respondents' year of study and their response to the above question, with a trend across years (a trend of increased awareness) (Gamma = 0.659, df = 4, $p < 0.001$). Table 1 shows a cross tabulation between respondents' year of study and their understanding of the permission of the slaughter of animals without stunning under UK animal welfare regulations.

Table 1. Cross tabulation of the year of study of respondent and awareness that slaughter without stunning for religious purposes was permissible in the UK.

Year of Study	UK Welfare Regulations Do Permit the Slaughter of Animals without Stunning, but Only for Religious Slaughter. Were You Aware of This?		
	No	Yes	Total
1	29% (20)	71% (48)	100% (68)
2	14% (14)	86% (83)	100% (97)
3	5% (5)	95% (99)	100% (104)
4	4% (4)	96% (101)	100% (105)
5	1% (1)	99% (83)	100% (84)
Totals	44	414	(458)

In a separate question, respondents were asked to share their opinion on the use of pre-slaughter stunning during meat production; 95.2% (437) of respondents indicated that all animals must be stunned before slaughter, including during religious slaughter, 2.2% (10) selected 'other' with the option to leave comments (Table 2 shows the comments left by these 10 respondents), 1.5% (7) indicated that they did not have an opinion on stunning and 1.1% (5) indicated that religious slaughter should be exempt from stunning to comply with traditional religious values. Respondents were also asked for their views on whether there is a need for meat to be labelled according to the method of slaughter (i.e., whether meat is from an animal that has been stunned or not). A total of 97.2% (446) indicated that meat should be labelled according to whether it was derived from stunned or nonstun animals, 1.3% (6) indicated that there is no need to label meat, whilst 1.5% (7) indicated they did not have an opinion. To gauge respondents' acceptability of meat derived from animals that had been effectively stunned during Halal slaughter, respondents were asked: 'If animals are effectively stunned before Halal slaughter, as an ordinary consumer, would you wittingly purchase and consume this type of Halal meat?'. Two separate analysis were made, first, with all respondents (including vegans and vegetarians) and a second, excluding vegans and vegetarians. In the first analysis, one respondent did not answer the question, 79.0% (362) answered 'Yes' and 21% (96) indicated 'No'. The results showed a significant trend with respondents' year of study and their willingness to purchase and consume Halal meat from effectively stunned animals (Gamma = 0.210, df = 4, $p = 0.009$), an increasing proportion answering 'yes', with increasing year of study (Table 3). In the second analysis, which excluded vegans and vegetarians, the majority of respondents, 88.4% (327) indicated that they would wittingly purchase Halal meat from stunned animals, whilst 11.6% (43) indicated that they would not purchase Halal meat from stunned animals. This still indicated a trend of increased acceptance of Halal meat from stunned animals as respondents progressed in their years of study, but this time, the trend was lesser and not statistically significant (Gamma = 0.147, df = 4, $p = 0.203$). Data were then further analysed to examine the attitudes of vegetarians and vegans alone towards Halal meat. The results showed a significant increase in the percentage of vegetarians and vegans who would wittingly purchase Halal meat from stunned animals as year of study increased (Gamma = 0.311, df = 4, p = 0.042).

Table 2. Comments by respondents who choose the option 'other' to the question regarding their opinion on the use of pre-slaughter stunning for meat animals.

There is no such thing as humane slaughter, no animal wants to die, therefore it is not humane
I do not agree with the exemption of stunning pre-slaughter and believe that stunning should be performed. This being said, I respect the choices and beliefs of religions other than my own.
All animals should be stunned before slaughter if is the most humane way—it is insanity that religion could come before the welfare of sentient beings.
I think a method should be employed to ensure the animals can't feel the pain of the true cause of death. I was told by a veterinarian that stunning renders the animal essentially 'brain-dead' and so prevents the feeling of pain during slaughter—any other method that achieves the same effect without causing further pain or discomfort to the animal would also be appropriate.
Conflicting between the two I have an opinion, but it changes whether I draw from my cultural upbringing or my veterinary knowledge.
Animals should not be slaughtered for meat at all, however whilst this continues to happen, they should all be stunned including for religious slaughter.
If animals aren't stunned, meat in stores should be labelled accordingly so that people can be informed/choose not to eat non-stunned meat.
I think it is difficult as people want to follow what their religion says, and I think they have that right, but I also think the welfare of the animal is important so I'm not sure if religious slaughter should be exempt from stunning or not.
Stunning makes us feel better at the animal doesn't display a classic pain response. It's impossible for us to actually know the level of pain the animal feels after stunning. If stunning does eliminate the pain, then it should be done in all cases of slaughter regardless of religious beliefs.
I agree most closely with the first option however I would rather non recoverable stun / killing methods were used to eliminate pain from recovery.

Table 3. Cross tabulation of the year of study of respondent and willingness to buy and consume Halal meat derived from animals that have been stunned before slaughter.

Year of Study	If Animals Are Effectively Stunned before Halal Slaughter, as an Ordinary Consumer, Would You Wittingly Purchase and Consume This Type of Halal Meat?		
	Yes	No	Total
1	69% (47)	31% (21)	100% (68)
2	75% (73)	25% (24)	100% (97)
3	82% (85)	18% (19)	100% (104)
4	81% (85)	19% (20)	100% (105)
5	86% (72)	14% (12)	100% (84)
Total	362	96	458

4. Discussion

The results of this study give an insight into the perception and understanding of religious slaughter issues by veterinary students at various levels of their studies from four universities in England. The results suggest some lack of a clear understanding of Halal slaughter with regard to the regulations and animal welfare issues surrounding the two main methods of slaughter, stun and nonstun. An understanding of these issues does appear to improve as they progress in their studies. The importance of the role of an independent, official veterinary surgeon (OV) in protecting animal welfare and public health cannot be underestimated. Due to the significance of their role, they require a better understanding of the burning issues around slaughter, particularly religious slaughter. The slaughter of animals under religious rites continues to attract public interest because of the insistence by a section of the religious communities that animals be slaughtered by severance of major blood vessels whilst they are fully conscious. There is scientific evidence to suggest that the slaughter of animals without stunning is painful [12,23], and loss of consciousness may be protracted [24], especially in the case of cattle were the vertebral artery is able to maintain blood supply to the brain [25]. However, a minority of the respondents in this study did not agree that stunning was necessary, by indicating that they believed 'stunning of animals prior to slaughter cannot abolish the pain associated with the neck-cut during slaughter'. The majority of the respondents 90.4% (413), however, indicated that they understood the slaughter of animals without stunning to be painful and that stunning is capable of abolishing the pain associated with neck-cutting. There was a trend of increased awareness of respondents' responses to whether stunning is capable of abolishing the pain associated with the neck-cut. This suggests that first-year students may not have yet undertaken any lectures on the science of stunning and slaughter and the rest of the years may have varying degrees of teaching, understanding and retention of the concept of stunning. Main [26] suggested that variation in veterinary curriculum and the way veterinary students are taught may account to variation in the level of understanding of students in the ever-evolving animal welfare module.

On the situation with regard to stunning of animals before Halal slaughter, the responses generally showed some lack of understanding of the facts surrounding Halal slaughter in the UK and suggest that this material needs to be better presented to students. This corroborates the conclusion made by Main [26], who noted that there is a need for veterinary institutions to include some core components of animal welfare in their curriculum in order to offer students a better appreciation of welfare science, ethics and standards. There is sufficient evidence from animal welfare surveys to show that the majority of animals are stunned during Halal slaughter in the England and Wales [27] and also within the EU [28]. There were, however, 23.1% (106) of respondents who indicated that they thought the majority of animals are stunned before Halal slaughter; this is consistent with the current situation in the England and Wales and parts of Europe [27,28]. Respondents' understanding on this issue (situation with regard to stunning of animals before Halal slaughter) tends to improve in later years of their study, with students describing the situation in line with current practices. On the acceptability to veterinary students of Halal meat from animals that have been stunned, the majority of respondents indicated that they would buy and consume Halal meat from stunned animals. Respondents who indicated that they would avoid stunned Halal meat may have done so with the belief that Halal stunning is not as humane as conventional stunning, or they may have done so for reasons not related to animal welfare. Levine and colleagues [29] observed differences in the level of understanding of humane procedures between US veterinary students with aspirations to work with food animals from those aspiring to work with companion animals; this may account for why some students consciously avoided meat derived from animals stunned with stunning methods they may perceive to be inhumane. Similarly, Mariti and others [30] observed that veterinary students in Italy gave more consideration to the welfare of companion animals than that of food animals; this may affect their perception and understanding of animal welfare issues around food animals. It must be noted that there is no real procedural difference between stunned Halal and conventional slaughter (with the exception of a short prayer during Halal slaughter). Therefore, the humaneness of stunned Halal slaughter is not inferior to the humaneness of conventional slaughter, and one cannot therefore use humaneness (or the lack of it) as a reason to avoid meat from effectively stunned Halal slaughter. Interestingly, the results also showed a significant increase in the percentage of vegans and vegetarians with year of study who would wittingly buy Halal meat from stunned animals. Although, presumably, vegans and vegetarians will usually avoid purchasing meat, their responses may have been for a number of reasons: (i) Some respondents may have been vegetarians who would consume meat if the humaneness of slaughter was guaranteed, (ii) the level of understanding of vegetarians and vegans on animal welfare issues (particularly humane slaughter) may have improved as they progressed in the level of study or (iii) they may just have given hypothetical answers to the question as they were only given the 'yes' or 'no' alternatives and did not feel they could miss a section of the questionnaire. In retrospect, more thought should have gone into the questionnaire to avoid ambiguity in interpretation of the responses made by the vegetarians and vegans to this question. Beardsworth and Keil [31], in a study on vegan and vegetarian trends, reported that ethics and welfare were the main reasons why some consumers avoided meat. One may argue that if some vegans and vegetarians can be assured of the highest welfare of animals, it may change their consumption pattern, which may explain this finding.

In line with scientific opinion on the welfare aspects of slaughter without stunning [12–14], the majority of respondents indicated that all animals must be stunned before slaughter. This is consistent with the observation made by Broom [32], who reported that there is increased awareness around animal welfare at slaughter which has led to an increasing number of consumers demanding humanely slaughtered products or avoiding those associated with poor welfare. A small proportion of respondents indicated that they did not have an opinion on stunning. With the recent rise in campaigning for restrictions or a ban on nonstun slaughter by the veterinary profession and other animal welfare charities (e.g., the British Veterinary Association, Royal Society for the Prevention of Cruelty to Animals), one would expect veterinary students to be well informed and hold an opinion on pre-slaughter stunning. The issue is further highlighted by the ten respondents who selected 'other' and left comments to support their choice (see Table 2). The comments showed variation in the opinion of veterinary students with regard to stunning and slaughter; some respondents questioned the humaneness of stunning and others called for the current exemption of religious slaughter from stunning to be withdrawn. One respondent cited their cultural upbringing as a factor that conflicts with their profession and opinion on stunning; the influence of the culture of students on their perception of animal welfare issues has been discussed by Philip and McCulloch [33], who reported that students' attitudes to animal welfare were influenced by their cultural upbringing, with students from Europe and the USA less likely to 'condone cruelty to animals'. A minority of respondents indicated that in their opinion, religious slaughter should be exempt from stunning to comply with traditional religious values. This is the case in some but not all EU member states (e.g., the UK, France, Germany and others), where a derogation is applied to permit the slaughter of animals without stunning for religious rites.

The results showed a higher proportion of vegans, 3.9% (18), compared with the general UK population, which is estimated at around 1.16% [34]. This may be in part due to an increased empathy for animals by veterinary students due to their close association and everyday contact with animals. A minority of the respondents (less than 1%) indicated that they did not eat meat due to 'dietary exclusion by choice. These respondents neither identified themselves as vegans, vegetarians, nor meat eaters. The assumption is that they probably had an intolerance, allergic or medical reason for not consuming meat.

5. Conclusions

As future enforcers of the law at the time of slaughter (including Halal), it is important for veterinary institutions in the UK to introduce students to the science and politics surrounding religious slaughter at all stages of their veterinary education so that students will be well informed about these issues on qualifying. It appears that for the majority of veterinary students, the debate surrounding Halal meat production is not concerned with religious ideas, but animal welfare, with the majority of respondents indicating that they would consciously consume Halal meat if it was obtained from animals that have been effectively stunned. Vegetarianism and veganism are slightly increased among veterinary students in comparison with the general UK population. It is recommended that future studies on this topic should consider evaluating the curriculum of different universities to examine whether there are disparities in teaching on religious slaughter, and how it might be improved in general.

Author Contributions: A.F. conceived and designed the study, analysed the data and produced the manuscript with T.G.K. A.G. assisted in data collection and a review of the paper before submission. All authors reviewed and approved the submitted version of the manuscript.

Animals **2019**, *9*, 293

Funding: This research received no external funding.

Acknowledgments: A.F. acknowledges the support of the Humane Slaughter Association (HSA) through an Animal Welfare Research Training (PhD) Scholarship and also the support of AHDB Beef and Lamb.

Conflicts of Interest: The authors declare no conflict of interest.

References

1. Tew, T.E.; Macdonald, D.W.; Rands, M.R.W. Herbicide application affects microhabitat use by arable wood mice (*Apodemus sylvaticus*). *J. Appl. Ethol.* **1992**, *29*, 352–359. [CrossRef]
2. Johnson, I.P.; Flowerdew, J.R.; Hare, R. Effect of broadcasting and drilling methiocarbs molluscicide pellets on field population of wood mice (*Apodemus sylvaticus*). *Bull. Environ. Contam. Toxicol.* **1991**, *46*, 84–91. [CrossRef] [PubMed]
3. Feber, R.E.; Firbank, L.G.; Johnson, P.J.; Macdonald, D.W. The effects of organic farming on pest and non-pest butterfly abundance. *Agric. Ecosyst. Environ.* **1997**, *64*, 133–139. [CrossRef]
4. Dale, V.H.; Polasky, S. Measures of the effect of agricultural practices on ecosystem services. *Ecol. Econ.* **2007**, *64*, 286–296. [CrossRef]
5. Edge, W.D. Wildlife of agriculture, pastures and mixed environs. In *Wildlife-Habitats Relationships in Oregon and Washington*; Johnson, D.H., O'Neil, T.A., Eds.; Oregon State University Press: Corvallis, OR, USA, 2000; pp. 342–360.
6. Regan, T. *A Case for Animal Rights*; University of California Press: Berkeley, CA, USA, 1983; pp. 226–329.
7. Fuseini, A.; Sulemana, I. An exploratory study of the influence of attitudes towards animal welfare on meat consumption in Ghana. *Food Ethics* **2018**, *2*, 57–75. [CrossRef]
8. Kauffmann, R.G. Meat composition. In *Meat Science and Applications*; Hui, Y.H., Nip, W.K., Rogers, R.W., Young, O.A., Eds.; Marcel Dekker, Inc.: New York, NY, USA, 2001; pp. 1–19.
9. Font-i-Furnols, M.; Guerrero, L. Consumer preference, behavior and perception about meat and meat products: An overview. *Meat Sci.* **2014**, *98*, 361–371. [CrossRef]
10. Council Regulation (EC) No 1099/2009 of 24 September 2009 on the protection of animals at the time of killing. *Off. J. Eur. Union* **2009**, *L301*, 1–30.
11. Welfare of Animals at the Time of Slaughter Regulations (England) 2015 (SI No. 1782). HMSO. Available online: http://www.legislation.gov.uk/uksi/2015/1782/contents (accessed on 15 March 2019).
12. Gibson, T.J.; Johnson, C.B.; Murrell, J.C.; Hulls, C.M.; Mitchinson, S.L.; Stafford, K.J.; Johnstone, A.C.; Mellor, D.J. Electroencephalographic responses of halothane-anaesthetised calves to slaughter by ventral-neck incision without prior stunning. *N. Z. Vet. J.* **2009**, *57*, 77–83. [CrossRef]
13. Mellor, D.J.; Littin, K.E. Using science to support ethical decisions promoting humane livestock slaughter and vertebrae pest control. *Anim. Welf.* **2004**, *13*, 127–132.
14. Gregory, N.G.; von Wenzlawowicz, M.; von Hollenben, K.; Fielding, H.R.; Gibson, T.J.; Mirabito, L.; Kolesar, R. Complications during shechita and halal slaughter without stunning in cattle. *Anim. Welf.* **2012**, *21*, 81–86. [CrossRef]
15. Grandin, T.; Regenstein, J.M. Religious slaughter and animal welfare: A discussion for meat scientists. *Meat Focus Int.* **1994**, *2*, 115–123.
16. Welfare of Animals at the Time of Slaughter Regulations (Scotland) 2012 (SI No. 321). HMSO. Available online: http://www.legislation.gov.uk/ssi/2012/321/contents/made (accessed on 15 March 2019).
17. Welfare of Animals at the Time of Slaughter Regulations (Wales) 2014 (SI No. 951 (W.92)). HMSO. Available online: http://www.legislation.gov.uk/wsi/2014/951/contents/made (accessed on 15 March 2019).
18. Animal Aid. Slaughter. 2017. Available online: https://www.animalaid.org.uk/the-issues/our-campaigns/slaughter/ (accessed on 1 March 2019).
19. Animal Aid. Veganism. 2018. Available online: https://www.animalaid.org.uk/veganism/ (accessed on 1 March 2019).
20. Harper, G.C.; Makatouni, A. Consumer perception of organic food production and farm animal welfare. *Br. Food J.* **2002**, *104*, 287–299. [CrossRef]
21. Wall, P. One health and the food chain: Maintaining safety in a globalised food industry. *Vet. Rec.* **2014**, *174*, 189–192. [CrossRef]

22. Spinka, M. AWARE Project: Background, objectives and expected outcomes. In *EAAP Scientific Committee Book of Abstracts of the 63rd Annual General Meeting of the European Federation of Animal Science*; Wageningen Academic Publishers: Bratislava, Slovakia, 2012.

23. Gibson, T.J.; Johnson, C.B.; Murrell, J.C.; Chambers, P.J.; Stafford, K.J.; Mellor, D.J. Components of electroencephalographic responses to slaughter in halothane-anaesthetised calves: Effects of cutting neck tissues compared to major blood vessels. *N. Z. Vet. J.* **2009**, *57*, 84–89. [CrossRef]

24. Gregory, N.G.; Fielding, H.R.; von Wenzlawowicz, M.; von Holleben, K. Time to collapse following slaughter without stunning in cattle. *Meat Sci.* **2010**, *85*, 66–69. [CrossRef]

25. Gregory, N.G.; Shaw, F.D.; Whitford, J.C.; Patterson-Kane, J.C. Prevalence of ballooning of the severed carotid arteries at slaughter in cattle, calves and sheep. *Meat Sci.* **2006**, *74*, 655–657. [CrossRef]

26. Main, D.C.J. Evolution of animal welfare education for veterinary students. *Vet. Med. Educ.* **2010**, *37*, 30–35. [CrossRef]

27. Food Standards Agency (FSA). *Results of the 2018 FSA Survey into Slaughter Methods in England and Wales*; Department for Environment Food and Rural Affairs: London, UK, 2019.

28. Dialrel. Report on Good and Adverse Practices: Animal Welfare Concerns in Relation to Slaughter Practices from the Viewpoint of Veterinary Sciences. 2010. Available online: http://www.dialrel.eu/images/veterinary-concerns.pdf (accessed on 13 March 2019).

29. Levine, E.D.; Mills, D.S.; Houpt, K.A. Attitudes of veterinary students at one US college towards factors relating to farm animal welfare. *Vetrinary Med. Educ.* **2015**, *32*, 481–490. [CrossRef]

30. Mariti, C.; Pirrone, F.; Albertini, M.; Gazzano, A. Familiarity and interest in working with livestock decreases the odds of having positive attitudes towards non-human animals and their welfare among veterinary students in Italy. *Animals* **2018**, *8*, 150. [CrossRef]

31. Beardsworth, A.D.; Keil, E.T. Vegetarianism, veganism and meat avoidance: Recent trends and findings. *Br. Food J.* **1991**, *93*, 19–24. [CrossRef]

32. Broom, D.M. Fish welfare and perception of farmed fish. In *Proceedings of Aquavision*; Aquavision: Stavenger, Norway, 1999; pp. 1–6.

33. Phillips, C.J.C.; McCulloch, S. Students attitudes on animal sentience and use of animals in society. *J. Biol. Educ.* **2010**, *40*, 17–24. [CrossRef]

34. The Vegan Society. Survey. 2018. Available online: https://www.vegansociety.com/my-account/the-vegan/issue-3-2018/survey (accessed on 30 March 2019).

Article

Animal Ethical Views and Perception of Animal Pain in Veterinary Students

Anna Valros † and **Laura Hänninen** *,†

Research Centre for Animal Welfare, Department of Production Animal Medicine, Faculty of Veterinary Medicine, University of Helsinki, 00014 Helsinki, Finland; anna.valros@helsinki.fi
*   Correspondence: laura.hanninen@helsinki.fi; Tel.: +358-50-415-1180
†   These authors contributed equally to this work.

Received: 12 November 2018; Accepted: 20 November 2018; Published: 23 November 2018

Simple Summary: Veterinary students face several ethical challenges during their curriculum. We surveyed the animal ethical views of Finnish veterinary students, and also asked them to score the level of pain perception in different animal species. We found that the appreciation of pain perception of different animal species, and especially of those taxonomically further away from humans, appeared to increase during Finnish veterinary education. This implies that knowledge is important in improving views towards capacities of animals of varying taxa. Finnish veterinary students have a clear domination of utilitarian views in animal ethics, and veterinary education appeared to influence their views only to a small degree. We suggest that understanding the ethical views of veterinary students enables better planning of educational activities, to ensure that the students gain a good understanding of the potential variety of, and differences in, ethical views they will encounter as future professionals.

Abstract: Veterinary students face several ethical challenges during their curriculum. We used the Animal Ethics Dilemma to study animal ethical views of Finnish veterinary students, and also asked them to score the level of pain perception in 13 different species. Based on the 218 respondents, the utilitarian view was the dominating ethical view. Mammals were given higher pain scores than other animals. The proportion of the respect for nature view correlated negatively, and that of the animal rights view positively, with most animal pain scores. Fifth year students had a higher percentage of contractarian views, as compared to 1st and 3rd year students, but this might have been confounded by their age. Several pain perception scores increased with increasing study years. We conclude that the utilitarian view was clearly dominating, and that ethical views differed only slightly between students at different stages of their studies. Higher pain perception scores in students at a later stage of their studies might reflect an increased knowledge of animal capacities.

Keywords: veterinary student; animal ethics; pain perception; animal; animal welfare

---

1. Introduction

Veterinary students face several ethical challenges during their curriculum [1,2]. Distinguishing their own view on animal ethics and how it develops over time may help veterinarians to cope with difficult situations and emotions [3]. Ethically motivated decision-making is also important for client communication, for improving job satisfaction and for maintaining a positive public profile [4].

The professional development during veterinary education may lead to contrasting changes in empathy towards, and pain assessment of non-human animals. As part of the veterinary curriculum, the students are taught about non-human animal cognition and pain perception. They also gain insights in the complexity of animal-related issues in the society, such as those related to the food production chain and to animal use in research.

The effect of experience and age on the perception of animal pain is not straightforward. Younger veterinarians have been shown to rate pain higher, and to treat pain more [5]; to show no difference in pain ratings [6]; or to show lower pain ratings [7], as compared to older veterinarians. Further, empathy scores have been shown to decline during the course of education in both human doctors [8,9] and veterinarians [10] or not to change [6]. Despite these contradictory results, in general, knowledge is a prerequisite for a proper pain management [7,11–13] and thus, to improve veterinary education, we believe it is important to understand how veterinary students develop their views of animal pain perception during their studies.

In many countries there is no faculty-wide agreement on animal ethics, beyond legislation [14], which is also true for Finland. In addition, there is no internationally agreed view on which contents are best suited for veterinary ethics [14,15]. At the University of Helsinki, we aim at equipping students with both a systematic understanding of their own animal ethics views, and knowledge of the variation of ethical views, which they may encounter, rather than teaching a single ethical standard. Our approach is thus similar to that recommended by Clarkeburn [16]. At the University of Helsinki, which is the only University in Finland with a Veterinary Faculty, veterinary ethics is taught during the preclinical studies, which includes the first three years of the veterinary curriculum. Veterinary ethics is integrated into other subjects including several courses with elements of importance for the development of animal ethics. These include subjects such as pharmacology, where students learn about animal pain perception; animal welfare; and physiology, including e.g. stress physiology. During the clinical studies a one-day workshop on animal ethics and client communication is held as a part of clinical rounds in year 5. Teaching of veterinary and animal legislation is integrated into the studies throughout the entire 6-year period, but most teaching is provided during years 4 to 6.

Animal Ethics Dilemma ([17], www.aedilemma.net, AED) is an internet tool developed to test the proportion (%) of common animal ethical views one may have. AED has questions for utilitarian (roughly defined as ethical judgements being based on a strive to maximize the human and non-human animal welfare), animal rights (animals have inherent value that should be respected), respect for nature (a duty to protect the integrity of each species), contractarian (only indirect ethical obligations towards animals, because they can matter to other humans) and relational views (the moral status of an animal is defined by the relationship to the animal).

As a part of developing our teaching on animal ethics, we collected data on student views on animal ethics with AED and on perception of pain in different animal species. Our aim was to study which animal ethical views dominate among Finnish veterinary students. Further, we explored how students rate pain perception in non-human animals from different taxa, and if these views and ratings change during the veterinary education.

2. Materials and Methods

First, 3rd and 5th year students (2014–2016) at the Veterinary Faculty, University of Helsinki, Finland, were asked to evaluate their animal ethical views using the AED and to report these via an electronic questionnaire. In the questionnaire, students were further asked to score the level of pain perception using a Likert-scale from 0 (pain causes only reflexes) to 10 (pain is subjectively experienced) in 13 different animals, representing wild and domestic animals and vertebrate and invertebrate animals of a wide range (see S1 for the questionnaire).

Background factors collected via the electronic questionnaire included gender, study year and year of birth. The replies were anonymous, and participation was voluntary.

The investigations were carried out following the rules of the Declaration of Helsinki of 1975. The work is in compliance with the guidelines of the Finnish national board of research integrity (TENK; http://www.tenk.fi/en/tenk-guidelines) according to which no ethical review was required.

2.1. Student Demographics

We obtained a total of 218 answers, which represents a response rate of approximately 35 %. The distributions of respondents of different study years over the three response years are presented in Table 1. The data included students from at least two response years for each study year, and each student replied to the questionnaire only once. The age of the students ranged from 19 to 44 years (on average 25 years). In total 21 of the respondents were male (9.6 %) which is equal to the proportion of male students in the faculty in 2016. Due to the small number of male students the effect of gender was not analyzed.

Table 1. Distribution of students replying to the survey (218 in total), according to year of study and response year.

Response Year	Year of Study		
	1	3	5
2014	66	19	0
2015	25	19	45
2016	28	0	16
Total	119	38	61

2.2. Data Handling and Statistical Analyses

For statistical analyses, the students were classified into two more or less equal age groups: 'older' (24 years or older, $n = 104$) and 'younger' (under 24 years, $n = 112$). Two respondents had not indicated their age. Age and study year were clearly associated with each other: 98% (60/61) of the 5th year students (average age 27.6 years) fell into the 'older' age class. The corresponding numbers were 40% (15/38) and 32% (37/117) for 3rd (average age 24.3 years) and 1st year students (average age 23.6 years), respectively.

Due to non-normal distribution of the data we tested differences between study years with Kruskall-Wallis tests, followed by Bonferroni-corrected pairwise tests when relevant, and between age classes ('older' and 'younger') with Mann-Whitney U tests. Correlations within and between the proportions of different ethical views from AED and animal pain scores were tested with Spearman rank correlations.

3. Results

3.1. Overall Results: Animal Ethical Views

In general, based on the replies to the AED, the utilitarian view was the dominating ethical view. Both the utilitarian and the respect for nature views were represented on some level in almost all the answers, while the utilitarian view got a much higher median percentage. Animal rights, relational and contractarian views were represented in a lower proportion of the answers but did occur to a significant degree in some of these (see Table 2).

Table 2. Overall animal ethical views of Finnish veterinary students based on the Animal Ethics Dilemma ($n = 218$).

Animal Ethical Views	Median	Min	Max	Answers with 0%, % [1]
Utilitarian	50	0	92	4
Respect for nature	25	0	67	3
Animal rights	17	0	92	15
Relational view	2	0	42	49
Contractarianism	0	0	67	86

[1] percentage of respondents for which the respective animal ethical view was not represented at all in the AED (Animal Ethics Dilemma) results.

The proportion of utilitarian view in the AED results correlated negatively with the proportion of all other ethical views ($r_s = -0.26$ to -0.54, $p < 0.001$ for all). In addition, the proportion of animal rights view correlated negatively with the proportion of the respect for nature view ($r_s = -0.26$, $p < 0.001$). No further correlations between the proportions of different animal ethical views were found ($p > 0.05$ for all).

3.2. Overall Results: Animal Pain Perception Scores

Overall descriptive results for the pain perception scores given by the students are presented in Table 3. Chimpanzees were scored the highest, while earthworms were given the lowest score. Pain perception scores could be numerically classified into three groups: the highest scoring group included all mammals as well as the budgerigar (median score 10), the intermediately scoring group included the fish and the invertebrate octopus (7) and the lowest scoring group included all other invertebrates (3–4).

Table 3. Animal pain perception scores given by veterinary students ($n = 213$), overall, and in study years 1, 3 and 5, based on a scale from 0: pain causes only reflexes–10: pain is subjectively experienced. Results are given as median (min-max). P-values indicate a difference between study years (n.s. for $p > 0.1$ and #, *, **, for $p < 0.1$, $p < 0.05$ and $p < 0.01$, respectively).

Animal	Overall Median (Min–Max)	1st Year $n = 119$	3rd Year $n = 38$	5th Year $n = 61$	Sign. Level
Chimpanzee	10 (7–10)	10 (8–10)	10 (7–10)	10 (8–10)	n.s.
Dolphin	10 (5–10)	10 (5–10)	10 (5–10)	10 (8–10)	n.s.
Cat	10 (5–10)	10 (5–10)	10 (7–10)	10 (8–10)	#
Pig	10 (6–10)	10 (6–10)	10 (6–10)	10 (8–10)	n.s.
Lion	10 (5–10)	10 (5–10)	10 (7–10)	10 (8–10)	n.s.
Cattle	10 (5–10)	10(5–10) [a]	10 (5–10) [ab]	10(7–10) [b]	*
Budgerigar	10 (2–10)	9 (2–10) [a]	10 (3–10) [b]	10(5–10) [b]	**
Octopus	7 (1–10)	7 (1–10)	8 (1–10)	7 (2–10)	n.s.
Fish	7 (1–10)	6 (1–10) [a]	8 (1–10) [b]	7 (2–10) [b]	**
Bee	4 (1–10)	3 (1–10) [a]	5 (1–10) [ab]	6 (1–10) [b]	*
Spider	4 (0–10)	3 (0–10) [a]	4 (1–10) [ab]	5 (0–10) [b]	*
Fly	3 (0–10)	3 (0–10) [a]	4 (1–10) [ab]	4 (1–10) [b]	*
Earthworm	3 (0–10)	2 (0–10) [a]	4.5 (1–10) [b]	4 (1–10) [b]	*
Average *	7.5 (3.3–10)	7.2 (3.3–10) [a]	7.7 (4.8–10) [ab]	7.8 (5.5–10) [b]	**

* average (min-max) pain perception score for all animals; [a, b] Values within rows lacking common superscript letters differ.

Pain scores of almost all different animals correlated positively with each other ($r_s = 0.18$ to 0.93, $p < 0.01$ for all), however, the chimpanzee pain score did not correlate with the pain score given to the fly or the earthworm ($p > 0.1$ for both).

3.3. Correlations between Animal Ethical Views and Pain Perception Scores

We found several statistically significant, though rather low correlation coefficients: The proportion of the respect for nature view in the AED correlated negatively ($r_s = -0.16$ to -0.25, $p < 0.05$ for all) and that of the animal rights view positively ($r_s = 0.15$ to 0.36, $p < 0.05$ for all) with all animal pain scores, except for the chimpanzee score ($p > 0.1$ for both). The proportion of these two ethical views also correlated with the average pain perception score (respect for nature; $r_s = -0.3$ and animal rights; $r_s = 0.3$, $p < 0.001$ for both). There was no correlation between the proportion of utilitarian, contractarian or relational view and any of the pain perception scores ($p > 0.1$ for all).

The proportion of contractarian view, according to the AED, differed between students of different study years (Kruskall–Wallis $\chi^2(2) = 11.9$, $n = 213$, $p < 0.01$). Fifth year students had a higher percentage of contractarian view in AED (Median: 0; Min–max: 0–67), as compared to 1st (0; 0–17, $p < 0.01$)

and 3rd year (0; 0–17, $p < 0.05$) students. Correspondingly, 'older' students tended to have a higher percentage of contractarian view in AED than 'younger' students (0; 0–67 and 0; 0–33, respectively, Mann-Whitney U: 6000, $n = 211$, $p < 0.1$).

Several pain perception scores, as well as the average pain perception score, increased with increasing study year (Table 3). Of the pain perception scores, only the budgerigar score was affected by age, with 'younger' respondents scoring lower scores than 'older' ones (9; 2–10 vs. 10; 3–10, Mann-Whitney U = 6363, $n = 211$, $p < 0.05$). In addition, the dolphin score tended to be lower for 'younger' than 'older' students (10; 5–10 vs. 10; 6–10, Mann-Whitney U = 6215, $n = 212$, $p = 0.05$)

4. Discussion

The animal ethical view estimated using the Animal Ethics Dilemma of Finnish veterinary students, was dominated by a utilitarian view throughout the study years. The veterinary students scored animal pain perception higher the further they had advanced in their studies, which appeared to be rather independent of their age.

The strong utilitarian view domination in this study might be due to the utilitarian view being more easily combined with a veterinary education, where some level of animal instrumentalization is unavoidable. Lund and others (2016) [18] studied ethical views of a general selection of meat eaters, vegetarians and vegans in the UK, and showed a more even distribution between the animal rights view and the utilitarian view than we report in this study. In addition, they showed that the animal rights view dominated in the vegetarians and vegans, while meat eaters were more utilitarian. As far as we know, however, there are no similar studies on animal ethical views on broader samples of the Finnish population, so we cannot exclude that the currently presented results are not merely a reflection of the attitudes of the general Finnish population. In a Finnish nationwide study Kupsala et al (2015) [19] reported an average rating of animal instrumentalization as 2.25 (SD 1.18) on a scale from 1–5, where 5 meant that the respondents totally agreed with the statement 'An animal should be seen primarily as a means of production'. Thus, a certain degree of utilitarian animal ethics was apparent also in the general public, but with quite a large variation. It is also worth mentioning that even though some EU-countries have implemented a non-kill-policy regarding healthy companion animals [20], this is not the case in Finland. As allowing euthanasia of healthy animals is typically against the view of animal rights advocates [21], this suggests that also in general the animal rights view might not be as strong in Finnish society as in at least some other European countries.

The animal ethics dilemma was originally developed for teaching [17] while components of it has been used also in research [18], even though it has not been formally validated for research purposes. It is probable that the tool simplifies the evaluation of animal ethical views, partly due to it´s forced-choice design [18] and as the respondents can chose only one answer, representing one ethical view, per question. The results regarding animal ethical views should thus be considered with some care.

Pain perception was scored the highest in mammals, followed by other vertebrates, and the lowest in invertebrates. This corresponds to pain ratings in the nationwide survey performed in Finland by Kupsala and others (2015) [19] where mammals were assessed to experience pain much more often than salmon and shrimp. Also, Phillips and McCulloch (2005) [22] found that students of different nationalities ascribed mammals as having higher levels of sentience than chicken and fish. The respondents in our study showed a higher variation in their rating of pain perception for non-mammals than mammals. A similarly higher uncertainty of pain rating for non-mammalian (chicken, salmon and shrimp), as compared to mammalian animals, was seen in the study by Kupsala and others (2015) [19].

The fact that the animal rights view correlated positively with pain perception scorings is not surprising. Previous studies have similarly shown that animal sentience is given a higher attribute by students applying more constrains on animal use [22], and that a less instrumental view on animals is correlated to a higher belief in animal mind [19]. We find it more difficult to speculate why the respect for nature view correlated negatively with pain perception scores, but this might be related to accepting pain as a part of 'natural life'. Lund and others (2016) [18] reported that meat eaters had a higher level of the respect for nature view as compared to vegans and vegetarians, which might indicate a generally less critical view of animal use. This correlation might, however, merely be due to the fact that the proportion of animal rights and respect for nature views correlated negatively in our study.

Pain perception scoring for chimpanzees appeared to be high irrespective of ethical view and slightly independent of pain perception in other species. Interestingly, Phillips and McCulloch (2005) [22] found that even though nationality had an effect on several attitudes of students towards animals, students of different nationalities did not differ in how they evaluated the level of sentience of monkey, dog and fox. It might therefore be that the perception of animals perceived to be culturally closer to humans, are less affected by other attitudes. Phillips and McCulloch (2005) [22] included 'new-born baby' in their list of animals to be assessed for level of sentience, and the students in their study scored the sentience level of a baby lower than that of a monkey or a dog. We only included non-human animals in our questionnaire, which does not allow us to estimate how our students would have rated human pain perception. Also, it is possible that the results would have changed if the students had related their ratings to humans.

It should be noted that the pain perception scale we used was a commonly used simple 11-point Numerical Rating Scale [6,7,23] and could have been somewhat confusing, as it assumed the students somehow understood the link between concepts of pain perception and the level of consciousness or mental abilities of the animals. This link, however, is a rather accepted issue in science. It has been suggested that while nociception is a mere sensory ability, pain perception is primarily a subjective feeling (for a review, see eg Sneddon and others [19]).

Students in later years of their studies (esp. 5th year) appeared to have more knowledge about animal pain perception, especially regarding species that were scored lower in general, such as most invertebrates. These are species often regarded as having a lower moral standing in society [19], which is reflected also in some pieces of legislation. One example is the EU directive (2010/63/EU) on the protection of animals used for scientific purposes, which does not include most invertebrates: *'In addition to vertebrate animals including cyclostomes, cephalopods should also be included … as there is scientific evidence of their ability to experience pain, suffering, distress and lasting harm'*. The students therefore might enter the veterinary curriculum with a more 'popular' view of animal pain, and then develop this by gaining more information during their education. Also, the scoring of fish and birds was higher in students a few years into their curriculum, which might reflect that these taxa are still somewhat underestimated regarding their cognitive abilities. This could be explained by the fact that the empathic response is amplified by similarity and familiarity [24–26]. Fish in particular are often treated in ways that would never be accepted for mammals; some species are, for example, slaughtered slowly by suffocating to death or bled without stunning [27], while an increasing body of evidence shows that fish actually perceive pain at a level comparable to mammals [28].

Further, as pain perception scores were affected by age of the students to a very small degree (only the score for budgerigar), the difference in the scoring between students at different stage in their curriculum, might reflect an improvement in the level of knowledge of pain perception, acquired during their veterinary education. Surprisingly, also the pain perception score of cattle was higher in 5th year students than in 1st year students, and production animals (cattle and pigs) scored more or less as high as cats. Related findings from North American Veterinary students showed, that the students were likely to assess farm animals to be less sentient or having lower cognitive processes than dogs and cats [29]. Also, a current study showed that Italian veterinary students estimated the freedom to express normal species-specific behaviors and the freedom from fear and distress less important for the welfare of livestock animals than to pets [30]. Further, Levine and others (2005) [29] and Kupsala and others (2015) [19] showed that dogs were more frequently assessed to be able to feel pain than farm animals, such as cows and pigs.

We could not see indications of the commonly reported effect of reduced perception of pain [31] and empathy in the course of medical education [9,10]. Instead, our results are in line with previous findings among Finnish veterinary students and graduated veterinarians on animal empathy and attitudes towards treating disbudding pain in calves [6]. In this study however, we did not test empathy as such, and it is possible, that the rating of pain perception used in this study merely tested the level of biological knowledge of other species. This study would be interesting to repeat in different countries, as nationality appears to affect attitudes to animals [22]. It would also be interesting to be able to compare multiple veterinary faculties within one country to evaluate a possible effect of faculty attitudes on the students´ perceptions.

Students in their 5th study year were somewhat more contractarian in their ethical views than students in earlier years. This, however, might be confounded by the fact that students of older age have more experience of life in general. We suggest this could be due to students usually entering the veterinary faculty with a rather low knowledge level regarding food producing animals, and animal use in general (*personal observation by the authors*), and as they learn more about the regulations and commonly accepted practices affecting and constraining animal husbandry, this might be reflected in increased contractarian views. In line with our findings, senior Swedish veterinary students were more likely to accept the use of animals in experiments than younger ones [32]. The Swedish veterinary students explained their attitude change to be due to greater knowledge, a better understanding of the necessity, more accepting and tolerant views, and a less prejudiced view of the use of animals.

5. Conclusions

The animal ethics views of Finnish Veterinary students are typically dominated by a utilitarian view, and that students at a later stage of their curriculum have a larger proportion of contractarian view than students in earlier years. The increased appreciation of animal pain perceptions in students further on in their education probably reflects increased knowledge of animal capacities, gained through the ongoing veterinary education.

Supplementary Materials: The following are available online at http://www.mdpi.com/2076-2615/8/12/220/s1, S1: The questionnaire. (Due to the data containing data, such as year of birth and gender, based on which individuals might be identifiable, the data is only supplied on request).

Author Contributions: Conceptualization, L.H.; formal analysis, A.V.; investigation, L.H. and A.V.; writing—original draft preparation, L.H. and A.V.; writing—review and editing, L.H. and A.V.

Funding: This research received no external funding.

Acknowledgments: The authors wish to thank all the students who filled in the questionnaire.

Conflicts of Interest: The authors declare no conflict of interest.

References

1. Herzog, H.A.; Vore, T.L.; New, J.C. Conversations with Veterinary Students: Attitudes, Ethics, and Animals. *Anthrozoos* **1989**, *2*, 181–188. [CrossRef]
2. Verrinder, J.M.; Phillips, C.J.C. Identifying Veterinary Students' Capacity for Moral Behavior Concerning Animal Ethics Issues. *J. Vet. Med. Educ.* **2014**, *41*, 358–370. [CrossRef] [PubMed]
3. Magalhães-Sant'Ana, M. A theoretical framework for human and veterinary medical ethics education. *Adv. Health Sci. Educ.* **2016**, *21*, 1123–1136. [CrossRef] [PubMed]
4. Mullan, S.; Main, D. Principles of ethical decision-making in veterinary practice. *In Pract.* **2001**, *23*, 394. [CrossRef]
5. Raekallio, M.; Heinonen, K.M.; Kuussaari, J.; Vainio, O. Pain Alleviation in Animals: Attitudes and Practices of Finnish Veterinarians. *Vet. J.* **2003**, *165*, 131–135. [CrossRef]
6. Norring, M.; Wikman, I.; Hokkanen, A.; Kujala, M.V.; Hänninen, L. Empathic veterinarians score cattle pain higher. *Vet. J.* **2014**, *200*, 186–190. [CrossRef] [PubMed]
7. Huxley, J.N.; Whay, H.R. Current attitudes of cattle practitioners to pain and the use of analgesics in cattle. *Vet. Rec.* **2006**, *159*, 662–668. [CrossRef] [PubMed]

8. Hojat, M.; Mangione, S.; Nasca, T.J.; Rattner, S.; Erdmann, J.B.; Gonnella, J.S.; Magee, M. An empirical study of decline in empathy in medical school. *Med. educ.* **2004**, *38*, 934–941. [CrossRef] [PubMed]
9. Neumann, M.; Edelhäuser, F.; Tauschel, D.; Fischer, M.R.; Wirtz, M.; Woopen, C.; Haramati, A.; Scheffer, C. Empathy Decline and Its Reasons: A Systematic Review of Studies With Medical Students and Residents. *Acad. Med.* **2011**, *86*, 996–1009. [CrossRef] [PubMed]
10. Hazel, S.J.; Signal, T.D.; Taylor, N. Can Teaching Veterinary and Animal-Science Students about Animal Welfare Affect Their Attitude toward Animals and Human-Related Empathy? *J. Vet. Med. Educ.* **2011**, *38*, 74–83. [CrossRef]
11. Paul, E.S.; Podberscek, A.L. Veterinary education and students' attitudes towards animal welfare. *Vet. Rec.* **2000**, *146*, 269–272. [CrossRef] [PubMed]
12. Hewson, C.J.; Dohoo, I.R.; Lemke, K.A.; Barkema, H.W. Factors affecting Canadian veterinarians' use of analgesics when dehorning beef and dairy calves. *Can. Vet. J.* **2007**, *48*, 1129–1136. [PubMed]
13. Mich, P.M.; Hellyer, P.W.; Kogan, L.; Schoenfeld-Tacher, R. Effects of a Pilot Training Program on Veterinary Students' Pain Knowledge, Attitude, and Assessment Skills. *J. Vet. Med. Educ* **2010**, *37*, 358–368. [CrossRef] [PubMed]
14. Magalhães-Sant'Ana, M.; Lassen, J.; Millar, K.M.; Sandøe, P.; Olsson, I.A. Examining Why Ethics Is Taught to Veterinary Students: A Qualitative Study of Veterinary Educators' Perspectives. *J. Vet. Med. Educ.* **2014**, *41*, 350–357. [CrossRef] [PubMed]
15. Magalhães-Sant'Ana, M. Ethics teaching in European veterinary schools: A qualitative case study. *Vet. Rec.* **2014**, *175*, 592. [CrossRef] [PubMed]
16. Clarkeburn, H. The aims and practice of ethics education in an undergraduate curriculum: Reasons for choosing a skills approach. *J. Furth. High. Educ.* **2002**, *26*, 307–315. [CrossRef]
17. Hanlon, A.J.; Algers, A.; Dich, T.; Hansen, T.; Loor, H.; Sandøe, P. Animal Ethics Dilemma: An interactive learning tool for university and professional training. *Anim. Welf.* **2007**, *16*, 155–158.
18. Lund, T.B.; McKeegan, D.E.F.; Cribbin, C.; Sandøe, P. Animal Ethics Profiling of Vegetarians, Vegans and Meat-Eaters. *Anthrozoös* **2016**, *29*, 89–106. [CrossRef]
19. Kupsala, S.; Vinnari, M.; Jokinen, P.; Räisänen, P. Citizen Attitudes to Farm Animals in Finland: A Population-Based Study. *J. Agric. Environ. Ethics* **2015**, *28*, 601–620. [CrossRef]
20. EU Animal Welfare Platform. In Proceedings of the 2nd Meeting, Brussels, Belgium, 10 November 2017. Available online: https://ec.europa.eu/food/animals/welfare/eu-platform-animal-welfare/meetings_en (accessed on 21 November 2018).
21. Sandøe, P.; Christiansen, S.B. *Ethics of Animal Use*; Wiley-Blackwell: Ames, IA, USA, 2008.
22. Phillips, C.J.C.; McCulloch, S. Student attitudes on animal sentience and use of animals in society. *J. Biol. Educ.* **2005**, *40*, 17–24. [CrossRef]
23. Wikman, I.; Hokkanen, A.-H.; Pastell, M.; Kauppinen, T.; Valros, A.; Hänninen, L. Dairy producer attitudes to pain in cattle in relation to disbudding calves. *J. Dairy. Sci.* **2013**, *96*, 6894–6903. [CrossRef] [PubMed]
24. de Waal, F.B. Putting the altruism back into altruism: The evolution of empathy. *Annu. Rev. Psycholo.* **2008**, *59*, 279–300. [CrossRef] [PubMed]
25. Drwecki, B.B.; FAU, M.C.; FAU, W.S.; Prkachin, K.M. Reducing racial disparities in pain treatment: The role of empathy and perspective-taking. *Pain* **2011**, *152*, 1001–1006. [CrossRef] [PubMed]
26. Rae Westbury, H.; Neumann, D.L. Empathy-related responses to moving film stimuli depicting human and non-human animal targets in negative circumstances. *Biol. Psychol.* **2008**, *78*, 66–74. [CrossRef] [PubMed]
27. Kestin, S.C.; Robb, D.H.F.; van de Vis, J.W. Protocol for assessing brain function in fish and the effectiveness of methods used to stun and kill them. *Vet. Rec.* **2002**, *150*, 302–307. [CrossRef] [PubMed]
28. Sneddon, L.U.; Elwood, R.W.; Adamo, S.A.; Leach, M.C. Defining and assessing animal pain. *Anim. Behav.* **2014**, *97*, 201–212. [CrossRef]
29. Levine, F.D.; Mills, D.S.; Houpt, K.A. Attitudes of veterinary students at one US college toward factors relating to farm animal welfare. *J. Vet. Med. Educ.* **2005**, *32*, 481–490. [CrossRef] [PubMed]
30. Mariti, C.; Pirrone, F.; Albertini, M.; Gazzano, A.; Diverio, S. Familiarity and Interest in Working with Livestock Decreases the Odds of Having Positive Attitudes towards Non-Human Animals and Their Welfare among Veterinary Students in Italy. *Animals* **2018**, *8*, 150. [CrossRef] [PubMed]

31. Cheng, Y.; Lin, C.; Liu, H.; Hsu, Y.; Lim, K.; Hung, D.; Decety, J. Expertise Modulates the Perception of Pain in Others. *Curr. Biol.* **2007**, *17*, 1708–1713. [CrossRef] [PubMed]

32. Hagelin, J.; Hau, J.; Carlsson, H.E. Attitude of Swedish veterinary and medical students to animal experimentation. *Vet. Rec.* **2000**, *146*, 757–760. [CrossRef] [PubMed]

Commentary

US Farm Animal Welfare: An Economic Perspective

Glynn T. Tonsor [1,*] and Christopher A. Wolf [2]

[1] Department of Agricultural Economics, Kansas State University, Manhattan, KS 66506, USA
[2] Department of Agricultural, Food and Resource Economics, Michigan State University,
     East Lansing, MI 48824, USA; wolfch@msu.edu
* Correspondence: gtonsor@ksu.edu

Received: 16 April 2019; Accepted: 10 June 2019; Published: 18 June 2019

Simple Summary: Farm animal welfare is one of the more controversial and complicated topics facing modern agriculture. To proceed forward, a sound understanding of multiple economic aspects must be appreciated. This article provides a short summary of these economic perspectives with a focus on the US situation.

Abstract: The topic of farm animal welfare (FAW) is both complex and controversial, and inherently involves expertise and views from multiple disciplines. This article provides a summary of economic perspectives on FAW issues in the United States. Practices related to FAW can occur through legal, market or voluntary programs. FAW is not a primary driver of US food demand but negative press has industry-wide effects. Aligning FAW supply and demand can be facilitated through labeling, education, and voluntary programs, but all have pros and cons.

Keywords: consumer demand; economics; farm animal welfare; producer perspective

---

1. Introduction

US livestock, dairy and poultry producers have faced increasing scrutiny of production practices perceived to be related to farm animal welfare (FAW) in recent years [1–4]. FAW is no longer a fringe issue—but it is also not the primary driver of US food demand. Consumers expect a safe, reliable, and affordable food supply, which US farmers have and will continue to provide. The general US population has little connection to modern commercial agriculture and, therefore, essentially no context to accurately understand production practices. Surveys consistently reveal a high level of trust and respect for farmers and maintaining that trust is of utmost importance to ensure market access and avoid unnecessary regulation.

Livestock producers are the primary caretakers of farm animals and their decisions regarding housing systems, feed type, and health management directly impact FAW [5]. In economic terms, livestock producers are the suppliers of farm animals, providing meat, milk, egg, and other products demanded by consumers [6]. Production economics can help us understand the producer's situation, reflecting the complex set of economic incentives and constraints they face. Moreover, recognizing behavioral aspects of how livestock producers view FAW and what attitudes they hold regarding FAW further this understanding [7].

The primary motivation for producers to recognize and respond to public and consumer concerns, perceptions and demands for FAW-related production practice changes is maintaining the social license to produce. Social license is a concept that relates to the public trust in the production system. Without social license, regulation of some sort would follow. In general, producers wish to avoid regulation as it tends to increase costs—particularly in heterogeneous industries when a rigid standard is applied.

Given a great deal of research on consumer perceptions and demand for FAW has been carried out, there are many general conclusions and lessons that can be drawn with implications for farm practices [8]. This article provides a summary of economic perspectives on FAW issues in the United States. Understanding the economic situation facing livestock producers and reflecting on the myriad of formal and informal signals they receive regarding production practices can elevate the overall understanding regarding current and future FAW [6]. This article is structured as a series of observations supported by the economics literature to spur more efficient decision-making and subsequently improve FAW management and policy outcomes.

2. Methods of FAW Practice Change

Over time, FAW-related production practices adjust as livestock producers respond to an array of signals which can broadly be viewed as two primary methods of change: Market and legal [9]. In the market method, FAW adjustments occur from livestock, dairy or poultry producers seeking to fulfill a perceived market opportunity. Market-driven changes occur as retailers, such as grocers and restaurant chains, insist on practices that they can market at the farm level. In some cases, the retailers are responding to pressure from consumer groups, but often they are seeking to increase corporate image. One example is voluntary enrollment in the 5-step Global Animal Partnership program, widely associated with Whole Foods Market (a US-based grocer primarily selling natural and organic products) [10]. A second example is enrollment in a United States Department of Agriculture (USDA) Process Verified Program [11]. Other examples include partial shifts toward swine production without use of gestation stalls and egg production without use of laying hen cages. In each case, voluntary adjustment by producers reflects a segment of livestock being produced in a different way, corresponding with private efforts to align the supply of livestock products carrying a FAW claim with higher livestock prices offered by upstream livestock buyers.

Perhaps the most contentious method of farm animal welfare-related practice change is legal. Changes in legislation or ballot initiatives result in legally binding consequences for how animals are reared in a specific geographic region [12]. A high-profile ballot example is the 2008 passage of Proposition 2 in California [13], while the US 28-h transportation rule [14] is an example of binding legislation. A common concern with the legal avenue is that "unfunded mandates" follow—producers are required to adjust practices without corresponding monetary benefits that help offset associated adjustment costs [15]. This "vote-buy" gap is core to contention of the legal avenue [16]. A main benefit of the legal avenue is establishment of clear minimum standards impacting all covered producers. Ballot initiatives can be attractive to activist groups. These initiatives are often written generically about farm animal space and natural behavior allowances. Ballot initiatives do not specify detailed system practices and the method to which they are implemented can be very important when determining actual farm effects. This lack of detail is also a source of great uncertainty for farmers who must make investment decisions. In some cases, legislation is used to avoid a ballot initiative and may be a useful alternative if the animal industries can be involved in the process.

Either the market or legal method can be the impetus for voluntary programs adopted by producers. These programs might realize premiums for farmers—or at least avoid market discounts. It is important to note that these programs are often motivated by the desire to avoid legal mandates or maintain market access. The primary advantage of a voluntary program can be input from farmers, industry and consumers to arrive at a set of standards that allow for a flexible farm response. As there is a great deal of variation in farm size, methods and practices, there are a number of ways for farmers to achieve FAW standards, which voluntary programs can enable.

3. A Minority of Residents Will Truly Pay More for FAW Change

In an examination of 11 different food values, Lister et al. (2018) found the importance of FAW was very low for US consumers [17]. Specifically, shares of importance on four different products were consistently 4–5%: Ground beef (5.2%), beef steak (4.6%), chicken breast (4.1%), and milk products (4.8%). Conversely, the relative importance of price was consistently around 20%, conveying that, on average, price is four-times as important to US consumers.

Consistently, the vast majority of consumers have rated price as the most important factor. Consumers that profess to higher willingness-to-pay (WTP) for animal welfare-related products tend to be older, female, and have higher incomes. However, all hypothetical WTP estimates should be interpreted in the context of hypothetical bias. That is, talk is cheap and saying you would pay more without the requirement to follow through is costless to the research subject. Further, when respondents believe that a specific answer reflects favorably upon them, they may choose the most flattering rather than most correct answer. This is referred to as social desirability bias and likely increases the stated desire for FAW change [18].

When making a purchase of almost any product, most consumers primarily focus on price. The reality is that budget constraints make concern over production practices a luxury for many consumers. A relatively small percentage of the US public cares a great deal about animal welfare to the extent it alters their actual food purchases. However, small, passionate groups drive discussions and often practice changes in many areas. Consumers that are most concerned with animal welfare when making food purchases tend to be older, wealthier, and female [15,19]. This is not to say that others do not consider animal welfare aspects, but they tend to be of secondary importance to most consumers.

4. FAW Is Confounded with Other Food Topics

With respect to FAW, public perceptions vary depending on species and the way in which questions are asked. For example, framing a practice in positive light will result in different responses than when it is framed in a negative light. However, some general patterns hold. First, food safety is an issue that overwhelms all others when it is present or perceived. Consumers have no interest in food products if safety is in question. This makes claims or labels that imply food safety issues particularly damaging. US agriculture has a stellar record of supplying safe food and maintaining this trust is paramount. Food safety is often conflated with technology, farm size, and environmental issues. Tonsor, Olynk, and Wolf (2009) found farm size attributes to be a substitute—from the perspective of US pork consumers—for FAW-oriented production practices [20].

5. FAW Negative Press Matters

When undercover videos or other evidence of animal welfare issues on commercial farms surface, they are widely viewed and affect public perceptions of the entire industry. Tonsor and Olynk (2011) found media attention to animal welfare had a small, but statistically significant impact on US pork and poultry demand [21]. The main economic impact was consumer expenditure reallocated to non-meat products, rather than spillover to competing meats. Contesting ballot initiatives has also been shown to have negative market repercussions, as it increases public attention on perceived farm animal welfare issues [19,22]. More recently, disputes over "ag gag" laws—that seek to limit the ability of opponents to take undercover videos and pictures—have led some to infer that animal agriculture has something to hide. Because negative media affects the entire industry, there is reason for programs that maintain social trust.

6. Why Don't Producers Maximize FAW?

Ultimately, a basic question arises in FAW discussions—why livestock producers simply do not "do more". A situational summary provided by Lusk and Norwood is useful in considering this question [6]. At the extreme, some ask why producers do not make a focused effort to maximize FAW. While improving FAW is a worthwhile goal, it is critical to appreciate that livestock producers are no different than other individuals—they face trade-offs. Any investment or adjustment in effort designed to improve FAW comes at the expense of an alternative investment or effort that is forgone [23]. Stated differently, difficult decisions are regularly made in rearing livestock to meet multiple goals, including, but certainly not limited to, FAW, safety of animal agriculture products, environmental impact, livestock disease resistance, and firm profitability. The key point is that while improvement in FAW is certainly worth discussing and a noble goal, there is no free lunch and tough decisions always must be made.

7. Ways Forward for US Animal Agriculture

So, what are some solutions for the farm animal industry to maintain social trust with respect to perceived animal welfare issues? Assuming that farmers would prefer to avoid formal regulation, options include labeling, education, and voluntary certification programs. Process labels describe how an animal was raised, crops were grown, or ingredients were transformed. Often these labels focus on what was not done or used, as is the case for "hormone-free" and "GMO-free" labels. Advantages of process labels include that they provide information, may enhance trust, and can assist in segmenting the market [24]. Disadvantages include potential information overload (when additional information can be distracting and complicate decisions), confusion, and elevated food safety and risk perceptions [25]. Labels communicating a production technology often induce a negative reaction, resulting in decreased demand for safe products. For example, "contains" has a negative connotation while "free of" a positive connotation. There are also labeling costs to consider, which results in less support for labeling [26].

Information changes consumer preferences. This is particularly true when evaluating food attribute trade-offs [3]. Educational programs are often put forward as a solution because few people have context for production agriculture (~2% of US citizens are farmers). There is some evidence that educational programming, such as Breakfast on the Farm, works to enhance public perception and trust. However, people must want to be educated. There is also some evidence that educational programs can work against intended goals with lack of context and understanding of production agriculture [27]. Educational programs can highlight that many valued practices are already supplied by current commercial agriculture [28]. Another aspect of education to consider is that consumer FAW concerns are about the production process, rather than the product itself. That is, the concern is with the animal rather than the meat, milk, or eggs. Lack of recognition about this often leads to producers and consumers talking past each other in an unproductive manner. Education should not be conflated with activism or it runs the risk of inducing a negative reaction.

Finally, voluntary programs can assist in maintaining public trust. Program effectiveness depends largely on certification. Research has shown that certification or verification is fundamental to public acceptance of animal welfare programs [18]. While people often profess to lack trust in the government, research reveals that consumers often do not trust industries to police themselves and prefer USDA certification [29]. Consumers also have more faith in programs where violations result in real consequences.

8. Conclusions

Viable food animal agriculture requires public trust. Social concerns related to production agriculture are unlikely to diminish in future years. Awareness and incorporation of these concerns in production practices and marketing will help ensure social trust, limit restrictive regulation, and ultimately improve the well-being of animals and people alike. The current literature suggests that FAW, while not the key driver of demand, has important market effects. The US livestock, dairy and poultry industries should continue to move to drive the conversation and be involved in any policy discussions related to potential regulation. Education, labeling, and voluntary programs can all be part of the discussion and ways forward to address public and consumer concerns related to FAW. Ultimately, understanding the economic situation facing livestock producers and reflecting on the formal and informal signals they receive regarding production practices can elevate the overall understanding regarding current and future FAW issues.

Funding: This research received no external funding.

Conflicts of Interest: The authors declare no conflict of interest.

References

1. Lusk, J.L.; Thompson, N.; Weimer, S.L. The Cost and Market Impacts of Slow Growth Broilers. *J. Agric. Resour. Econ. Forthcom.* **2018**. Available online: https://static1.squarespace.com/static/502c267524aca01df475f9ec/t/5bdaf60e562fa73fc1584bef/1541076494784/slow+growth+costs+paper+3.pdf/ (accessed on 26 February 2019).
2. Mullally, C.; Lusk, J.L. The Impact of Restrictions on Farm Animal Housing on Egg Prices, Consumer Welfare, and Production in California. *Am. J. Agric. Econ.* **2018**, *100*, 649–669. [CrossRef]
3. Ochs, D.; Wolf, C.A.; Widmar, N.O.; Bir, C. Is there a "cage-free" lunch in US egg production? Public views of laying hen housing attributes. *J. Agric. Resour. Econ.* **2019**, *44*, 345–361.
4. Widmar, N.J.O.; Morgan, C.J.; Croney, C.C. Perceptions of Social Responsibility of Prominent Animal Welfare Groups. *J. Appl. Anim. Welf. Sci.* **2018**, *21*, 27–39. [CrossRef] [PubMed]
5. Henningsen, A.; Czekaj, T.G.; Forkman, B.; Lund, M.; Nielsen, A.S. The Relationship between Animal Welfare and Economic Performance at Farm Level: A Quantitative Study of Danish Pig Producers. *J. Agric. Econ.* **2018**, *69*, 142–162. [CrossRef]
6. Lusk, J.L.; Norwood, F.B. Animal welfare economics. *Appl. Econ. Perspect. Policy* **2011**, *33*, 463–483. [CrossRef]
7. Hansson, H.; Lagerkvist, C.J. Defining and Measuring Farmers' Attitudes to Farm Animal Welfare. *Anim. Welf.* **2014**, *23*, 47–56. [CrossRef]
8. Lagerkvist, C.J.; Hess, S. A meta-analysis of consumer willingness to pay for farm animal Welfare. *Eur. Rev. Agric. Econ.* **2011**, *38*, 55–78. [CrossRef]
9. Harvey, D.; Hubbard, C. Reconsidering the political economy of farm animal welfare: An anatomy of market failure. *Food Policy* **2013**, *38*, 105–114. [CrossRef]
10. Global Animal Partnership (GAP). Available online: https://globalanimalpartnership.org/ (accessed on 26 February 2019).
11. United States Department of Agriculture, Process Verified Program (USDA PVP). Available online: https://www.ams.usda.gov/services/auditing/process-verified-programs (accessed on 26 February 2019).
12. Schulz, L.L.; Tonsor, G.T.; The, U.S. Gestation Stall Debate. *Choices* **2015**, *30*, 1–7.
13. Malone, T.; Lusk, J.L. Putting the Chicken before the Egg Price: An Ex Post Analysis of California's Battery Cage Ban. *J. Agric. Resour. Econ.* **2016**, *41*, 518–532.
14. United States Department of Agriculture, National Agricultural Library (USDA NAL). Available online: https://www.nal.usda.gov/awic/twenty-eight-hour-law (accessed on 26 February 2019).
15. Norwood, B.; Tonsor, G.T.; Lusk, J.L. I Will Give You My Vote but not My Money: Preferences for Public versus Private Action in Addressing Social Issues. *Appl. Econ. Perspect. Policy* **2018**, *41*, 96–132. [CrossRef]
16. Paul, A.S.; Lusk, J.L.; Norwood, F.B.; Tonsor, G.T. An Experiment on the Vote-Buy Gap with Application to Cage-Free Eggs. *J. Behav. Exp. Econ.* **2019**, *79*, 102–109. [CrossRef]
17. Lister, G.; Tonsor, G.T.; Brix, M.; Schroeder, T.C.; Yang, C. Food Values Applied to Livestock Products. *J. Food Prod. Mark.* **2017**, *23*, 326–341. [CrossRef]
18. Olynk, N.J.; Tonsor, G.T.; Wolf, C.A. Verifying Credence Attributes in Livestock Production. *J. Agric. Appl. Econ.* **2010**, *42*, 439–452. [CrossRef]
19. Smithson, K.; Corbin, M.; Lusk, J.L.; Norwood, F.B. Predicting State-Wide Votes on Ballot Initiatives to Ban Battery Cages and Gestation Crates. *J. Agric. Appl. Econ.* **2014**, *46*, 107–124. [CrossRef]
20. Tonsor, G.T.; Olynk, N.J.; Wolf, C.A. Consumer Preferences for Animal Welfare Attributes: The Case of Gestation Crates. *J. Agric. Appl. Econ.* **2009**, *41*, 713–730. [CrossRef]
21. Tonsor, G.T.; Olynk, N.J. Impacts of Animal Well-Being and Welfare Media on Meat Demand. *J. Agric. Econ.* **2011**, *62*, 59–72. [CrossRef]
22. Richards, T.; Allender, W.J.; Fang, D. Media Advertising and Ballot Initiatives: The Case of Animal Welfare Regulation. *Contemp. Econ. Policy* **2013**, *31*, 145–162. [CrossRef]
23. Nocella, G.; Hubbard, L.; Scarpa, R. Farm Animal Welfare, Consumer Willingness to Pay, and Trust: Results of a Cross-National Survey. *Appl. Econ. Perspect. Policy* **2010**, *32*, 275–297. [CrossRef]
24. Britton, L.; Tonsor, G.T. Consumers' Willingness to Pay for Beef Products Derived from RNA Interference Technology. *Food Qual. Prefer.* **2019**, *75*, 187–197. [CrossRef]
25. Messer, K.D.; Costanigro, M.; Kaiser, H.M. Labeling Food Processes: The Good, the Bad and the Ugly. *Appl. Econ. Perspect. Policy* **2017**, *39*, 407–427. [CrossRef]

26. Tonsor, G.T.; Wolf, C.A. On Mandatory Labeling of Animal Welfare Attributes. *Food Policy* **2011**, *36*, 430–437. [CrossRef]
27. Cummins, A.M.; Olynk Widmar, N.J.; Croney, C.C.; Fulton, J.R. Exploring Agritourism Experience and Perceptions of Pork Production. *Agric. Sci.* **2016**, *7*, 239–249. [CrossRef]
28. Wolf, C.A.; Tonsor, G.T. Cow Welfare in the U.S. Dairy Industry: Willingness-to-Pay and Willingness-to-Supply. *J. Agric. Resour. Econ.* **2017**, *42*, 164–179.
29. Wolf, C.A.; Tonsor, G.T.; Olynk, N.J. Understanding US Consumer Demand for Milk Production Attributes. *J. Agric. Resour. Econ.* **2011**, *36*, 326–342.

Article

Taiwanese Consumers' Willingness to Pay for Broiler Welfare Improvement

Yu-Chen Yang [1] and Cheng-Yih Hong [2,*]

[1]   Department of Applied Economics, National Chung-Hsing University, Taichung 402, Taiwan;
     ycyang@dragon.nchu.edu.tw
[2]   Department of Finance, Chaoyang University of Technology, Taichung 413, Taiwan
*   Correspondence: hcyih@cyut.edu.tw

Received: 21 March 2019; Accepted: 7 May 2019; Published: 10 May 2019

Simple Summary: In Taiwan, the development of farm animal welfare practice is in its beginning stage. Consumers' attitude toward farm animal welfare products is important for the development of this practice. The main goal of this research is to explore the consumers' willingness to pay for broilers' welfare improvement and to identify the factors that affect this willingness to pay. The results of this study showed that consumers' food safety concerns combined with farm animal welfare can influence consumers' willingness to pay. The more consumers believed that they could make a difference in the improvement of animal welfare, the more they were willing to pay. Consumers who felt that farm animal welfare was the producers' responsibility were less willing to pay. The results of this study can be used to evaluate whether or not farm animal welfare practice is market viable. Moreover, the results can be used to develop marketing strategies for high welfare broilers.

Abstract: In this study, we explored the willingness to pay (WTP) for broilers raised under the high welfare system. The interval data model and the ordered probit model were used to investigate the factors that affect consumers' WTP for broiler meat produced by farm animal welfare (FAW), practice. Our results from both methods suggest that socioeconomic characteristics such as education level, income level, gender, and age significantly affect consumers' WTP. The food safety concerns of consumers and perceived consumer effectiveness also influence consumers' WTP. Using the interval data method, we computed the mean and median of the estimated WTP from our survey sample. The mean was 46.7745 New Taiwanese dollar per kilogram. The marginal effects of the different variables are also presented.

Keywords: farm animal welfare (FAW); willingness to pay; food safety concerns; ethical concerns; perceived consumer effectiveness; broiler

---

1. Introduction

In response to public pressure regarding farm animal welfare (FAW) in industrial livestock farming, the Brambell report claimed that farm animals are sentient beings capable of displaying fear, anger, and thirst. It inspired strong legislation to protect farm animals [1]. After that, the farm animal welfare council in the U.K. was established and proposed the concept of the Five Freedoms [2,3]. The Five Freedoms have gained global acceptance and commendation and have been incorporated in national regulations and food marketing schemes [4]. The ethical concerns associated with animal welfare are related to the animal's quality of life [5,6].

Intensive animal farming has raised public concern from ethical, public health, environmental, and food safety perspectives [7–12]. Recently, in the European Union and the U.K., citizens have been paying more attention to the issue of farm animal welfare and these concerns have become a major force, pushing the government to implement higher standards [13,14].

Consumer attitudes toward FAW play an important role in promoting animal welfare practice. Without the support from consumers, the appropriate FAW standards cannot be implemented through the supply chain [15]. Generally, legislation on animal welfare originated as a result of pressure from the public. However, the animal welfare standards in food retail companies have a greater effect on animal welfare than do government regulations [3]. Consumer attitudes toward FAW products act as the main driving force behind food retail companies implementing appropriate FAW standards [16]. Like other agricultural products, livestock products need a marketing system to fulfill the final demand of consumers. A marketing system ranges from the slaughtering-house to food retailers, such as supermarkets. Therefore, we can say that the farm level demand for broilers is the derived demand from consumers demand for chicken. To attract consumers caring about FAW, supermarkets can use over-compliance with government regulation as a marketing strategy [3]. Under this circumstance, they will require their contract animal farm to comply with stricter standards than that promulgated by the government—a stricter FAW standard would be more effective in improving FAW than a relaxed one. For example, Carrefour in Taiwan will not accept eggs from production processes that do not comply with FAW standards, even though such a regulation has not been issued by the Taiwan government.

In Taiwan, the concept of the Five Freedoms has spread gradually among the public, and this is demonstrated by the increase in market share of farm animal welfare eggs from 1% to 7% between 2012 and 2018 [17]. From the results of this study, more than 77% of respondents were willing to pay more for high welfare broiler. The certification system in Taiwan for high welfare products is private. The system was established in 2007, providing a basic market segmentation of labeled and unlabeled goods. According to Taiwan's society of agricultural standards, a third-party certification body, the maximum premium for high welfare products can be nine times larger than those for conventional products. This label system is guided by the principles of the Five Freedoms. So far, most of the farms certified by this label are shell eggs and indigenous chickens. For the broiler industry, only a few farms are able to fulfill the standards of this label system [18].

In order to improve the welfare of farm animals, it is important to create market segmentation and implement a market-driven strategy, particularly in countries where no legislation on high welfare standard is available. Improving the welfare of broilers will increase the cost of broiler production. If consumers are willing to pay the extra cost of high welfare broiler production, market segmentation should be applied, and a market-driven strategy should be implemented. However, if consumers do not want to pay the extra cost associated with high welfare broiler production, then government regulation is needed to improve the welfare of farm animals, and to prevent externality costs and other issues caused by intensive animal farming.

Retailers sponsoring the use of chicken and eggs produced under proper high welfare practice can appeal to more poultry farmers to participate in humane animal farming. The transformation of consumers' ethical intentions into actual ethical buying serves as one of the driving forces influencing food retailers to develop appropriate high welfare standards. The gap between ethical intention and ethical buying remains a crucial issue [19].

Currently, in Taiwan, minimum animal welfare standards have not been passed to improve the welfare of farm animals. The opinion of food supply chain members and consumers' perception toward the welfare of animals are important driving forces. For consumers, willingness to pay (WTP) is the amount of money a consumer wants to pay for achieving an attribute. In other words, WTP is the amount of money a consumer wants to give up in order to keep him in the same satisfaction status while obtaining an attribute [20]. The goal of this research is to investigate consumers' willingness to pay (WTP) for broilers produced by high welfare practice and to determine the factors that affect this willingness to pay in Taiwan. The results of the study can be used to evaluate whether or not high welfare practice is market viable.

2. Materials and Methods

2.1. The Survey

We used a structured questionnaire to elicit information about consumers' WTP for high welfare broiler meat in Taiwan. We employed a pilot survey before conducting the final survey. The survey period ranged from May to November 2018. Both online and hard copy surveys were distributed to respondents, with a total of 480 questionnaires returned to us. However, only 441 questionnaires were completed. During the survey, most of our respondents demonstrated a lack of familiarity with farm animal welfare issues. This lack of familiarity made it difficult for them to assign a financial value to high welfare products. In this study, convenient sampling was used. We sent our survey conductors to the exit of the Carrefour supermarket, which is one of the major channels that sell high welfare products in Taiwan. For the online survey, we selected respondents who frequently purchased high welfare or organic products. Generally, for both the online and hard copy survey, the Five Freedoms principle of high welfare was clearly explained on the cover page of our questionnaire. About 80% of people we selected rejected our invitation to answer the questionnaire.

The questionnaire contained three components. In the first section, we asked respondents about their purchasing habits and knowledge of high welfare products, including their purchasing concerns. Respondents' food safety and ethical concerns in regard to high welfare products were also incorporated into this section. Moreover, the role of the government in supporting high welfare and perceived consumer effectiveness (PCE) regarding high welfare products were also contained in this part. For WTP, a list of options consisting of the various price intervals that respondents had to choose from was included in the second section. In the third section, respondents' socioeconomics conditions were incorporated. The questionnaire didn't reveal any personal information of the respondents so in this case ethical approval was not needed.

2.2. Variables

The variables included in the model to explain the WTP for high welfare broiler meat were categorized into three groups. The first group of variables consisted of perception variables, such as the respondent's food safety concerns towards high welfare broiler meat, the respondent's ethical perception towards broiler meat produced under high welfare practice, the respondent's belief that they could make a difference, and the respondent's belief that it was the producers' responsibility to care about the welfare of farm animals that they keep. In previous literature, age has been suggested as one variable that can affect ethical consumption [21–23]. Besides age, the second group of factors included socioeconomic characteristics, such as education and income levels [24,25]. The third group consisted of other variables, such as whether or not the respondent had heard about high welfare practice, the concerns of the respondent while purchasing broiler meat, and the frequency of eating broiler meat (Table 1).

Table 1. Descriptive Statistics.

Variable	Definition	Mean	Std. Dev
heard	Respondents who have heard about farm animal welfare	0.421769	0.494403
male	Respondents who are male	0.462585	0.499164
eth1	Respondents who very strongly agree that high welfare practice is ethical	0.492064	0.500505
eth2	Respondents who strongly agree that high welfare practice is ethical	0.328798	0.47031
fs1	Respondents who very strongly agree that products produced by high welfare practice are healthier	0.560091	0.49694
fs2	Respondents who strongly agree that products produced by high welfare practice are healthier	0.285714	0.452267
pce1	Respondents whose belief that they can make a difference in solving animal welfare problem is very strong	0.253968	0.435774
pce2	Respondents whose belief that they can make a difference in solving animal welfare problem is strong	0.356009	0.479362
ag5	Respondents' age is between 55 and 65 years old	0.102041	0.303046
ag6	Respondents' age is between 65 and 75 years old	0.040816	0.198089
ed4	Respondents have College degree	0.544218	0.498607
ed5	Respondents have Master's degree	0.14966	0.357143
ed6	Respondents have Ph.D.	0.036281	0.187201
inc4	Respondents' income is between 55000–70000 NTD	0.090703	0.287512
inc5	Respondents' income is more than 70000 NTD	0.131519	0.338351
nonp	Price is not the respondents' only concern	0.950113	0.217958
freq_c	Frequency of eating chicken meat	3.011338	1.601805
prodh	Respondents think that farm animal welfare is the responsibility of producers	0.938776	0.240014

2.2.1. Socioeconomic Conditions

In our survey, respondents were divided into five different groups based on their ages. However, none of the respondents who completed the questionnaires were older than 64. In the WTP equation, we added dummy variables for age. The dummy variables were denoted *ag4*, which represented the group of respondents aged between 45 and 55 years; and *ag5*, which represented the group of respondents aged between 55 and 64. The remaining respondents were incorporated into the reference group(For ag4 and ag5, the reference group covers respondents less than 45 years old). Gender was also incorporated, using a dummy variable (*gender* (Reference group is female)) set as 1 for male respondents and 0 for female ones. Regarding income, the dummy variable *inc4* was set as 1 for the group of respondents whose monthly incomes ranged between NTD 55000 and 70000; otherwise, it was set to 0. The dummy variable *inc5* represented the group of respondents with incomes more than NTD 70000 as 1; otherwise, *inc5* was set to 0. The remainder of the respondents were incorporated into the reference group. For respondents' education, we used *ed4*, *ed5*, and *ed6* for respondents who had a College degree, Master's degree, or Ph.D., respectively.

2.2.2. Respondents' Food Safety Concern

In the survey, respondents were asked about whether or not they agreed with the following statement: "The food safety quality can be improved if an animal is raised by farm animal welfare practice." We used a Likert scale to measure the degree to which a respondent agreed with the statement. The Likert scale ranged from 1 to 5. If a respondent agreed strongly with the statement, the dummy variable, denoted *fs1*, was set to 1; otherwise, it was set to 0. If a respondent agreed with the statement, a dummy variable, denoted *fs2*, was set to 1; otherwise, it was set to 0. The reference group for dummy variables *fs1* and *fs2* represented the group that included respondents whose agreement scores for the statement were 3, 4, and 5.

2.2.3. Respondents' Ethical Concern

Similarly, in the survey, we used the degree to which a respondent agreed with the statement: "Paying courtesy to farm animals' welfare is an appearance of the harmonious coexistence of people, animals, and the natural environment and is a symbol of a progressive society." The extent to which respondents agreed ranged from 1 to 5. To measure respondent's ethical concern [26], a dummy variable, eth1, was employed for the group of respondents who agreed very strongly with the statement. Similarly, a dummy variable, eth2, was employed for the group of respondents who agreed with the statement. The reference group for dummy variables eth1 and eth2 represented the groups that included respondents who scored 3, 4, and 5 in terms of agreement with the statement.

2.2.4. Perceived Consumer Effectiveness

Perceived consumer effectiveness is an important factor that determines the transformation of ethical intention to ethical behavior [27]. In this study, we used the degree to which a respondent agreed with the following statement: "Your personal support for animal welfare can solve the problem of farm animals being abused" to measure the extent to which a respondent believes that his or her effort could make a difference. To measure respondent's perceived consumer effectiveness, a dummy variable, denoted *pce1*, was used for the group of respondents who highly agreed with the statement. Similarly, a dummy variable, *pce2*, was used for the group of respondents who agreed with the statement. The reference group for dummy variables *pce1* and *pce2* is the group including the respondents whose agreement scores for the statement were 3, 4, and 5.

2.2.5. Producers' Responsibility

In this research, we used the degree to which a respondent agreed with the statement: "The industry should offer sufficient space for extension and appropriate facilities for farm animals to have an appropriate environment." We used a dummy variable, denoted *prodh*, for the group of respondents whose agreement scores for the statement were 1, 2, and 3; otherwise, it was set to be 0. Therefore, the reference group included the group made up of respondents whose agreement scores for the statement were 4 and 5.

Research Methodology

To evaluate consumer WTP for broilers produced using high welfare practice in Taiwan, a contingent valuation survey was carried out to collect data. In this structured questionnaire, five WTP intervals (represented as percentage increase) were listed and respondents were asked to select the interval that they believed corresponded with their WTP. The intervals were numbered 1, 2, 3, 4, and 5, valued respectively as follows: (0–25%), (25–50%), (50%–75%), (75%–100%), and (100–∞%). The percentage increase was calculated as the increase in the price of broiler meat as a result of changing from conventional methods to high welfare methods, divided by the price of broilers produced under conventional methods. If the respondent's WTP for high welfare broiler meat fell within Interval 1, then we considered the WTP to be very weak. If the WTP fell within Interval 2, we considered WTP to be weak. The WTP was considered moderate if it fell within Interval 3. Willingness to pay was considered strong if it fell within Interval 4, and willingness to pay was considered to be very strong if it fell within Interval 5. (Table 2).

Table 2. Strength of willingness to pay for broilers produced by high welfare practice.

Strength of Willingness to Pay	Interval Label
Between 0 and 25 percent (Very weak)	Interval 1
Between 25 and 50 percent (Weak)	Interval 2
Between 50 and 75 percent (Modest)	Interval 3
Between 75 and 100 percent (Strong)	Interval 4
More than 100 percent (Very Strong)	Interval 5

We assumed that the willingness to pay (WTP) for high welfare broiler was expressed by the following equation:

$$WTP^* = \sum_i \beta_i z_i + \varepsilon, \tag{1}$$

where ε is assumed to be normally distributed with a mean of 0 and a variance σ^2, z_i are the explanatory variables. Willingness to pay (WTP^*) of consumers is unobservable.

We informed the respondents that the price of broiler chicken raised under the conventional farming method was 50 NTD/kg (NTD is new Taiwanese dollar). Accordingly, the corresponding monetary values for the various intervals were as follows: Interval 1, 0–12.5 NTD; Interval 2, 12.5–25 NTD; Interval 3, 25–37.5 NTD; Interval 4, 37.5–50 NTD; and Interval 5, 50–∞ NTD. The survey elicited both the strength and intervals of respondents' willingness to pay for high welfare broilers. Therefore, both the ordered probit model and interval data method can be used. The factors that affect the WTP for high welfare broiler can be verified by both methods.

Unlike other contingent valuation methods that use bidding approaches, especially double bound dichotomous choices, this method allowed us to avoid the problem of bias resulting from changes in the incentive structure [28–30]. The interval data model includes more information and can improve the efficiency of estimation [31].

For this study, we denote an observable variable WTP. If Respondent i indicates that his willingness to pay is very weak, then $WTP_i = 1$; similarly, if the WTP is weak, then $WTP_i = 2$; if moderate, $WTPi = 3$; if strong, $WTP_i = 4$; and if very strong, $WTP_i = 5$.

The corresponding probability for $WTP_i = j-1$ is expressed as the following equation:

$$\Phi\left(\gamma_j - \sum_i \beta_i z_i\right) - \Phi\left(\gamma_{j-1} - \sum_i \beta_i z_i\right), \tag{2}$$

where γ_j are unknown category threshold parameters that can be estimated, and Φ is the distribution function of standard normal distribution.

Therefore, Equation (2) can be used to estimate the value of the coefficient of z_i. Ordered qualitative response models were employed. The estimation results of β_i cannot be used as the marginal impact of z_i. To estimate the marginal impact of z_i, we used the interval data method. In the interval data regression, the interval boundaries are known. Therefore, the probability that WTP^* falls into a specific interval (α_{j-1}, α_j) is $F\left(\frac{\alpha_j - \sum_i \beta_i z_i}{\sigma}\right) - F\left(\frac{\alpha_{j-1} - \sum_i \beta_i z_i}{\sigma}\right)$, where F(.) is the distribution function of the random variable ε in WTP^*. If we assume the probability distribution of the random variable ε is normal, then the probability that the WTP^* falls into a specific interval (α_{j-1}, α_j) is

$$\Phi\left(\frac{\alpha_j - \sum_i \beta_i z_i}{\sigma}\right) - \Phi\left(\frac{\alpha_{j-1} - \sum_i \beta_i z_i}{\sigma}\right). \tag{3}$$

3. Results and Discussion

3.1. Descriptions of Results from Two Methods

The empirical results from the ordered probit and interval data model were similar. The sign of the coefficients of variables were the same for both models, however, the explanations of the coefficients differed. The coefficients in the ordered probit model do not represent the marginal impacts of variables on the respondents' WTP. The dependent variables in the ordered probit model were categorical ordinal variables. Hence, we could only evaluate the impact of variables on the probability of the occurrence of a specified category. Furthermore, the assumptions underlying the error term also differed. In the ordered probit model, the standard normal distribution was used. Four threshold parameters were estimated. In the interval data model, the random variables were normalized by assuming a scale parameter sigma (σ). Hence, only the coefficients in the interval data model could be used to represent the impact of the variables on WTP. From Table 3, it can be seen that out of 19 variables, 12 have coefficients that are significant. For a dummy variable, the coefficient represents the difference in the impact between respondents in a specific dummy group and those in the reference group. For a continuous variable, the coefficient represents its marginal impact.

Table 3. Estimation Results.

Variables	Interval Data Regression	Ordered Probit Model
heard	1.6354	0.1521
male	−2.745**	−0.2418**
eth1	2.4394	0.2128
eth2	−3.0056	−0.2591
fs1	5.5652*	0.5332*
fs2	5.3528*	0.5088*
pce1	5.0128***	0.4408***
pce2	1.0305	0.08418
ag5	3.9727*	0.3570*
ag6	0.2342	−0.006
ed4	3.4320**	0.3081**
ed5	7.2680***	0.6344***
ed6	7.6075**	0.6584**
inc4	2.4710	0.2259
inc5	6.7881***	0.5937***
nonp	4.8749*	0.4751*
freq_c	0.5389	0.0486
Prodh	−8.8927***	−0.8176***
Constant	19.4765***	
/lnsigma	2.4357***	
sigma	11.4243	
threshold parameter1		−0.3368
threshold parameter 2		0.4404
threshold parameter3		1.7015
threshold parameter 4		2.6773
	$\chi^2 = 128.86$	$\chi^2 = 127.45$
	p-value=0	p-value=0

* is significant at 10% confidence level. ** is significant at 5% confidence level. *** is significant at 1% confidence level.

3.2. Results from the Interval Data Method

From the econometric results of the interval data method, it can be seen that the coefficient of the gender variable is significant and negative. This implies that gender has a significant effect on WTP. The female respondents are more willing to pay for high welfare products. This result is consistent with previous literature. The coefficient of the variable, *nonp*, was significant and positive, which implies that a consumer who does not only consider price is willing to pay more for high welfare products. The coefficient of ag5 was positive and significant. This result implies that the group of respondents aged between 45 and 55 years old are more willing to pay than respondents in the reference group.

Regarding the respondents' income, the coefficient of variable inc5 was both significant and positive. This implies that respondents in the group with incomes more than 70000 NTD are more willing to pay than respondents in the reference group. For education, the coefficients of ed4, ed5, and ed6 were all significant and positive. This result suggests that respondents with a College degree, Master's degree, or Ph.D. are more willing to pay for high welfare products than respondents in the reference group.

Regarding consumers' attitude toward high welfare products, the coefficients of variable *sf1* and *sf2* represent the differences in WTP between respondents in the dummy groups and the reference group. Both of the coefficients for variables *sf1* and *sf2* were significant and positive. This result suggests that respondents strongly agree with the statement that farm animal products are healthier and are inclined to pay more for high welfare products. Also, respondents strongly agreed with the statement that farm animal products were healthier and that they were inclined to pay more for these products. The results of this research indicate that consumer food safety concerns in regard to intensive farming significantly affect their WTP for high welfare products. The more a consumer agrees that high welfare products are healthier, the more he or she is willing to pay for high welfare products.

Regarding consumer's ethical concerns, the coefficients of variables *eth1* and *eth2* represent the differences in WTP between respondents in the dummy groups and respondents in the reference group. Both of the coefficients for variables *eth1* and *eth2* were insignificant but positive. This indicates that consumer's ethical concern of high welfare has no impact on their WTP for high welfare products.

The coefficient of variable *prodh* represents the difference in WTP between respondents who agreed with the statement that producers are responsible for animal welfare and should provide a decent environment for farm animals, and those who did not believe so. The coefficient of variable *prodh* was significant and negative. This result implies that respondents who agree that producers are responsible for animal welfare are less willing to pay for high welfare products.

The coefficient of variable *pce1* represents the difference in WTP between respondents who strongly believe that they can have an effect in improving the welfare of farm animals and those who believe that they can have no effect. The coefficient of variable *pce1* was significant and positive. This implies that a respondent who strongly believes that he or she can have an effect in improving the welfare of farm animals is more willing to pay for high welfare products than one who believes that he or she has no effect. The result was similar for variable *pce2*—with a coefficient that was both significant and positive.

After estimating the coefficients of variables in Equation (1), we were able to estimate the WTP for respondents with varying socioeconomic characteristics and varying degrees of agreement in terms of the link between food safety, ethical quality, and high welfare broiler products. By substituting the coefficients in Equation (1) with the estimations of the coefficients in Table 3 from the interval data model, we obtained the sample mean and median of the estimated WTP for high welfare broiler meat in Taiwan. The estimates of the WTP are shown in Table 4. The sample mean of consumer WTP was 28.0648 NTD/kg, while the sample median of WTP was 28.4762 NTD/kg.

Table 4. Estimated willingness to pay for broiler produced by high welfare practice.

	Mean	Maximum	Median	Minimum
Willingness to pay	46.7745	81.0495	47.4603	16.6580

3.3. Results from the Ordered Probit Model

Since the meaning of the coefficients for the variables in the ordered probit model differed from those in the interval data model, it cannot represent the marginal impact. We computed changes in the probability of a specific category outcome occurring. In the ordered probit model, we used the following equation:

$$\frac{\Delta \Pr(WTP = J - 1)}{\Delta x_i} = \Phi(WTP = J - 1|x_i = 1) - \Phi(WTP = J - 1|x_i = 0). \tag{4}$$

The marginal effects are shown in Table 5. It can be seen that males are more likely to have lower WTP compared to females. The probability that categories specify very strong, strong, and moderate groups both diminish. However, the probabilities for weak and very weak groups increase. This result implies that, in general, female respondents have a higher likelihood of willingness to pay more.

For age, the marginal impact of age5 was negative for very weak WTP and weak WTP. The marginal effects were positive for modest, strong, and very strong WTP. This implies that respondents aged between 55 and 65 years had a higher probability of modest, strong, and very strong WTP compared with those in the reference group.

Table 5. Marginal effects.

Variables	Very Weak	Weak	Moderate	Strong	Very Strong
heard	−0.0322	−0.0168	0.0103	0.02400	0.0147
male	0.0512**	0.0267**	−0.0163**	−0.0381**	−0.0234**
eth1	−0.0450	−0.0235	0.01437	0.0336	0.0206
eth2	0.0548	0.0286	−0.0175	−0.0409	−0.0250
fs1	−0.1128*	−0.0588*	0.0360*	0.0841*	0.0515*
fs2	−0.1077*	−0.0561*	0.0344*	0.0803*	0.0492*
pce1	−0.0933***	−0.0486***	0.0298***	0.0695***	0.0426***
pce2	−0.0178	−0.0093	0.0057	0.0133	0.008
age5	−0.0756*	−0.0394*	0.0241*	0.0563*	0.0345*
age6	0.0014	0.0007	−0.0004	−0.0010	−0.0006
ed4	−0.0652**	−0.0340**	0.02081**	0.0486**	0.0298**
ed5	−0.1342***	−0.0700***	0.0428***	0.1001***	0.0613***
ed6	−0.1393**	−0.0726**	0.0444**	0.10381**	0.0636**
in4	−0.0478	−0.0249	0.01525	0.03561	0.0218
in5	−0.1256***	−0.0655***	0.0400***	0.09361***	0.0574***
nonp	−0.1005**	−0.0524**	0.0321**	0.0749**	0.0459**
freq_c	−0.0103	−0.0054	0.0033	0.0077	0.0047
peodh	0.1723***	0.0902***	−0.0552***	−0.1290***	−0.0790***

* is significant at 10% confidence level. ** is significant at 5% confidence level. *** is significant at 1% confidence level.

For education levels, in comparison with those in the reference group, respondents with a College degree, Master's degree, or Ph.D. had a higher probability of having modest, strong, and very strong WTP. For weak and very weak WTP, the marginal effects were negative.

The marginal effects of *fs1* were negative for very weak WTP and weak WTP. For modest, strong, and very strong WTP, the marginal effects were positive. This indicates that a consumer who had very strong recognition that high welfare products are healthier had a higher probability of having modest, strong, and very strong WTP compared with a consumer in the reference group. For weak and very weak WTP, the marginal effects were negative. The marginal effects of *fs2* on each of the categories were similar.

Regarding perceived consumers effectiveness, the marginal impacts of *pce1* were negative for outcomes of weak and very weak WTP; however, for outcomes of strong and very strong WTP, marginal effects were all positive. Respondents who had a very strong belief that their contribution could make a difference in improving the welfare of farm animals had a higher likelihood to have modest, strong, and very strong WTP than those in the reference group.

For consumers' purchasing concern for broiler, the marginal effect of *nonp* was negative for outcomes of weak and very weak WTP; and was positive for outcomes of modest, strong, and very strong WTP. This indicates that if respondents did not only care about pricing, they were more likely to pay more in comparison with those who only cared about broiler meat price.

Concerning consumers' altitude toward producers' responsibility for high welfare, the marginal effect of *prodh* was positive for outcomes of weak and very weak WTP; and was negative for those of modest, strong, and very strong WTP. This result implies that respondents who agree that it is the responsibility of the farmer to provide a decent environment for farm animals have a higher likelihood to have weak and very weak WTP than those who do not.

3.4. Marketing Strategies for High Welfare Products

The results of this research demonstrate that if consumers recognize that animal welfare products are healthier, they are willing to pay premium prices for high welfare products. Such a result implies that adopting high welfare practices will not result in businesses losing competitive advantage due to increased production costs [32]. Producers adopting high welfare methods could gain competitive advantage as a result of market segmentation by providing healthier products to consumers.

Food safety concerns over intensive animal farming practices are the driving force for consumers to purchase ethically products produced by high welfare methods. Consequently, in order to provide consumers more links to high welfare practices, one suggested strategy is to assist consumers to completely comprehend the ethical high welfare production processes. Additionally, another strategy is helping consumers realize that their purchasing of high welfare products is beneficial. The payoff for ethical buying is the improvement in consumers' food safety.

Regarding perceived consumer effectiveness (PCE), our results demonstrate that PCE is one factor that can inspire consumers' buying intentions [33–36]. Therefore, in order to encourage consumers' intention to buy high welfare products, their PCE should be motivated. A number of researchers have discussed how to activate consumer PCE [37–39]. Accumulation of knowledge in high welfare practices and awareness of individuals' effort will make a difference in solving the problem of weak consumer PCE. Therefore, animal welfare education should begin from the early childhood stage. In addition, government and non-profit organizations should campaign on behalf of improved and humane animal welfare practice to demonstrate to individual consumers that they are not alone.

3.5. The Gap between Ethical Intention and Ethical Buying

Our results reveal that some consumers with high ethical concerns for farm animal welfare issues are not actually willing to pay the premium for high welfare products. There exists a gap between ethical concern and ethical buying. This attitude–behavior gap is an example of so-called consumer dualism, discussed by Verbeke (2009) and Grunert (2006) [40,41]. One factor possibly discouraging customers from participating in dollar voting is their lack of trust in the ability of producers to fulfill all the requirements of high welfare. Therefore, they feel that their efforts in that regard are impractical. This ambiguity may also result from the fact that consumers want to support high welfare practice, but they worry that the price is too high. As mentioned earlier, farm animal welfare is deemed a public good. This is a problem related to the phenomenon of free riding [42–44].

Consumers who think that producers are responsible for high welfare are less willing to pay for farm animal products. This result could serve as an explanation of the gap between ethical concern and ethical buying. Consumers are concerned with the issue of animal welfare; however, they feel disengaged from animal welfare issues and believe that it is not their personal responsibility.

4. Conclusions and Limitations of Research

One relevant result emerging from this study is the agreement between interval data methods and the ordered probit method. In addition, the results of this study indicated that consumers' food safety concern regarding high welfare was one factor that influenced consumers' WTP for high welfare broiler chicken meat. Socioeconomic characteristics such as education, income level, gender, and age also influence consumers' willingness to pay. Females were more willing to pay than males. Consumers with a College degree, Master's degree, or Ph.D. were more willing to pay compared with those with an educational attainment below the college level. Consumers with monthly income levels ranging between 55000 and 700000NTD were also more willing to pay than the others. Respondents who fell within the age range of 55-65 were more willing to pay than others. Regarding respondents perceived consumer effectiveness, our results demonstrated that the more consumers believe that they could make a difference in solving the animal welfare problem; the more they were willing to pay for high welfare products. However, consumers who felt that farm animal welfare was the producers' responsibility were less willing to pay.

Although the survey question should be more concise and less subjective, the results of this study showed that an individual's moral intensity is irrelevant to WTP for improvement in broiler welfare.

During the survey, some respondents felt that it was hard to assign a financial value to the broiler chicken produced by high welfare practice. This may be one of the reasons why some respondents gave up answering the questionnaire completely.

Author Contributions: Y.-C.Y. and C.-Y.H. conceived and designed the experiments; Y.-C.Y. and C.-Y.H. performed the experiments; Y.-C.Y. and C.-Y.H. analyzed the data; Y.-C.Y. and C.-Y.H. contributed reagents/materials/analysis tools; Y.-C.Y. and C.-Y.H. wrote the paper.

Funding: This research received no external funding.

Conflicts of Interest: The authors declare no conflict of interest.

References

1. Rogers Brambell, F.W. *Committee. Report of the technical committee to enquire into the welfare of animals kept under intensive livestock husbandry systems*; Report No.: 2836; Her Majesty's Stationery Office: London, UK, 1965.
2. Botreau, R.; Veissier, I.; Butterworth, A.; Bracke, M.B.M.; Keeling, L.J. Definition of criteria for overall assessment of animal welfare. *Anim. Welfare* **2007**, *16*, 225–228.
3. Veissier, I.; Butterworth, A.; Bock, B.; Roe, E. European approaches to ensure good animal welfare. *Appl. Anim. Behav. Sci.* **2008**, *113*, 279–297. [CrossRef]
4. Buller, H.; Blokhuis, H.; Jensen, P.; Keeling, L. Towards farm animal welfare and sustainability. *Animal* **2018**, *8*, 81. [CrossRef]
5. Broom, D.M. Quality of life means welfare: How is it related to other concepts and assessed? *Anim. Welfare* **2007**, *16*, 45–53.
6. Broom, D.M. Welfare assessment and relevant ethical decisions: Key concepts. *Annu. Rev. Biomed. Sci.* **2008**, *10*, 79–90. [CrossRef]
7. Li, P.J. Exponential growth, animal welfare, environmental and food safety impact: The case of China's livestock production. *J. Agric. Environ. Ethics* **2009**, *22*, 217–240. [CrossRef]
8. de Passillé, A.M.; Rushen, J. Food safety and environmental issues in animal welfare. *Rev. Sci. Tech. OIE* **2005**, *24*, 757–766.
9. Akhtar, A. The need to include animal protection in public health policies. *J. Public Health Pol.* **2013**, *34*, 549–559. [CrossRef]
10. Harper, G.C.; Makatouni, A. Consumer perception of organic food production and farm animal welfare. *Br. Food J.* **2002**, *104*, 287–299. [CrossRef]
11. Evans, A.; Miele, M. Consumers' Views about Farm Animal Welfare. Part II: European Comparative Report Based on Focus Group Research. In *Welfare Quality Reports*; School of City and Regional Planning, Cardiff University: Cardiff, UK, 2008.
12. Rostagno, M.H. Can stress in farm animals increase food safety risk? *Foodborne Pathog. Dis.* **2009**, *6*, 767–776. [CrossRef]
13. Broom, D.M. Does present legislation help animal welfare? *Landbauforsch Volk.* **2002**, *227*, 63–69.
14. Broom, D.M. Animal Welfare and Legislation. In *Welfare of Production Animals: Assessment and Management of Risks*; Smulders, F., Algers, B.O., Eds.; Wageningen Press: Wageningen, The Netherlands, 2009; pp. 341–354.
15. Harvey, D.; Hubbard, C. The supply chain's role in improving animal welfare. *Animal* **2013**, *3*, 767–785. [CrossRef]
16. Broom, D.M. Animal welfare: An aspect of care, sustainability, and food quality required by the public. *J. Vet. Med. Educ.* **2009**, *37*, 83–88. [CrossRef]
17. The Environment & Animal Society of Taiwan. Available online: https://www.east.org.tw (accessed on 6 April 2019).
18. Taiwan society of agricultural standards. Available online: http://www.tsas.tw (accessed on 6 April 2019).
19. Carrington, M.J.; Neville, B.A.; Whitwell, G.J. Why ethical consumers don't walk their talk: Towards a framework for understanding the gap between the ethical purchase intentions and actual buying behaviour of ethically minded consumers. *J. Bus. Ethics* **2010**, *97*, 139–158. [CrossRef]
20. Hanemann, W.M. Willingness to pay and willingness to accept: How much can they differ? *Am. Econ. Rev.* **1991**, *81*, 635–647. [CrossRef]
21. Spain, C.; Freund, D.; Mohan-Gibbons, H.; Meadow, R.; Beacham, L. Are They Buying It? United States Consumers' Changing Attitudes toward More Humanely Raised Meat, Eggs, and Dairy. *Animals* **2018**, *8*, 128. [CrossRef]

22. Makdisi, F.; Marggraf, R. Consumer willingness-to-pay for farm animal welfare in Germany—the case of broiler. In Proceedings of the German Association of Agricultural Economists 51st Annual Conference, Halle, Germany, 28–30 September 2011.

23. Lagerkvist, C.J.; Hess, S. A meta-analysis of consumer willingness to pay for farm animal welfare. *Eur. Rev. Agric. Econ.* **2011**, *38*, 55–78. [CrossRef]

24. Mulder, M.; Zomer, S. Dutch Consumers' Willingness to Pay for Broiler Welfare. *J. Appl. Anim. Welf. Sci.* **2017**, *20*, 137–154. [CrossRef]

25. Toma, L.; McVittie, A.; Hubbard, C.; Stott, A.W. A structural equation model of the factors influencing British consumers' behaviour toward animal welfare. *J. Food Prod. Market.* **2011**, *17*, 261–278. [CrossRef]

26. Mann, S. Ethological farm programs and the "market" for animal welfare. *J. Agric. Environ. Ethics* **2005**, *18*, 369–382. [CrossRef]

27. Frey, U.J.; Pirscher, F. Willingness to pay and moral stance: The case of farm animal welfare in Germany. *PLoS ONE* **2018**, *13*, e0205551. [CrossRef] [PubMed]

28. Alberini, A. Efficiency vs bias of willingness-to-pay estimates: Bivariate and interval-data models. *J. Environ. Econ. Manag.* **1995**, *29*, 169–180. [CrossRef]

29. Cameron, T.A.; Quiggin, J. Estimation using contingent valuation data from a Dichotomous Choice with Follow-Up. *J. Environ. Econ. Manag.* **1994**, *27*, 218–234. [CrossRef]

30. Herriges, J.A.; Shogren, J.F. Starting point bias in dichotomous choice valuation with follow-up questioning. *J. Environ. Econ. Manag.* **1996**, *30*, 112–131. [CrossRef]

31. Hanemann, M.; Loomis, J.; Kanninen, B. Statistical efficiency of double-bounded dichotomous choice contingent valuation. *Am. J. Agric. Econ.* **1991**, *73*, 1255–1263. [CrossRef]

32. Blandford, D.; Bureau, J.C.; Fulponi, L.; Henson, S. Potential implications of animal welfare concerns and public policies in industrialized countries for international trade. In *Global Food Trade and Consumer Demand for Quality*; Springer: Boston, MA, US, 2002; pp. 77–99.

33. Vermeir, I.; Verbeke, W. Sustainable food consumption: Exploring the consumer "attitude–behavioral intention" gap. *J. Agric. Environ. Ethics* **2006**, *19*, 169–194. [CrossRef]

34. Ellen, P.S.; Wiener, J.L.; Cobb-Walgren, C. The role of perceived consumer effectiveness in motivating environmentally conscious behaviors. *J. Public Policy Mark.* **1991**, *10*, 102–117. [CrossRef]

35. Fishbein, M.; Ajzen, I. *Belief, Attitude, Intention and Behavior: An Introduction to Theory and Research*; Addison-Wesley, Reading: Boston, MA, USA, 1975.

36. Vanhonacker, F.; Verbeke, W. Buying higher welfare poultry products? Profiling Flemish consumers who do and do not. *Poult. Sci.* **2009**, *88*, 2702–2711. [CrossRef]

37. Engel, J.F.; Blackwell, R.D.; Miniard, P.W. *Consumer Behaviour*, 9th ed.; Harcourt College Publisher: New York, NY, USA, 2001.

38. Why Consumers Buy Green; Why They Don't. A Barrier/Motivation Inventory: The Basis of Community-Based Social Marketing. Available online: http://www.acetiassociates.com/pubs/greenbuying.pdf (accessed on 1 February 2019).

39. Mills, J.; Margaret, S.C. Exchange and communal relationships. In *Review of Personal and Social Psychology*; Wheeler, L., Ed.; Sage: Beverly Hills, CA, USA, 1982; Volume 3, pp. 121–144.

40. Verbeke, W. Stakeholder, citizen and consumer interests in farm animal welfare. *Anim. Welfare* **2009**, *18*, 325–333.

41. Grunert, K.G. Future trends and consumer lifestyles with regard to meat consumption. *Meat Sci.* **2006**, *74*, 149–160. [CrossRef]

42. Bennett, R. The value of farm animal welfare. *J. Agric. Econ.* **1995**, *46*, 46–60. [CrossRef]

43. Bennett, R.; Larson, D. Contingent valuation of the perceived benefits of farm animal welfare legislation: An exploratory survey. *J. Agric. Econ.* **1996**, *47*, 224–235. [CrossRef]

44. Bennett, R.M. Farm animal welfare and food policy. *Food Policy* **1997**, *22*, 281–288. [CrossRef]

animals

Article

Towards the Abandonment of Surgical Castration in Pigs: How is Immunocastration Perceived by Italian Consumers?

Jorgelina Di Pasquale [1], Eleonora Nannoni [2], Luca Sardi [2,*], Giulia Rubini [2], Renato Salvatore [3], Luca Bartoli [3], Felice Adinolfi [2] and Giovanna Martelli [2]

[1] Faculty of Veterinary Medicine, University of Teramo, 64100 Piano D'Accio, Teramo, Italy; jdipasquale@unite.it

[2] Department of Veterinary Medical Sciences, University of Bologna, Via Tolara di Sopra 50, 40064 Ozzano Emilia (BO), Italy; eleonora.nannoni2@unibo.it (E.N.); giulia.rubini8@unibo.it (G.R.); felice.adinolfi@unibo.it (F.A.); giovanna.martelli@unibo.it (G.M.)

[3] Department of Economics and Law, University of Cassino and Southern Lazio, Viale dell'Università, 03043 Cassino (FR), Italy; rsalvatore@unicas.it (R.S.); bartoli@unicas.it (L.B.)

* Correspondence: luca.sardi@unibo.it

Received: 19 March 2019; Accepted: 23 April 2019; Published: 26 April 2019

Simple Summary: The European Declaration on alternatives to surgical castration of pigs was aimed at abandoning surgical castration and switching to alternative techniques. Immunocastration (a vaccination against Gonadotropin Releasing Hormone) can be a viable alternative method. This technique offers some advantages in terms of animal welfare compared to surgical castration. Nevertheless, the main obstacle to the diffusion of immunocastration seems to be related to consumers' acceptance, since the use of new technologies in the food chain often generates mistrust. The objective of this research was to assess how immunocastration is perceived by Italian consumers, and how complex and complete information (on advantages and disadvantages of the technique) can influence their perception. The results show that immunocastration is perceived in a predominantly positive manner (54.5%), with a relatively low level of risk perception (34.2%) and a good willingness to pay more for meat from immunocastrated pigs (+18.7%). However, there were no statistically significant differences between the control group (receiving only a neutral technical information) and groups to which complete and complex information was provided.

Abstract: Immunocastration of pigs represents an alternative method to surgical castration, being more respectful of animal welfare. However, this new technology may not be accepted by consumers due to their perception of possible risks tied to the use of the product, thus representing a concern for the production sector. The study aimed at verifying the attitude of Italian consumers towards immunocastration and to assess whether their perception can be affected by science-based information on advantages and disadvantages of immunocastration. A total of 969 consumers (divided in three groups representative of the Italian population) were contacted and asked to complete an online questionnaire. Only technical (neutral) information on immunocastration was provided to the first group; the second and the third group received information on the advantages (+) and disadvantages (-) of the technique, shown in reverse order (+/- and -/+, respectively). The level of information did not affect consumers' perception of immunocastration. Overall, immunocastration is perceived in a predominantly positive manner (54.5%), with a relatively low level of risk perception (34.2%), and a good willingness to pay more for meat deriving from immunocastrated pigs (+18.7%).

Keywords: animal welfare; consumer; willingness to pay; pig; castration; immunocastration; information; survey

1. Introduction

Since 3000–4000 BC, male piglets have been surgically castrated for diverse reasons [1]: first of all, to reduce the occurrence of boar taint, which has an objectionable odor and flavor of meat deriving from entire males. The boar taint is associated with androstenone (a testicular steroid) and skatole (which is bacterially produced from tryptophan degradation in the hindgut of the pig) [2]. The second reason for castration is to reduce aggressive and sexual behaviors [3]. As a side effect of castration, it favours a higher fatness degree [4], which is appreciated for peculiar production schemes.

At present, surgical castration is the most frequently applied method in piglets [5]. In agreement with Council Directive 2008/120/EC, this procedure is usually carried out within the first 7 days of life, with minimal (or not at all) pain relief or anesthesia. Surgical castration is an obviously painful and stressful procedure that undermines piglet's welfare, possibly resulting in detrimental effects on growth and the immune system, and hence on the health of animals [6]. In 2010, the European Declaration on alternatives to surgical castration of pigs [7] recommended to switch to alternatives (such as castration with anesthesia or analgesia, raising entire males, sperm sexing, and immunocastration), with the aim to abandon castration by 2018, with the exception of pork meat for products under "Traditional Speciality Guaranteed -TSG", "Protected Geographical Indication PGI" or "Protected Designation of Origin -PDO" labels, for which castration is deemed to be unavoidable to meet the current quality standards [7]. Although to date the target set by the Declaration has not been met by most European countries, the use of local anesthesia during surgical castration has been mandatory in Norway since 2002, and in Switzerland but by using general anesthesia [5].

Immunocastration is a viable alternative to surgical castration. It consists in a vaccination against GnRH (Gonadotropin Releasing Hormone), through the administration of a protein-conjugate analogue of GnRH that results in the production of antibodies against the animal's own GnRH. This subsequently stops the synthesis of LH (Luteinizing Hormone) and FSH (Follicle Stimulating Hormone) with consequent testis regression and reduced production and accumulation of steroid hormones, including boar-taint-causing androstenone [8]. Immunocastration therefore controls boar taint and aggressive and sexual behavior [3,9]. This minimally invasive technique offers some advantages compared to surgical castration: absence of acute pain, reduced stress [10], and simplified handling (it consists of a subcutaneous injection at the base of the neck, just behind the ear). Some studies have also shown that immunocastrated pigs have a better feed conversion ratio and their carcasses have a higher percentage of lean meat than surgically castrated animals [4,5,11]. This could be due to the fact that two administrations of the vaccine are needed for a full response in pigs: the first one is aimed to prime the animals' immune system and the second one is applied when animals approach their sexual maturity [8]. Until the second injection, the immunocastrated males are physiologically more similar to entire males than to surgically castrated animals, with consequently higher lean meat yield, potentially lower environmental impact, and higher cost efficiency [2,12–14]. Other studies found no differences in terms of carcasses and meat quality [1].

According to the recommendations for vaccine administration, the first injection should be given at week 17–18 or earlier, while the second should be administered at 21–22 weeks old if the pigs are slaughtered at 26 weeks old [15]. For pigs slaughtered at higher age and body weight (such as Italian heavy pigs intended for PDO dry-cured hams), a third dose becomes necessary to control boar taint [8].

The drawbacks of immunocastration are tied to the possibly increased costs for the farmer (purchase of the product and workforce for its administration), the risk of accidental self-injection by the farm workers, and the uncertain consumer's attitude towards meat from pharmacologically castrated animals. From the economic point of view, surgical castration with local anaesthesia could represent a less expensive option in comparison with immunocastration. On the other hand, immunocastration seems to result in better feed conversion rate which can compensate the costs of vaccination, particularly in Italy where pigs are slaughtered at a very high body weight (about 170 kg) [16]. As regards the risks for farm workers, accidental self- injection may produce similar effects in people to those seen in pigs (temporary reduction in sexual hormones and reproductive functions in both men and women, adverse effect on pregnancy), with increased risk after a second or subsequent accidental injection. Therefore, according to specifications, the vaccine must only be administered with a safety vaccinator having both a needle guard and a mechanism to prevent accidental operation of the trigger [15].

One of the issues that most concerns the primary sector seems to be related to consumer acceptance. Some studies have investigated consumer's attitude towards immunocastration [17–26]. These surveys identified two main aspects: concerns for food safety and sensitivity towards animal welfare. Based on their attitudes towards these topics, EU consumers can be divided into two main groups: one broadly in favor and other one broadly against immunocastration [17]. Across the mentioned studies, participants expressed mainly favorable attitudes towards the abandonment of castration without anesthesia and analgesia and its substitution with alternative methods, although in some cases they expressed some apprehensions about immunocastration [18,26]. In a Norwegian study, despite their considerable trust in national control authorities, respondents were skeptical towards immunocastration, due to concerns about possible residuals in meat and unpredictable long-term consequences for consumers' health. On the other hand, these consumers categorically refused surgical castration without anaesthesia [18]. Similarly, a study carried out in Italy confirms consumers' skepticism about the use of immunocastration in pigs intended for the production of traditional products (PDO and PGI) with doubts similar to those indicated by Norwegians [17]. Only few studies focus on the role that information plays on the consumer's acceptance of immunocastration. The field of information is a topic of extreme importance for the primary sector. The way and the type of information provided to the consumer can result in the acceptance or refusal of an innovative production technique. The study carried out by Vanhonacker et al. [5], concludes that information concerning the potential benefits and risks of immunocastration does not affect much consumers' attitudes. Tuyttens et al. [20], stressed the role of information type: audiovisual information revealed a more marked effect than basic and detailed written information: students were more in favor of immunocastration after viewing videos showing the different methods of castration. Consumers' attitudes towards immunocastration change across countries, differs between citizens and stakeholders, and between different stakeholder categories (individuals vs. organizations). For example, scientists tend to consider immunocastration more favorably than producers, which tend to express worries about operator safety and public acceptance [19], this latter one in particular when the PDO/PGI supply chains are involved [17]. An exhaustive list of concerns expressed by the stakeholders involved in the pork chains across Europe is detailed in the final report of the CASTRUM project [27]. With respect to studies on consumers' acceptance, a common limitation is the very little knowledge about boar taint, castration of male piglets, or alternative strategies to reduce the occurrence of boar taint [20]. Therefore, in order to study consumer's attitude and perception, many variables such as education, social background, gender and age of the respondent need to be accounted for.

Despite the increasing attention toward animal welfare, studies carried out on consumers led to conflicting results on their WTP (Willingness To Pay) for immunocastration: a study carried out in 10 countries in 2013 [28] estimated the WTP in 0.04€/kg of pig meat. Vanhonacker et al. [5] examined Belgian consumers and found a 5% WTP, despite a very positive attitude towards immunocastration. Heid and Hamm [29] found among German consumers a negative WTP for immunocastrated pork compared to both castration with pain relief and fattening of entire males, but a positive WTP (+12%) for immunocastrated pork compared with castration without pain relief, as the result of the fact that all the alternatives have (perceived) drawbacks that force consumers to make trade-offs among different aspects. Lagerkvist et al. [30] found WTP for immunocastration to be 21% higher than that for surgical castration among Swedish consumers, who perceived immunocastration to be a socially viable alternative to castration without anaesthesia.

Given the Italian scenario (castration is necessary because pigs are slaughtered at a very high body weight, i.e., after sexual maturity, and they are intended for PDO, i.e., high quality, products), and considering the advantages of immunocastration in terms of animal welfare, the aim of the present study is to assess the attitude of Italian consumers towards immunocastration, and how this attitude is influenced by the extent in the detail of the information, and by the order in which information is provided.

2. Materials and Methods

2.1. Questionnaire and Consumers Sample

A questionnaire was formulated and submitted to a sample of 969 Italian consumers (supplemental material). The survey was carried out in Italy, between December 2018 and January 2019. Interviewees were contacted by a specialized agency (DemetraOpinioni.net S.r.l., Venice, Italy), with CAWI (Computer Assisted Web Interview) methodology. Participation quotas were identified in order to obtain three representative samples of the Italian population for gender, age (over 18 years), and geographic area (Northwestern, Northeastern, Center, South, Islands) [31]. People below 18 years of age, people who do not consume swine meat (or cured products), and people exceeding quotas were excluded from the survey. A total of 1463 invitations were sent and 1062 answers to the survey were received. Of these, 56 interviews were screened-out (people excluded from the survey because they were vegans, vegetarians or non-consumers of pork). The size of the remaining sample was 1006 consumers. However, after a quality control, 37 interviews were excluded from the sample (e.g., incomplete forms, partial answers). The final sample was of 969 Italian respondents. The average completion time was 15 min.

After filling the socio-demographic section (gender, age, occupation, education, household size and income, area of residence: rural vs. urban), all consumers were asked to respond to a *first part* of the questionnaire (14 questions) focusing on:

- consumer background (meat consumption habits, direct visual experience through visits of animal farms, attitude and perception towards the welfare of farmed animals);
- consumer knowledge (on animal-friendly foods, on swine castration).

Consumers were then asked to read attentively a short paragraph (approximately 12 lines) containing general information on the reasons why pigs are castrated, on how this procedure is at present carried out mainly surgically, and on immunocastration as a possible alternative to surgical castration. The paragraph contained technically neutral information. The information was preliminarily evaluated by a group of experts in the field of swine science and pre-tested on a small group of consumers (15 people), and was therefore modified in order to eliminate all the words that could bias the perception of the interviewees. For example, words unfamiliar to the general public or which may generate a greater sensitivity (e.g., "piglet") have been accurately avoided, together with words suggesting advantages or disadvantages of one technique respect to the other.

All consumers were then asked to express their level of agreement with:

- the need to abandon surgical castration without pain relief and/or anaesthesia;
- the use of immunocastration.

For the *second part* of the questionnaire, the three groups of consumers previously identified were asked to answer to three different questionnaires, in order to study whether the information provided could affect their attitude towards immunocastration, their WTP and propensity to consume pork obtained with this technique and their risk perception.

The three groups filled the questionnaire as follows:

1. "Neutral information" group (N) (n = 319). This group, after reading the general paragraph described above, was asked to answer a short group of questions (8) on:

   - their preference in purchasing pork obtained either with the different methods of castration (surgical; surgical with analgesia and/or anesthesia, immunocastration, or meat from entire pigs or from pigs selected—genetic improvement—for low boar taint);

- their WTP a premium price for these products;
- their perceived risk with respect to immunocastration.

2. "Positive-negative information" group (+/-) (n = 323). This group, immediately after the first part of the questionnaire, was asked to read attentively a short paragraph on the advantages of immunocastration ("positive information") in comparison with surgical castration (reduction in animal pain and discomfort, absence of negative effects on meat quality together with an improved feed efficiency in some cases). Immediately after, these consumers were asked to read a short paragraph ("negative information") on the disadvantages of immunocastration (increased production costs and accidental self-injection risks for farm workers). Lastly, they were asked to respond to a set of questions (11) on

    - their preference in purchasing pork obtained either with the different methods of castration (surgical; surgical with analgesia and/or anesthesia, immunocastration, or meat from entire pigs or from pigs selected—genetic improvement—for low boar taint);
    - their WTP a premium price for these products;
    - their perceived risk with respect to immunocastration.

3. "Negative-positive information" group (-/+) (n = 327). This group was presented with exactly the same information and questions than the +/- group, but information was presented in reverse order (first the disadvantages and then the advantages of immunocastration).

The reverse order of presentation of the +/- information to these latter two groups was aimed to avoid any influence due to the order of presentation itself.

Given that the additional information was provided in two subsequent steps, positive/negative or vice versa, we defined this information as "complex". Overall information consumers of groups 2 and 3 owned by the end of the questionnaire (neutral and positive/negative, regardless of the order in which this latter was received) is defined as "complete information".

2.2. Statistical Analysis

Statistical analysis was carried out using SPSS software (v. 25.0). The dataset was organized in three groups according to the sampling methodology and the kind of information provided to consumers, and a one-way ANOVA was used. In the "neutral information" group, the perception of immunocastration was tested after providing technically neutral information, whereas in the other two groups the difference in perception was tested after providing technically neutral information and after providing the complex and complete information. Similarly, ANOVA was applied to the variables related to the willingness to consume products obtained with the use of immunocastration, the WTP and the risk perception.

Brown and Forsythe [32] and the Welch [33] tests were carried out. Using absolute deviations from the group medians, the Brown–Forsythe is a robust test for data that potentially violate the assumption of normality. The test compares the variance within each group with the median value of the variance across groups. The Welch test is an alternative test for the one factor analysis of variance F-test. It is a parametric test for equal population means, to be used when we do not have equal population variances.

Moreover, pairwise comparisons, by using Tukey HSD [34], Duncan [35], and Scheffe [36] post hoc tests, were carried out to confirm the absence of significant differences between all possible pairs of averages. The Tukey test is a non-parametric test which in our case is particularly suitable since it is structured for data measured in ordinal scales. To confirm the validity of the test, two other post hoc tests were run: the Duncan test (commonly used in agronomy and in other agricultural economics research), which is more protective against the type II error (although it implies a greater risk of type I errors), and the Scheffe test, which is a more flexible test and was used only as confirmation and reinforcement of the goodness of the ANOVA analysis.

Statistical significance was set at $p < 0.05$ for all tests.

3. Results

3.1. Consumers Background and Knowledge On Animal Welfare

According to our results, the direct knowledge of Italian consumers regarding animal welfare can be defined as very limited. Only 12.6% of the interviewees gathered their knowledge through direct visiting of farms, and for about half of them this experience was sporadic (1 or 2 farm visits in their lifetime). About one fifth (21.2%) of the responders say to have no knowledge on issues related to animal welfare and 66.3% have received their information through the mass media.

Among those who visited a farm at least once, the most common species seen is swine (n = 84), followed shortly by beef cattle (n = 79), while the less observed species is sheep (n = 4) (multiple answers allowed–total number of answers received: 411).

In general, consumers believe that avian species are those having the worst welfare conditions. On a Likert scale with a score of 1 to 5 (where 1 = minimum welfare, 5 = maximum welfare), broilers have an average score of 2.54 and laying hens 2.70. Pigs do have a level of welfare comparable to that of laying hens (2.72), while dairy cows and beef cattle get relatively higher scores (3.2 and 3.0, respectively).

On a scale from 0 to 10 (with 0 = "not at all" and 10 = "extremely"), 82% of respondents attributed a value equal or higher than 6 to the importance of animal welfare during purchases. In particular, a quarter of the interviewees stated that they attribute a valor equal to 10 to animal welfare. The average value attributed by the whole group of respondents was 8.4.

The large majority of consumers (69.7%) declared to purchase food obtained respecting a level of animal protection higher than the minimum set up by legislation; out of them, 40% declare to do it always, while the remaining 60% only sometimes. Among consumers who purchased these products, 59% bought organic foods or products obtained from animals having an outdoor access. However, only one half of those giving very high importance to animal welfare at time of purchasing (score 10 = extremely) declare to buy "animal friendly" products always.

3.2. Consumers Knowledge and Perception on Swine Castration

Only about a quarter of respondents (n = 259) are aware that male pigs undergo castration within the first week of their life. On a total of 381 selections made (multiple answers were allowed), 198 answers indicated that this practice is aimed to improve meat quality or meat production and 106 indicated that castration is aimed to avoid boar taint.

After reading the short "neutral" paragraph, respondents were asked to express their level of agreement with the abandonment of surgical castration (without anesthesia and/or analgesia) in favor of alternative methods. Sixty-eight percent of respondents agreed (i.e., scores equal or above 6 on a 0-to-10 Likert scale) with the abandonment of surgical castration without anesthesia and analgesia and with the implementation of alternative castration techniques. Approximately two out of three respondents expressed a positive score (equal to or greater than 6) and more than a quarter (28%) was extremely in favor of immunocastration (scores 9 and 10).

When asked to choose among meat from surgically castrated pigs (with or without anesthesia and/or analgesia) or from animals subject to alternative methods (such as immunocastration, breeding of entire males, or animals genetically selected to not express the boar taint), consumers indicated a clear preference towards products obtained through the use of immunocastration (34%), followed equally by entire males and pigs surgically castrated with anesthesia (20.8 and 20.4%, respectively). Genetic selection was the penultimate choice (16%), followed only by surgical castration without anesthesia/analgesia.

However, at the end of the first part of the questionnaire, apprehension about immunocastration was expressed by 23% of the respondents; on the other hand, 19% had no fear of this technique, but the remaining 58% was undecided about whether the vaccine is harmless or not. Among the consumers who perceived risks or were undecided at the previous question (n = 782 consumers), 596 answers express concern about "possible unknown long-term risks", 478 about residues in meat, and 255 indicated apprehension for pigs' health (multiple answers allowed–total answers received: 1329).

3.3. Effect of Complex and Complete Information on Pig Immunocastration

Immediately after providing neutral information to all three groups, they were asked to indicate their degree of agreement with the use of immunocastration on a scale from 0 to 10. This answer will be defined as "starting point" in the results description below.

The same question was then repeated only to the second and the third groups ("positive-negative group" and "negative-positive group") after they had received the complex and complete information. This second answer to the question will be defined as "ending point" in the results description below. It is intuitive that, for the neutral group, the starting point and the ending point coincide.

Table 1 shows the results for the question "Please indicate, on a 0-to-10 scale, your degree of agreement/disagreement with the use of immunocastration (0 = completely disagree, 10 = completely agree)"

Table 1. One-way ANOVA results for consumers' answers to the question "Please indicate, on a 0-to-10 scale, your degree of agreement/disagreement with the use of immunocastration (0 = completely disagree, 10 = completely agree)" after receiving different levels and complexity of information.

Variables	Total	Information (Means)			ANOVA	
	Mean (n = 969)	Neutral (n = 319)	Complete +/− (n = 323)	Complete −/+ (n = 327)	F-Test	*p*-Value
Starting Point	6.18	6.40	6.14	6.00	1.321	0.267
Ending Point	6.36		6.31	6.38	0.09	0.914

The analysis shows no significant differences of variances among groups at the starting point. Therefore, the three groups can be considered homogeneous at the beginning of the questionnaire. Moreover, taking into account the order of the information presented, there are no significant differences at the end point after complete information has been given to the second and the third group, confirming the impossibility to reject the null hypoyhesis of equality of means. Although the sample is composed of a large number of cases, the two parametric tests have been run, assuming the possibility that the assumption of normality could be violated.

As shown in Table 2, the robust test of equality of means was carried out, with the results of Welch and Brown–Forsythe tests confirming the impossibility to reject the null hypothesis of equality of means. The post hoc tests of Tukey HSD, Duncan, and Scheffe confirmed the absence of statistically significant differences and therefore the presence of homogeneous subsets for alpha = 0.05.

Table 2. Robust and post hoc tests and results for consumers' agreement/disagreement on the use of immunocastration.

		Statistic [a]	Significance.
Starting Point	Welch	1.348	0.260
	Brown-Forsythe	1.322	0.267
Ending Point	Welch	0.092	0.912
	Brown-Forsythe	0.090	0.914
		Starting Point	**Ending Point**
Tukey HSD [b,c]	Significance	0.243	0.913
Duncan [b,c]	Significance	0.130	0.705
Scheffe [b,c]	Significance	0.275	0.920

Means for groups in homogeneous subsets are displayed. [a.] Asymptotically F distributed; [b.] Uses Harmonic Mean Sample Size = 322,967; [c.] The group sizes are unequal, the harmonic mean of the group sizes is used. Type I error levels are not guaranteed.

The one way-ANOVA analysis between the starting and ending point for each of the groups led to the same results as above, i.e., no significant differences were observed indicating that complex and complete information did not change consumers' perception of immunocastration.

The results on the effect of information on the willingness to consume products obtained with the use of immunocastration, the WTP for the same products and the level of risk perception related to the technique are shown in Table 3.

Table 3. One-way ANOVA results for consumers' willingness to consume, willingness to pay, and level of perception of risks for consumers' health tied to immunocastration (answers were expressed on a 0-to-100 scale) after receiving different levels and complexity of information.

Variables	Total		Information (Means)			ANOVA	
	Mean (n = 969)	Neutral (n = 319)	Complete +/− (n = 323)	Complete −/+ (n = 327)		F-Test	p-Value
Willingness to consume	54.54	54.80	54.29	54.54		0.020	0.980
Willingness to pay	18.74	18.04	18.58	19.58		0.489	0.613
Level of risk perception	34.23	33.273	32.31	37.06		2.775	0.063

Information did not affect willingness to pay or to consume products obtained from immunocastrated pigs, as confirmed also by the post hoc tests summarized in Table 4. The tests carried out confirmed the absence of significant differences between all possible pairs. A tendency ($p = 0.063$) towards a higher level of risk perception by consumers receiving at first the negative information was observed, confirmed also by the measures of Welch and Brown–Forsythe (used to confirm our results if the assumption of normality is violated) and the post hoc tests.

Table 4. Robust tests and post hoc tests results for consumers' willingness to pay, willingness to consume, and perception of risks tied to immunocastration.

		Statistic [a]	Significance
Willingness to consume	Welch	0.020	0.981
	Brown–Forsythe	0.020	0.980
Willingnes to pay	Welch	0.490	0.613
	Brown–Forsythe	0.490	0.613
Level of risk perception	Welch	2.747	0.065
	Brown-Forsythe	2.776	0.063

		Willingness to consume	Willingness to pay	Level of risk perception
Tukey HSD [b,c]	Significance	0.978	0.594	0.068
Duncan [b,c]	Significance	0.852	0.363	0.077
Scheffe [b,c]	Significance	0.980	0.623	0.085

Means for groups in homogeneous subsets are displayed. [a.] Asymptotically F distributed; [b.] Uses Harmonic Mean Sample Size = 322,967; [c] The group sizes are unequal, the harmonic mean of the group sizes is used. Type I error levels are not guaranteed.

In absence of statistically significant differences among the three groups, they can be considered as a single representative population of Italian consumers. From our study, it emerges that average WTP stands at 18.7%, the willingness to consume products obtained through the use of immunocastration at 54.5%, and the extent to which consumers perceived the presence of risks for their health due to immunocastration was on average 34.2%.

4. Discussion

The results of the present survey show that the Italian population does not have a high level of direct knowledge on livestock living conditions. Most of the sample gets information on this issue from the mass media (TV, internet, and newspapers). Similar results were previously observed by our research group on a smaller sample of Italian consumers [37]. It is therefore reasonable to hypothesize that consumers' perception of the reality of farm animals' welfare can be chronically distorted, or at least not realistic. Information provided by media is often either entirely negative (focus on scandals, inhumane practices, mistreatments, etc.) or extremely positive (description and videos showing animals kept under bucolic conditions such as grazing on green pastures), omitting the more debatable issues/practices, e.g., mutilations.

In agreement with the findings of the 2005 Eurobarometer survey on animal welfare [38], the present study confirms that Italian consumers still perceive a marked difference between species in the level of welfare attained under common farming conditions. In particular, the welfare level of avian species is believed to be worse than cattle welfare, while pig welfare is perceived as intermediate.

Results also indicate that the Italian population is not particularly aware about the practice of pig castration. However, those who are aware of pig castration are also aware of the reasons for this practice (improving meat quality and avoiding boar taint). This lack of knowledge about castration of the general public has already pointed out by several authors [20,25,26]. The majority of respondents declared to buy product obtained with high animal welfare standards (at least sometimes), and their self-assessment regarding the level of attention paid to animal welfare during purchases returns a high score (8.4 on a 0–10 scale). This result is not entirely surprising considering that in the 2016 Eurobarometer survey on animal welfare [39], 97% of EU consumers (and the same percentage of Italian consumers) declared to perceive the protection of farmed animals as an important topic; however, it should be noted that these two questions are not directly comparable. The high percent of people indicating organic products or products obtained from animals having outdoor access mirrors the positive trend of organic market and the understanding of a link between animal welfare and ethical value of food [37].

The high attention toward animal welfare is probably the reason why the majority of respondents fully agree with the abandonment of surgical castration without anesthesia. However, there is no full agreement within the population on the alternative methodology that should be used. In fact, one-third of the respondents would prefer immunocastration, but the other two-thirds expressed preferences for other methods.

The risks perceived by Italian consumes in this study ("possible unknown long-term risks", followed by "residues of the product in meat") are in agreement with those recorded by Fredriksen et al. [18], although Norwegian consumers indicated the fear of residuals in meat as the main reason. Nevertheless, Italian consumers declared a good willingness to pay a premium price for products obtained from immunocastrated pigs (+18.7%). In the light of specific literature, this result can be considered as particularly positive. The value is higher than those reported in other studies on animal-friendly foods [37] and on immunocastrated meat [5,29], and similar to the WTP observed among Swedish consumers [30], although in this last study WTP was measured through a choice experiment. This is of peculiar interest for the Italian market, since it has been hypothesized that for pigs slaughtered at high body weights the increase of costs associated with immunocastration would be offset by the increase in production [16,19] and therefore would not affect the final cost of the product. It should nonetheless be noted that all studies (including the present one) based on a declared (i.e., hypothetical) WTP might lead to its overestimation.

Also, the willingness to consume pork from immunocastrated animals was positive (54.5%), even if our result is only slightly higher than the central value. This outcome must be read in light of the results obtained on the level of risk perception, i.e., consumers are not completely convinced of the harmlessness of immunocastration, even if the level of risk perception expressed by consumers was relatively low on average (34%).

The values of willingness to pay, willingness to consume, and level of risk perception do not change after reading neutral or complete and complex information. Given the lack of direct knowledge of the consumer regarding the conditions of rearing and the technique of surgical castration without anesthesia, this remains a topic to be investigated further in future research. Some studies reported similar findings [5,20] and pointed out that it is likely that the written information does not have particularly evident effects, whereas images (videos, pictures, etc.) may produce a greater impact. However, it is also possible that, given the substantial naiveness of the general public with respect to this topic, simply providing consumers with the technically neutral information about castration might have been sufficient to negatively affect their perception, by letting them know about a mutilation of which they were completely unaware before. If this is the case, it could be argued that the information we provided as "neutral" might not have been perceived as actually neutral by consumers, and every subsequent piece of information they received might not have been able to change their initial opinion. To the best of our knowledge, this aspect has never been investigated before and could be of interest for future surveys. Although the experiment did not aim to assess the effect of the information order, an aspect worth mentioning is the tendential difference we observed in the level of risk perception, giving some evidence that probably the order in which the information was provided changed at least a little the level of risk perception by consumers: in one case (+/-) it slightly decreased the level of risk perception, whereas in the other one (-/+) it slightly increased it. It would be of interest to explore if such a tendency is confirmed or becomes stronger with larger groups of consumers.

5. Conclusions

The sensitivity on the European population towards animal welfare is progressively increasing and the demands for adaptation of the production systems are becoming increasingly urgent. However, it is not always easy to respond to consumer demands, as the introduction of new technologies (including vaccines) in the food chain often generates mistrust. Information can shift consumers' preference from organoleptic characteristics to intangible characteristics such as ethic attributes (e.g., animal welfare) [40]. Until a real influence of information in favor of immunocastration will be verified, the concerns of the livestock sector over the acceptance of immunocastration by consumers are obviously legitimate.

The results of this study confirm that in the Italian population the level of attention to animal welfare is increasing and that more and more consumers are looking for and buying animal-friendly products, also demonstrating their willingness to recognize a premium price to farmers for their efforts. In this study, immunocastration was perceived in a predominantly positive manner, with a relatively low level of risk perception and a good willingness to pay, showing that, once the consumer's trust is gained (by means of transparent information on production systems), immunocastration may become an acceptable way forward for Italian producers. However, from a practical standpoint, the numerical paucity of studies concerning the adoption of immunocastration in PDO production chains may represent a gap to be filled by future research.

Supplementary Materials: The supplementary materials are available online at http://www.mdpi.com/2076-2615/9/5/198/s1.

Author Contributions: Conceptualization, J.D.P., G.M., and E.N.; methodology, F.A., L.B., J.D.P., and R.S; formal analysis, L.B., J.D.P., and R.S.; data curation, L.B., J.D.P., and R.S.; writing—original draft preparation, J.D.P., E.N., G.R., and G.M.; writing—review and editing, J.D.P., G.M., E.N., and L.S.; visualization, J.D.P. and E.N.; project administration, J.D.P. and L.S.; funding acquisition, L.S.

Funding: This research received no external funding.

Acknowledgments: We would like to thank all those (experts, colleagues, family and friends) who helped us in pre-testing the questionnaire for the generous amount of time they spent analyzing the questionnaire with us and highlighting their perplexities on the preliminary draft. We would also like to thank the anonymous reviewers for their fundamental support in improving the quality of our manuscript.

Conflicts of Interest: The authors declare no conflict of interest.

References

1.   Zamaratskaia, G.; Rasmussen, M.K. Immunocastration of male pigs-situation today. *Procedia Food Sci.* **2015**, *5*, 324–327. [CrossRef]

2. Brunius, C.; Zamaratskaia, G.; Andersson, K.; Chen, G.; Norrby, M.; Madej, A.; Lundstrom, K. Early immunocastration of male pigs with Improvac—Effect on boar taint, hormones and reproductive organs. *Vaccine* **2011**, *29*, 9514–9520. [CrossRef] [PubMed]

3. Rydmher, L.; Lundstrom, K.; Andersson, K. Immunocastration reduces aggressive and sexual behavior in male pigs. *Animal* **2010**, *4*, 965–972. [CrossRef] [PubMed]

4. Poulsen Nautrup, B.; Van Vlaenderen, I.; Aldaz, A.; Mah, C.K. The effect of immunization against GnRh on growth performance, carcass characterics and boar taint relevant to pig producers and the pork packing industry: A meta analysis. *Res. Vet. Sci.* **2018**, *119*, 182–195. [CrossRef] [PubMed]

5. Vanhonacker, F.; Verbeke, W.; Tuyttens, F.A.M. Belgian's attitude towards surgical castration and immunocastration of piglets. *Anim. Welf.* **2009**, *18*, 371–380.

6. AHAW. Opinion of the Scientific Panel on Animal Health and Welfare on a request from the Commission related to welfare aspects of the castration of piglets. *EFSA J.* **2004**, *91*, 1–18.

7. European Declaration on Alternatives to Surgical Castration of Pigs. Available online: https://ec.europa.eu/food/sites/food/files/animals/docs/aw_prac_farm_pigs_cast-alt_declaration_en.pdf (accessed on 1 February 2019).

8. Pinna, A.; Schivazappa, C.; Virgili, R.; Parolari, G. Effect of vaccination against gonadotropin-releasing hormone (GnRH) in heavy male pigs for Italian typical dry-cured ham production. *Meat Sci.* **2015**, *110*, 153–159. [CrossRef] [PubMed]

9. Karaconji, B.; Lloyd, B.; Campbell, N.; Meaney, D.; Ahern, T. Effect of an anti-gonadotropin-releasing factor vaccine on sexual and aggressive behavior in male pigs during the finishing period under Australian field condition. *Aust. Vet. J.* **2015**, *93*, 121–123. [CrossRef]

10. Martins, P.C.; Albuquerque, M.D.; Machado, I.P.; Mesquita, A.A. Implicações da imunocastração na nutrição de suínos e nas características de carcaça. *Arch. Zootec.* **2013**, *62*, 105–118.

11. Aluwé, M.; Tuyttens, F.; Millet, S. Field experience with surgical castration with anaesthesia, analgesia, immunocastration and production of entire male pigs: Performance, carcass traits and boar taint prevalence. *Animal* **2015**, *9*, 500–508. [CrossRef]

12. Dunshea, F.R.; Colantoni, C.; Howard, K.; McCauley, I.; Jackson, P.; Long, K.A.; Lopaticki, S.; Nugent, E.A.; Simons, J.A.; Walker, J.; et al. Vaccination of boars with a GnRH vaccine (Improvac) eliminates boar taint and increases growth performance. *J. Anim. Sci.* **2001**, *79*, 2524–2535. [CrossRef] [PubMed]

13. Fàbrega, E.; Velarde, A.; Cros, J.; Gispert, M.; Suárez, P.; Tibau, J. Effect of vaccination against gonadotrophin-releasing hormone, using Improvac®, on growth performance, body composition, behaviour and acute phase proteins. *Livest. Sci.* **2010**, *132*, 53–59. [CrossRef]

14. Pauly, C.; Spring, P.; Odoherty, J.V.; Ampuero Kragten, S.; Bee, G. Growth performance, carcass characteristics and meat quality of group-penned surgically castrated, immunocastrated (Improvac) and entire male pigs and individually penned entire male pigs. *Animal* **2009**, *3*, 1057–1066. [CrossRef]

15. EC, Union Register of Veterinary Medicinal Products. Available online: http://ec.europa.eu/health/documents/community-register/2018/20181030142742/anx_142742_en.pdf (accessed on 1 February 2019).

16. De Roest, K.; Montanari, C.; Fowler, T.; Baltussen, W. Resource efficiency and economic implications of alternative to surgical castration without anaesthesia. *Animal* **2009**, *3*, 1522–1531. [CrossRef]

17. Mancini, M.C.; Menozzi, D.; Arfini, F.; Veneziani, M. How Do Firms Use Consumer Science to Target Consumer Communication? The case of Animal Welfare. In *Case Studies in the Traditional Food Sector*, 1st ed.; Cavicchi, A., Santini, C., Eds.; Woodhead Publishing: Kidlington, UK, 2018; pp. 337–357.

18. Fredriksen, B.; Johnsen Sibeko, A.M.; Skuterud, E. Consumer attitudes toward castration of piglets and alternatives to surgical castration. *Res. Vet. Sci.* **2011**, *90*, 352–357. [CrossRef]

19. Mancini, M.C.; Menozzi, D.; Arfini, F. Immunocastration: Economic implications for the pork supply chain and consumer perception. An assessment of existing research. *Livest. Sci.* **2017**, *203*, 10–20. [CrossRef]

20. Tuyttens, F.A.M.; Vanhonacker, F.; Langendries, K.; Aluwè, M.; Millet, S.; Bekaert, K.; Verbeke, W. Effect of information provisioning on attitude toward surgical castration of male piglets and alternative strategies for avoiding boar taint. *Res. Vet. Sci.* **2011**, *91*, 327–332. [CrossRef]

21. Allison, J.; Wright, N.; Martin, S.; Wilde, N.; Izumi, E. Consumer acceptance of the use of vaccination to control boar taint. In Proceedings of the 59th Annual Meeting of the European Association of Animal Production, Vilnius, Lithuania, 24–27 August 2008; p. 97.

22. Liljenstolpe, C. Evaluating animal welfare with choice experiments: An application to Swedish pig production. *Agribusiness* **2008**, *24*, 67–84. [CrossRef]
23. Huber-Eicher, B.; Spring, P. Attitudes of swiss consumers towards meat from entire or immunocastrated boars: A representative survey. *Res. Vet. Sci.* **2008**, *85*, 625–627. [CrossRef]
24. Kallas, Z.; Gil, J.M.; Panella-Riera, N.; Blanch, M.; Font-i-Furnols, M.G.; Chevillon, P.; De Roest, K.; Tacken, G.; Angels Oliver, M. Effect of tasting and information on consumer opinion about castration. *Meat Sci.* **2013**, *95*, 242–249. [CrossRef]
25. Vanhonacker, F.; Verbeke, W. Consumer response to the possible use of a vaccine method to control boar taint v. physical piglet castration with anaesthesia: A quantitative study in four european countries. *Animal* **2011**, *5*, 1107–1118. [CrossRef]
26. Heid, A.; Hamm, U. Consumer attitudes towards alternatives to piglet castration without pain relief in organic farming: Qualitative results from Germany. *J. Agric. Environ. Eth.* **2012**, *25*, 687–706. [CrossRef]
27. CASTRUM Consortium. Pig Castration: Methods of Anaesthesia and Analgesia for All Pigs and Other Alternatives for Pigs Used in Traditional Products. 2016. Available online: http://boars2018.com/wp-content/uploads/2017/02/Castrum-study.pdf (accessed on 10 April 2019).
28. FCEC–Food Chain Evaluation Consortium. *Study and Economic Analysis of the Costs and Benefits of Ending Surgical Castration of Pigs; Final Reports;* DG SANCO: Bruxelles, Belgium, 2013; Available online: https://ec.europa.eu/food/sites/food/files/animals/docs/aw_prac_farm_pigs_cast-alt_research_civic_pt1-synthesis_20131202.pdf (accessed on 1 February 2019).
29. Heid, A.; Hamm, U. Animal welfare versus food quality. Factors influencing organic consumer's preferences for alternatives to piglet castration without anaesthesia. *Meat Sci.* **2013**, *95*, 203–211. [CrossRef] [PubMed]
30. Lagerkvist, C.J.; Carlsson, F.; Viske, D. Swedish consumer preferences for animal welfare and biotech: A choice experiment. *AgBioForum* **2006**, *9*, 51–58.
31. ISTAT—National Institut of Statistic. 2018. Available online: https://www.istat.it/ (accessed on 10 April 2019).
32. Brown, M.B.; Forsythe, A.B. Robust tests for the equality of variances. *J. Am. Stat. Assoc.* **1974**, *69*, 364–367. [CrossRef]
33. Welch, B.L. The generalization of Student's problem when several different population variances are involved. *Biometrika* **1947**, *34*, 28–35. [CrossRef] [PubMed]
34. Tukey, J.W. Comparing individual means in the analysis of variance. *Biometrics* **1949**, *5*, 99–114. [CrossRef]
35. Duncan, D.B. Multiple range rests for correlated and heteroscedastic means. *Biometrics* **1957**, *13*, 164–176. [CrossRef]
36. Scheffé, H. A method for judging all contrasts in the analysis of variance. *Biometrika* **1953**, *40*, 87–104.
37. Di Pasquale, J.; Nannoni, E.; Del Duca, I.; Adinolfi, F.; Capitanio, F.; Sardi, L.; Vitali, M.; Martelli, G. What foods are identified as animal friendly by Italian consumers? *Ital. J. Anim. Sci.* **2014**, *13*, 782–789. [CrossRef]
38. EC. Attitudes of Consumers towards the Welfare of Farmed Animals. Special Eurobarometer 229. Wave 63.2. 2005. Available online: http://ec.europa.eu/public_opinion/archives/ebs/ebs_229_en.pdf (accessed on 25 February 2019).
39. EC. Attitudes of Europeans towards Animal Welfare. Special Eurobarometer 442. Wave EB84.4.—TNS Opinion & Social. 2016. Available online: https://ec.europa.eu/commfrontoffice/publicopinion/index.cfm/ResultDoc/download/DocumentKy/71348 (accessed on 25 February 2019).
40. Napolitano, F.; Braghieri, A.; Caroprese, M.; Marino, R.; Girolami, A.; Sevi, A. Effect of information about animal welfare, expressed in terms of rearing conditions, on lamb acceptability. *Meat Sci.* **2007**, *77*, 431–436. [CrossRef] [PubMed]

Article

The Development and Evaluation of 'Farm Animal Welfare': An Educational Computer Game for Children

Roxanne D. Hawkins [1,*]**, Gilly A. R. Mendes Ferreira** [2] **and Joanne M. Williams** [3]

[1] Psychology, School of Media, Culture and Society, University of West Scotland, Paisley PA1 2BE, UK
[2] Scottish SPCA, Kingseat Road, Halbeath, Dunfermline, Fife KY11 8RY, UK; gilly.ferreira@scottishspca.org
[3] Clinical and Health Psychology, University of Edinburgh, Edinburgh EH8 9AG, UK; jo.williams@ed.ac.uk
* Correspondence: Roxanne.hawkins@uws.ac.uk

Received: 9 February 2019; Accepted: 8 March 2019; Published: 13 March 2019

Simple Summary: The aim of this study was to design and evaluate a new digital game 'Farm Animal Welfare' to teach children about farm animal welfare. The game focuses on chickens and cows, and children played the game on touchscreen netbooks. To evaluate the game, we measured children's knowledge, attitudes, compassion, and beliefs about whether farm animals have emotions and feelings, both before and after the game, using a child-friendly questionnaire. We found that the new game led to increases in children's knowledge about animal welfare, knowledge about welfare in different farming systems (such as caged hens vs. free range), and children were more likely to believe that farm animals can feel emotions. The game did not seem to impact children's attitudes about cruelty or compassion towards farm animals. The new game shows promise, and to improve children's understanding of animal welfare, we recommend further research on digital animal welfare education interventions for children.

Abstract: Many children growing up in urban areas of Western countries have limited contact with and knowledge of farm animals and food production systems. Education can play an important role in children's understanding of farm animal welfare issues, however, most education provided focuses on pets. There is a need to develop new farm animal welfare interventions for young children. This study examines the process of designing, developing, and evaluating the effectiveness of a new theoretically-driven digital game to teach children, aged 6–13 years, about farm animal welfare. 'Farm Animal Welfare' aimed to promote children's knowledge about animal welfare, promote beliefs about animal sentience, and promote positive attitudes and compassion. A quasi-experimental design was carried out, using self-report questionnaires that children (n = 133, test = 69, control = 64) completed in the classroom. Test and control groups were from different schools and the control group did not engage in the intervention. Findings indicate a positive impact on beliefs about animal minds, knowledge about animal welfare needs, and knowledge about welfare in different farming systems, but there was no change in compassion or attitudes about cruelty. This study presents the first evaluation of a digital animal welfare education intervention for children, demonstrating the benefits of incorporating 'serious games' into farm animal welfare education. The findings will inform future practice around farm animal welfare education interventions for primary school children.

Keywords: children; farm animals; animal welfare; education; technology

---

1. Introduction

Public concern over the treatment of farm animals has increased over time, and production systems have faced public scrutiny, with natural living and humane treatment being central to public perceptions on what is considered good welfare [1]. However, the public lack sufficient knowledge about farm animal welfare, and often their concerns are misinformed [2]. Growing research into farm animal sentience (i.e., cognitive and emotional abilities) might be contributing to this increased moral concern over farm animal treatment, and the 'use of animals' [3,4], and might, in part, be contributing to the rise in vegetarians and vegans in European countries [5]. Surprisingly little research has examined children's knowledge of farm animal welfare needs and production systems, and their attitudes towards farm animals. It has been suggested by past studies that urban children are disconnected from rural life and agricultural systems and have low levels of food knowledge [6,7]. Yet, these cognitive and attitudinal factors can underpin food and consumer choices and impact farm animal welfare. Therefore, promoting positive orientations to farm animals and informed consumer choices, from an early age, might be beneficial [7,8]. Efforts are required to encourage children in developing a duty of care for farm animals [8] and to encourage children to value positive farm animal welfare, by preventing harm, minimizing suffering, and ensuring that farm animals 'have a good life' [9]. The UK Government has made recommendations about farm animal welfare for children [7], emphasizing the importance of school-based education interventions to inform children as future consumers, and impact upon future animal welfare standards and practices [7,9].

Targeting younger children for animal welfare education is common amongst animal welfare organizations [10]. This is because this age range is an important time for the development of empathy, morality, and receptivity to learning about animals, leading to conceptual changes in knowledge of biology [11–13]. We know from previous research that children rate farm animals lower on their ability to feel emotions and display intelligence, compared to other animals such as dogs and cats, which could impact upon children's compassion and attitudes towards farm animals [14]. Targeting children's beliefs about farm animal minds, as well as their knowledge of farm animal welfare needs, and the welfare implications of different food production systems are essential for educational programs.

Very little research currently exists on the benefits or effectiveness of farm animal welfare education for children. Most studies of farm welfare education have focused on training for veterinarians, and very little has examined education for children. Jamieson and colleagues [8] investigated the impact of a one-off poultry welfare education event and found positive but short-term improvements in children's knowledge and positive behavior towards poultry, and consideration of welfare needs, but not on the value afforded to animal life. In a previous study [10], we found that a one-hour educational workshop in school classes, focusing on farm animal welfare, significantly increased children's knowledge about farm animal welfare needs and increased children's beliefs about animal minds, which are related to compassion towards animals and acceptance of animal cruelty [14]. In-classroom animal welfare education (not focusing on farm animals specifically) has been shown to be effective for improving children's knowledge about animals, attitudes, and empathy towards animals [15–17]. However, the number of studies remains insufficient to make necessary assertions about the effectiveness of classroom-based education programs for improving children's knowledge about and attitudes towards farm animals. Written materials and in-class presentations can be valuable methods of animal welfare education [17] but more interactive and modern methods, such as the use of digital gaming interventions, might lead to a more effective and significant change [18].

There is an opportunity to utilize children's interest in technology to help them connect or re-connect with the natural world, and teach them about farm animals and agriculture [19]. 'Serious games' are increasingly being used by environmentalists, such as 'conservation games' to promote conservation attitudes, through interactions with virtual conservation landscapes [19–21]. There is scope to use these educational advances within animal welfare education, especially in relation to animals with which children have less direct contact, such as farm animals. Gamification of education can bring about a host of benefits by encompassing the 'key components' of effective learning [22], such as building upon the 'Science of Learning' or 'the pillars of learning' [23,24], which are important for stimulating learning and achieving learning outcomes. Moreover, 'serious games' can be more effective than conventional methods at promoting children's intrinsic motivation to learn, promoting 'active' learning, as well as facilitating 'stealth learning' (children do not realize they are learning embedded content for an enjoyable experience) [25]. The purpose of the present study was, therefore, to design and evaluate a digital educational game, focusing on farm animal welfare, to promote knowledge of and positive attitudes towards farm animals.

This study aimed to answer the following research question: Does the Farm Animal Welfare educational digital game have a significant impact on children's beliefs about farm animal minds, knowledge about farm animal welfare, compassion towards animals, and attitudes toward farm animal cruelty? It was hypothesized that there would be a significant pre- to post-test change for all target outcomes for the intervention group, but not the control group. The present study also took into consideration two factors that have been found to affect the baseline scores relating to children's knowledge and attitudes towards animals: gender and age [26,27].

2. Materials and Methods

2.1. The Development of 'Farm Animal Welfare' Digital Game

Farm Animal Welfare is an educational or 'serious game' designed to teach primary school children about farm animal welfare. A series of three interactive levels were developed for each of two types of farm animals (cows and chickens), incorporating text, images, and sound. Children received feedback throughout the game and viewed their scores. All images were purchased from photo stock websites. Once developed, the game was downloaded and played offline, via touchscreen notebook computers, in a school classroom. See Supplementary Materials (Figure S1) for example screenshots of the game.

All content and feedback provided within the game was based on current scientific research and confirmed by animal welfare experts to ensure accuracy and to avoid misinformation. Images were checked by animal behavior experts to ensure they accurately reflected the emotional and behavioral states they were representing in the game. Children received feedback after answering each question and gained points throughout the game for correct answers. There were three levels in the game for each of the two animals (Figure 1):

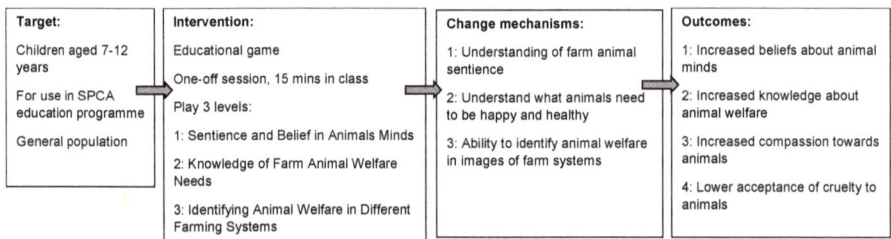

Target:	Intervention:	Change mechanisms:	Outcomes:
Children aged 7-12 years	Educational game	1: Understanding of farm animal sentience	1: Increased beliefs about animal minds
For use in SPCA education programme	One-off session, 15 mins in class	2: Understand what animals need to be happy and healthy	2: Increased knowledge about animal welfare
General population	Play 3 levels:	3: Ability to identify animal welfare in images of farm systems	3: Increased compassion towards animals
	1: Sentience and Belief in Animals Minds		4: Lower acceptance of cruelty to animals
	2: Knowledge of Farm Animal Welfare Needs		
	3: Identifying Animal Welfare in Different Farming Systems		

Figure 1. Logic Model for Farm Animal Welfare game.

Level 1—Sentience and Belief in Animals Minds: Level 1 targeted children's beliefs about farm animal minds. The aim of this level was to teach children that farm animals are sentient and to change their beliefs about farm animal minds. Children were provided with photographic images of animals in different emotional and welfare states (positive and negative) and were asked sentience questions relating to each image (e.g., "Is this chicken in pain?"), which the children would rate on a scale ('not at all', 'not really', 'maybe', 'yes', 'yes very'). The questions (see Supplementary Materials Table S1) focused on the items from the Children's Beliefs about Animal Minds measure, a key target variable, due to concerns highlighted in previous research about children's low beliefs in farm animal minds [14].

Level 2—Knowledge of Farm Animal Welfare Needs: Level 2 aimed to promote new knowledge of the welfare needs of farm animals, focusing on the five welfare needs/freedoms. This level focused on what farm animals need to be 'happy and healthy' through a 'drag and drop' game. For each animal, children had options of items (e.g., straw) and distractors (e.g., dog lead) to move on screen. Children were required to drag and drop items that were required for animal welfare to the target animal icon and distractor items to a bin icon. Incorrect answers 'bounced back' and so children had to keep trying until all items were on the correct location. Once finished, feedback was provided about the five welfare needs for each animal, to reinforce learning and provide further information on welfare needs.

Level 3—Identifying Animal Welfare in Different Farming Systems: Level 3 aimed to promote children's understanding of animal welfare in different farming systems, focusing on the balance between profit (through intensive systems) and positive welfare. This level also involved a 'drag and drop' game, where children had to move images onto an image of a balance-scale, depending on what they judged it was better for—'money' or 'happiness'—in the farming system. Once finished, feedback was provided to reinforce learning and provide context to the items.

2.2. Evaluation Method

2.2.1. Participants

Participants were 133 primary school children, 69 test and 64 control (53% boys, 47% girls) from three schools in West Lothian, Scotland, UK. Two schools were included in the test group, and one school was in the control group. Randomization was not possible for this study, so a quasi-experimental design was used. Children were aged between 6 and 13 years (M = 9.4, SD = 1.2) and were from two age classes, 6–9 years (42%) and 10–12 years (58%). The control group was age-matched to the test group.

2.2.2. Design

A mixed factorial design was used to evaluate the intervention. One variable was phase of testing (time), a repeated-measures variable with two conditions—pre-tests (day before intervention) and post-tests (two days after intervention). The between-subjects variable was the intervention condition (intervention vs. control).

2.2.3. Ethical Considerations

The ethical guidelines of the British Psychological Society, specifically relating to research with children, were adopted for this research, and ethical consent was granted by the University of Edinburgh Clinical and Health Psychology Ethics Committee. All information was treated confidentially and kept in a secure location at all times; child and school data were anonymized during data preparation by adopting identity numbers.

2.2.4. Intervention Materials and Procedure

The pre-tests and post-tests were conducted during three school days within one week. Children completed the pre-test on day one (Monday), played the game on day two (Tuesday), and completed the post-test on day four (Thursday). The control group followed the same pattern but did not play the game until immediately after completing the post-test questionnaire on the Thursday. The control group received their usual classes on the Tuesday, when the intervention group was engaging with the educational game. During the intervention, children took turns to play the game on the touch screen netbook computers provided by the research team. The two netbooks were set up in a quiet space within the school and children took turns throughout the day to play the game in pairs on a netbook. The researcher called two children at a time to come over from their classes to play the game on a netbook, while the rest of the class carried on with their usual school activities. The game took each child approximately 15 min to complete. A self-complete questionnaire was developed as the evaluation tool (see below). This was administered to children during class, as a pre-test and post-test. The questionnaire took each child approximately 20 min to complete and children could ask the researcher or their teacher for assistance but not for the answers. Each child received a gift as a thank you for participation (certificate and Scottish SPCA magnets and stickers). The school also received a gift for participation, including a sponsored space at a Scottish SPCA animal rescue center.

2.2.5. Pre-Test Questionnaires and Post-Test Questionnaires

A short, child-friendly, paper questionnaire served as the evaluation tool for this study. The questionnaire used appropriate language and terminology for 6- to 13-year-olds and was piloted before the study. The questionnaire included questions regarding age, gender, and school class, and the following measures.

Children's Compassion towards Animals (CCA): The Children's Compassion towards Animals measure [14] uses a 5-item scale asking the question, "What do you think about animals?" with five statements (e.g., "When I see an animal that is hurt or upset I feel upset" and "When I see an animal that is hurt or upset I want to help it"). The measure was scored on a 5-point Likert scale ("Strongly disagree" to "Strongly agree"). Total scores were calculated (range 5–25). This measure demonstrated good reliability within the current sample ($\alpha = 0.71$).

Children's Beliefs about Farm Animal Minds: An adapted version of the Children's Beliefs about Animal Minds measure [14] was created for the purpose of this evaluation, changing the animal types to focus on chickens and cows. Each scale (e.g., "Do you think the following animals are … ?") relates to a specific sentience item (clever/pain/happiness/sadness/fear). Each item is scored on a 5-point Likert scale ("Strongly disagree" to "Strongly agree"). Total scores were calculated for each animal (score range 5–25), as well as an overall beliefs about farm animal minds (BAM) score, across the two animal types (score range 10–50). The measure demonstrated a high reliability within the current sample ($\alpha = 0.84$).

Acceptance of Cruelty to Farm Animals (chickens and cows): An adapted version of the Children's Acceptance of Cruelty to Pets [18,28] measure was created for this evaluation to focus on attitudes towards cruelty to cows and chickens. This measure included two 9-item scales, with the question, "Do you think it is alright to...?" with nine statements (e.g., "Not give a chicken a comfortable place to live?"). The measure was based on animal sentience (e.g., "Make a chicken scared?" and "injure a cow"), and welfare needs (e.g., "Get treatment for an ill or injured chicken?"). The measure comprised of two separate scales, one for each animal type. Each item was scored on a 5-point Likert scale ("Strongly disagree" to "Strongly agree"). Total scores were calculated for each animal (score range 9–45), as well as an overall cruelty attitude score across the two animals (score range 18–90), where high scores indicate a high acceptance of animal cruelty. This measure showed reliability within the current sample ($\alpha = 0.83$).

Children's Knowledge about Farm Animals Welfare: This knowledge question asked, "What do cows/chickens need to be happy and healthy?" An image of each animal was provided with space around the image, for children to write freely. Answers were coded according to the five animal welfare needs/freedoms. For example, mentioning food and water would score two points for 'freedom from thirst, hunger, and malnutrition'. Total scores for each animal were calculated as well as a total knowledge score across animals. The measure demonstrated very good reliability within the current sample ($\alpha = 0.70$).

Knowledge about Animal Welfare in Farming Systems: Children were asked, "Which farm systems are better for animal welfare and which are better for making money?" They were presented with a series of images of different farming systems for both cows and chickens (battery farm, free range, organic, and crated), the same images as within the game. Children were asked to identify which images displayed animals in a better welfare state. The measure demonstrated very good reliability ($\alpha = 0.84$).

2.3. Analysis

For the purpose of this evaluation, total scores were summed for each key variable for each individual at each sample point, and data were analyzed at an individual level, using the Statistical package for the Social Sciences Statistics 22 (IBM Corp. Released 2013. IBM SPSS Statistics for Windows, Version 22.0. Armonk, NY: IBM Corp.), with a two-tailed significance of $p < 0.05$. A two-way repeated measures ANOVA using time (phase of testing—pre-test and post-test) as the within-subject variable, and group (two conditions—test, control) as the between-subject variable, tested main effects and interaction effects. The main focus of the results reported here is the interaction effects, which showed a difference in performance between the intervention group and the control group.

3. Results

3.1. Compassion towards Animals

Farm Animal Welfare did not significantly improve children's scores for compassion towards animals (Figure 2). No statistically significant interaction between the intervention condition and time was found ($F(1,121) = 0.32$, $p = 0.576$, $\eta^2 = 0.003$) (Tables 1 and 2). This result remained non-significant, when adjusting for pre-test scores and age and gender, using an analysis of covariance (ANCOVA) ($F(1,123) = 0.189$, $p = 0.665$, $\eta^2 = 0.002$), even though the control group scored significantly higher than the test group, at baseline ($p = 0.0001$). Independent t-test at pre-test found no significant age or gender differences in compassion scores ($p > 0.05$).

Table 1. Descriptive statistics for children's compassion towards animals.

Measure Item	Test				Control			
	Pre-Test		Post-Test		Pre-Test		Post-Test	
	M	SD	M	SD	M	SD	M	SD
When I see an animal that is hurt or upset I feel upset	4.28	0.8	4.26	0.9	4.11	0.8	4.19	0.9
When I see an animal that is hurt or upset I want to help it	4.42	0.8	4.46	0.7	4.48	0.6	4.43	0.9
When I see an animal that is hurt or upset I want to tell someone	4.54	0.6	4.55	0.7	4.58	0.5	4.52	0.9
When I see an animal that is hurt or upset I think it is my responsibility to help	3.65	1	3.62	1	4.03	0.8	4.02	1
I know what to do when I see an animal that is hurt or upset	3.83	1	4.22	0.8	3.81	0.9	3.91	1
TOTAL Compassion	20.71	3	21.12	3	21.02	3	21.06	4

Note: Scores for each item ranged from 1–5. Total compassion scores ranged from 5–25.

Table 2. Results from main effects analysis for each intervention, following insignificant interactions.

Main Effect of Time				Main Effect of Group			
df	F	p	η^2	df	F	p	η^2
Compassion towards animals							
1121	0.8	0.37	0.01	1121	0.04	0.85	0.00
Total attitudes towards animal cruelty							
1116	7.3	0.008	0.059	1116	5.87	0.017	0.048
Attitudes towards cruelty to chickens							
1119	6.36	0.013	0.051	1119	7.27	0.008	0.058
Attitudes towards cruelty to cows							
1117	3.97	0.049	0.033	1117	3.92	0.05	0.032
Total knowledge about the five freedoms							
1114	5.49	0.021	0.046	1114	36.6	0.0001	0.243
Knowledge about the five freedoms for cows							
1117	8.58	0.004	0.068	1117	36.7	0.0001	0.24

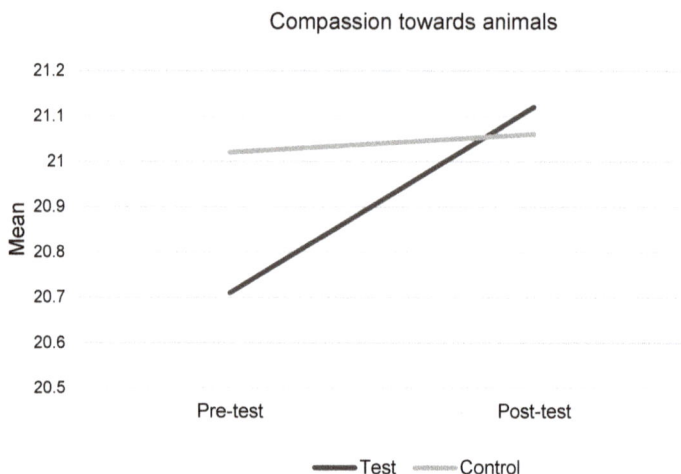

Figure 2. Children's scores for compassion towards animals.

3.2. Beliefs about Farm Animal Minds

Beliefs about farm animal minds scores improved significantly following the intervention (Tables 3 and 4 and Figure 3). There was a statistically significant interaction between intervention condition and time (F(1,116) = 20.92, *p* = 0.0001, η² = 0.15); the intervention group significantly improved at post-test, whereas the control group did not. The difference between the game intervention and control, at post-test, remained significant when adjusting for pre-test scores, age, and gender, using ANCOVA (F(1,118) = 20.79, *p* = 0.0001, η² = 0.16), as the control group scored significantly higher than the test group at baseline (*p* = 0.0001). Independent *t*-test at pre-test found no significant age or gender differences in BAM scores (*p* > 0.05).

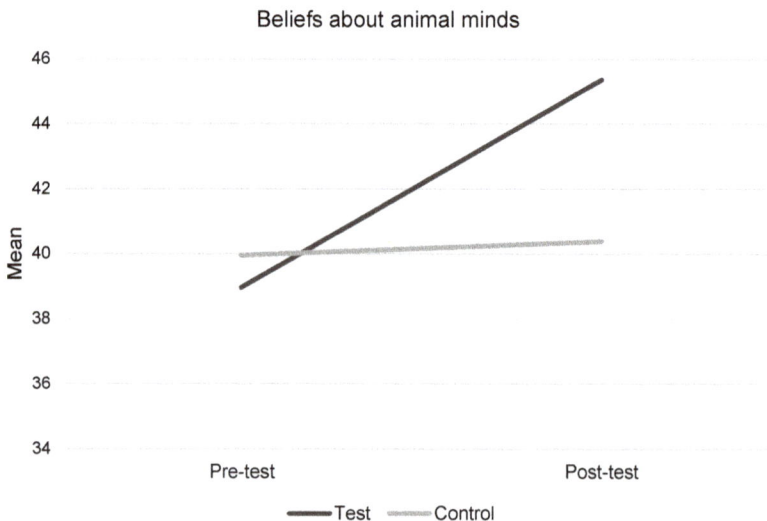

Figure 3. Children's beliefs about farm animal minds.

Table 3. Descriptive statistics for children's beliefs about farm animal minds (BAM).

Measure Item	Test				Control			
	Pre-Test		Post-Test		Pre-Test		Post-Test	
	M	SD	M	SD	M	SD	M	SD
Chicken Clever	3.06	0.9	4.39	0.7	3.11	1	3.48	1
Chicken Pain	4.36	0.9	4.67	0.7	4.13	1	4.06	1
Chicken Happy	4	0.9	4.68	0.6	4.28	0.8	4.33	0.9
Chicken Sad	3.99	0.9	4.49	0.8	4.02	1	4.12	1
Chicken Fear	4.28	1	4.57	0.8	4.33	1	4.19	1
Total Chicken Minds	19.68	3	22.80	3	19.86	3	20.17	5
Cow Clever	3.22	1	4.43	0.8	3.33	1	3.81	2
Cow Pain	4.06	1	4.57	0.8	4.15	1	4.06	1
Cow Happy	4.10	1	4.68	0.6	4.38	0.8	4.26	1
Cow Sad	3.77	1	4.41	1	4.02	1	4.06	1
Cow Fear	4.13	1	4.41	1	4.16	1	4.11	1
Total Cow Minds	19.28	4	22.50	3	20.03	4	20.30	5
TOTAL BAM	38.96	6	45.37	5	39.95	7	40.38	10

Note: Total scores for each animal type, ranged from 5–25. Overall BAM scores across all animal types, ranged from 10–50, where a high score indicated high BAM.

Table 4. Results for simple effects following significant interactions.

Test × Control at Pre-Test				Test × Control at Post-Test			
df	F	p	η^2	df	F	p	η^2
Total beliefs about farm animal minds							
130	0.759	0.385	0.006	120	13	0.0001	0.10
Beliefs about chicken minds							
131	0.11	0.743	0.001	119	15.1	0.0001	0.113
Beliefs about cow minds							
128	1.21	0.274	0.009	119	8.69	0.004	0.07
Knowledge about the five freedoms for chickens							
129	19.86	0.0001	0.133	119	40.3	0.0001	0.253
Total knowledge about farm welfare from farming system photos							
131	0.064	0.801	0.00	119	11.58	0.001	0.10
Knowledge about chicken welfare in farming system photos							
131	1.27	0.262	0.01	119	17.1	0.0001	0.128
Knowledge about cow welfare in farming system photos							
129	1.27	0.262	0.01	117	3.99	0.048	0.033

3.2.1. Beliefs about Chickens' Minds

There was a statistically significant interaction between intervention condition and time ($F(1,119) = 17.0$, $p = 0.0001$, $\eta^2 = 0.13$); the intervention group significantly improved at post-test, whereas the control group did not. The difference between game intervention and control at post-test remained significant, when adjusting for pre-test scores, age, and gender, using ANCOVA ($F(1,121) = 19.5$, $p = 0.0001$, $\eta^2 = 0.14$), as the control group scored significantly higher than the test group at baseline ($p = 0.0001$). Independent *t*-tests at pre-test found no significant differences in scores by age or gender ($p > 0.05$).

3.2.2. Beliefs about Cows' Minds

There was a statistically significant interaction between intervention condition and time ($F(1,117) = 17.7$, $p = 0.0001$, $\eta^2 = 0.13$); the intervention group significantly improved at post-test whereas the control group did not. The difference between game intervention and control at post-test remained significant when adjusting for pre-test scores, age, and gender, using ANCOVA (F(1,119) = 17.07, p = 0.0001, η^2 = 0.13), as the control group scored significantly higher than the test group at baseline (p = 0.0001). Independent t-tests at pre-test found no significant differences in scores by age or gender (p > 0.05).

3.3. Attitudes towards Cruelty to Farm Animals

There was no significant overall change in children's acceptance of cruelty to farm animals, following the intervention (Table 5, Figure 4). No statistically significant interaction between intervention condition and time was found (F(1,116) = 0.22, p = 0.641, η^2 = 0.002). This result remained non-significant when adjusting for pre-test scores, age, and gender, using ANCOVA (F(1,118) = 0.682, p = 0.411, η^2 = 0.006), even though the test group scored significantly higher than the control at baseline (p = 0.0001). Independent t-tests at pre-test found no significant differences in scores by age or gender (p > 0.05).

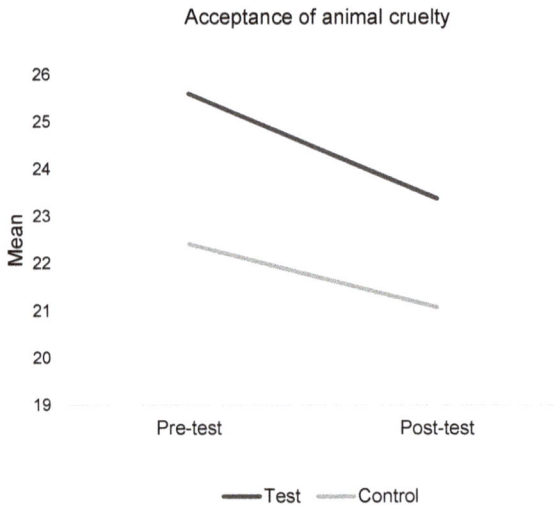

Figure 4. Children's acceptance of cruelty to farm animals.

Table 5. Descriptive statistics for children's attitudes towards cruelty to farm animals.

Measure Item	Test				Control			
	Pre-Test		Post-Test		Pre-Test		Post-Test	
	M	SD	M	SD	M	SD	M	SD
Attitudes towards cruelty to chickens								
Injure a chicken	1.28	0.8	1.19	0.5	1.08	0.3	1	0
Make a chicken scared	1.23	0.4	1.17	0.4	1.11	0.4	1.04	0.2
Make a chicken angry	1.28	0.5	1.16	0.4	1.11	0.4	1.02	0.1
Make a chicken sad	1.25	0.6	1.23	0.5	1.16	0.4	1.04	0.2
Make a chicken happy (r)	1.32	0.5	1.35	0.8	1.53	1	1.31	0.7
Not give a chicken food or water	1.33	0.8	1.32	0.9	1.14	0.5	1.08	0.6
Get treatment for an ill or injured chicken (r)	1.54	0.8	1.52	0.9	1.6	1	1.48	1
Not give a chicken a comfortable place to live	1.65	1	1.38	1	1.2	0.6	1.15	0.6
Not allow a chicken to perch or dig	1.80	1	1.65	1	1.27	0.6	1.21	0.7
Total cruelty to chickens	12.67	4	11.97	4	11.16	3	10.33	2

Table 5. *Cont.*

Measure Item	Test				Control			
	Pre-Test		Post-Test		Pre-Test		Post-Test	
	M	**SD**	**M**	**SD**	**M**	**SD**	**M**	**SD**
Attitudes towards cruelty to cows								
Injure a cow	1.17	0.5	1.15	0.6	1.23	0.8	1.21	0.7
Make a cow scared	1.29	0.5	1.12	0.5	1.08	0.3	1.04	0.3
Make a cow angry	1.23	0.5	1.13	0.5	1.11	0.4	1.04	0.3
Make a cow sad	1.33	0.6	1.28	0.8	1.46	0.8	1.04	1
Make a cow happy (r)	1.57	0.9	1.19	0.4	1.15	0.7	1.45	0.6
Not give a cow food or water	1.51	0.9	1.43	1	1.15	0.7	1.11	1.4
Get treatment for an ill or injured cow (r)	1.58	0.9	1.62	1	1.72	1	1.77	1
Not give a cow a comfortable place to live	1.71	1	1.29	0.9	1.23	0.7	1.11	0.6
Not allow a cow to move around or go outside	1.54	1	1.34	1	1.16	0.7	1.09	0.6
Total cruelty to cows	12.93	4	11.54	4	11.30	3	10.87	3
Total cruelty to farm animals	25.59	7	23.38	7	22.41	6	21.08	5

Note: Scores for each item ranged from 1–5. Total cruelty scores for each animal ranged from 9–45. Overall cruelty to farm animals scores, ranged from 18–90. High scores = high acceptance of cruelty.

3.3.1. Attitudes towards Cruelty to Chickens

No statistically significant interaction between the intervention condition and time was found ($F(1,119) = 0.21$, $p = 0.652$, $\eta^2 = 0.002$). This result remained nonsignificant when adjusting for pre-test scores, age, and gender, using ANCOVA ($F(1,121) = 3.08$, $p = 0.082$, $\eta^2 = 0.03$), even though the test group scored significantly higher than the control at pre-test ($p = 0.0001$). Independent *t*-tests at pre-test found no significant differences in scores by age or gender ($p > 0.05$).

3.3.2. Attitudes towards Cruelty to Cows

No statistically significant interaction between the intervention condition and time was found ($F(1,117) = 1.4$, $p = 0.239$, $\eta^2 = 0.012$). This result remained non-significant when adjusting for pre-test scores, age, and gender, using ANCOVA ($F(1,119) = 0.05$, $p = 0.82$, $\eta^2 = 0.001$), even though the test group scored significantly higher than the control at pre-test ($p = 0.001$). Independent *t*-tests at pre-test found no significant differences in scores by age or gender ($p > 0.05$).

3.4. Knowledge about Farm Animal Welfare Needs

Farm Animal Welfare improved children's scores for knowledge of welfare needs (Tables 6 and 7 and Figure 5). No statistically significant interaction between the intervention condition and time was found ($F(1,114) = 3.72$, $p = 0.056$, $\eta^2 = 0.032$), but a significant difference between the test and the control, was found at post-test, when adjusting for pre-test scores, age, and gender, using ANCOVA ($F(1,116) = 26.8$, $p = 0.0001$, $\eta^2 = 0.194$), as the test group scored significantly higher than the control at baseline ($p = 0.0001$). Independent *t*-tests at pre-test found no significant differences in scores by age ($p > 0.05$), but girls scored significantly higher than boys ($t(127) = -2.71$, $p = 0.008$).

Table 6. Descriptive statistics for knowledge about the five freedoms for animals, displaying the percentage of children in the intervention group who mentioned each freedom.

Measure Item	Pre-Test %		Post-Test %	
	Yes	No	Yes	No
Cows				
Free from discomfort due to suitable environment	87	13	89.6	10.4
Free from thirst, hunger and malnutrition through a suitable diet	91.3	8.7	97	3
Freedom to exhibit natural behavior patterns	68.1	31.9	62.7	37.3
Free from pain, injury and disease	33.3	66.7	29.9	70.1
Free from fear and distress	72.5	27.5	77.6	22.4
Chickens				
Free from discomfort due to suitable environment	84.1	15.9	86.8	13.2
Free from thirst, hunger and malnutrition through a suitable diet	88.4	11.6	98.5	1.5
Freedom to exhibit natural behavior patterns	37.7	62.3	57.4	42.6
Free from pain, injury and disease	23.2	76.8	30.9	69.1
Free from fear and distress	72.5	27.5	72.1	27.9

Table 7. Descriptive statistics for knowledge about the five freedoms for animals, displaying mean scores for each item for the intervention group.

Measure Item	Pre-Test		Post-Test	
	M	SD	M	SD
Cows				
Free from discomfort due to suitable environment	1.86	1.3	1.9	1.3
Free from thirst, hunger and malnutrition through a suitable diet	1.88	0.78	2	0.65
Freedom to exhibit natural behavior patterns	0.94	0.98	0.73	0.67
Free from pain, injury and disease	0.46	0.76	0.39	0.65
Free from fear and distress	1.43	1.3	1.37	1.2
Chickens				
Free from discomfort due to suitable environment	1.61	1.2	1.68	1.2
Free from thirst, hunger and malnutrition through a suitable diet	1.75	0.76	1.96	0.47
Freedom to exhibit natural behavior patterns	1.06	1.1	0.90	0.96
Free from pain, injury and disease	0.32	0.65	0.38	0.65
Free from fear and distress	1.25	1.1	1.34	1.2

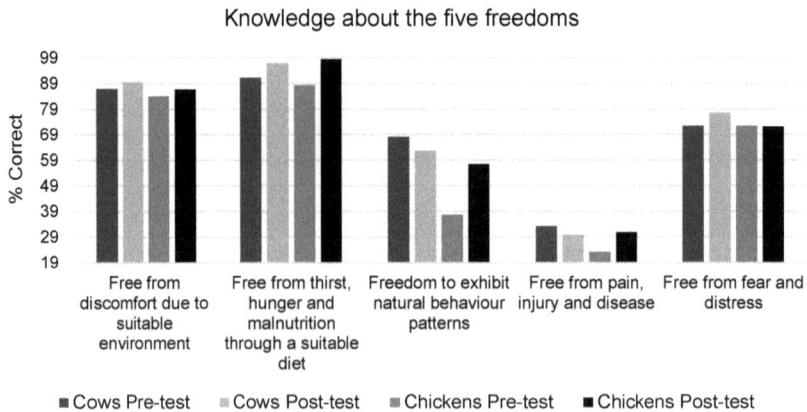

Figure 5. Children's knowledge about the five freedoms for animals.

3.4.1. Knowledge of Chicken Welfare Needs

There was a statistically significant interaction between the intervention condition and time (F(1,118) = 6.15, p = 0.015, η^2 = 0.05), the intervention group significantly improved at post-test, whereas the control group did not. The difference between the intervention and the control at post-test remained significant, when adjusting for pre-test scores, age, and gender, using ANCOVA (F(1,120) = 102, p = 0.0001, η^2 = 0.201), as the test group scored significantly higher than the control at baseline (p = 0.0001). Independent t-tests at pre-test found no significant differences in scores by age (p > 0.05), but girls scored significantly higher on chicken welfare needs knowledge than boys (t(129) = −2.65, p = 0.009).

3.4.2. Knowledge of Cow Welfare Needs

No statistically significant interaction between the intervention condition and time was found (F(1,117) = 1.26, p = 0.264, η^2 = 0.011), but a significant difference between the test and the control was found at post-test, when adjusting for pre-test scores, age, and gender, using ANCOVA (F(1,119) = 24.02, p = 0.0001, η^2 = 0.174), as the test group scored significantly higher than the control at baseline (p = 0.0001). Independent t-tests at pre-test found no significant differences in scores by age (p > 0.05), but girls scored significantly higher on cow welfare needs knowledge than boys (t(129) = −3.02, p = 0.003).

3.5. Knowledge of Animal Welfare in Farming Systems

There was a statistically significant interaction between the intervention condition and time (F(1,115) = 8.81, p = 0.004, η^2 = 0.071) for overall understanding of welfare in food production systems (Table 8, Figure 6). However, the significant difference between intervention and control groups at post-test was lost, after adjusting for pre-test scores, age, and gender, using ANCOVA (F(1,117) = 3.65, p = 0.059, η^2 = 0.032), as the control group scored significantly higher than the test group at baseline (p = 0.0001). Independent t-tests at pre-test found no significant differences in scores by age or gender (p > 0.05).

Table 8. Descriptive statistics for knowledge about animal welfare in photos of farming systems.

Measure Item	Test		Control	
	Pre-Test %	Post-Test %	Pre-Test %	Post-Test %
Cow a	85.5	95.5	93.5	92.5
Cow b	68.1	89.4	71	83
Cow c	59.4	81.8	59.7	64.2
Cow d	88.4	93.9	95.2	94.3
Cow e	78.3	97	79	77.4
Cow f	76.8	92.4	75.8	84.9
Cow g	82.6	89.4	93.5	94.3
Cow h	87	95.5	95.2	94.3
& of all cow pictures correct	36.2	66.7	48.4	52.8
Chicken a	84.1	95.5	74.2	77.4
Chicken b	79.7	89.4	90.3	83
Chicken c	53.6	57.4	43.5	43.4
Chicken d	84.1	94.1	75.8	73.6
Chicken e	76.8	94.1	77.4	73.6
Chicken f	78.3	92.6	74.2	69.8
Chicken g	81.2	85.3	90.3	90.6
Chicken h	84.1	91.2	74.2	67.9
& of all chicken pictures correct	23.2	43.9	21	20.8
All photos correct	10.1	34.8	16.1	15.1

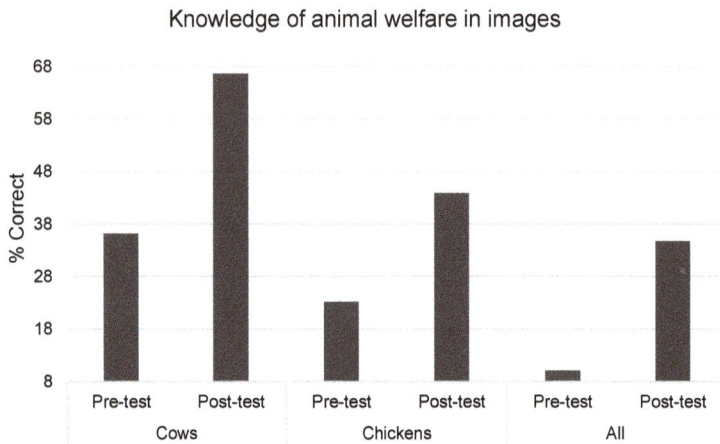

Figure 6. Children's knowledge about animal welfare in farming systems.

3.5.1. Knowledge of Chicken Welfare and Farming Systems

There was a statistically significant interaction between the intervention condition and time ($F(1,115) = 10.31$, $p = 0.002$, $\eta^2 = 0.082$); the intervention group significantly improved at post-test, whereas the control group did not. The difference between intervention and control at post-test remained significant, when adjusting for pre-test scores, age, and gender, using ANCOVA ($F(1,117) = 5.33$, $p = 0.023$, $\eta^2 = 0.05$), as the test group scored significantly higher than control at baseline ($p = 0.0001$). Independent t-tests at pre-test found no significant differences in scores for age ($p > 0.05$), but boys scored significantly higher than girls ($t(129) = -2.794$, $p = 0.006$).

3.5.2. Knowledge of Cow Welfare and Farming Systems

There was a statistically significant interaction between the intervention condition and time ($F(1,115) = 8.81$, $p = 0.004$, $\eta^2 = 0.07$); the intervention group significantly improved at post-test whereas the control group did not. The difference between intervention and control at post-test remained significant when adjusting for pre-test scores, age, and gender, using ANCOVA ($F(1,117) = 5.33$, $p = 0.023$, $\eta^2 = 0.05$), as the control group scored significantly higher at pre-test ($p = 0.000$). Independent t-tests at pre-test found no significant differences in scores by age or gender ($p > 0.05$).

4. Discussion

The purpose of this study was to evaluate the effectiveness of a novel digital farm animal welfare educational intervention named 'Farm Animal Welfare'. The game had a significant impact on children's beliefs about farm animal minds, children's knowledge about farm animal welfare needs, and knowledge about farm animal welfare in different farming systems. The game did not however, have an impact on children's compassion towards farm animals or their acceptance of cruelty to farm animals. The results, therefore, only partially supported the hypotheses.

The present study demonstrated that a one-off gaming intervention can lead to positive increases in children's knowledge about animal welfare. This is in line with previous research on short-term animal welfare education interventions and might be because knowledge is the easiest variable to change through education [10,28]. Although no overall difference in scores for knowledge about animal welfare in farming systems was found, when analyzing the scores separately for each type of farm animal, significant improvements were found. In combination, these findings are promising, given that knowledge about animal welfare needs in school-aged children is generally low, especially for farm animals [8,28]. Although an overall significant change was found for knowledge about farm animal welfare needs, when examining the percentages of children who mentioned each welfare need/freedom, changes were only seen for some of the welfare needs/freedoms. For example, very few children mentioned freedom from pain, injury, and disease for either the cow and the chicken, and this percentage did not improve at post-test for cows, but did so for chickens. Children were more likely to mention the dietary needs of farm animals and were less likely to mention the health needs or the freedom to exhibit natural behavior patterns, this is consistent with previous findings for children's knowledge about the welfare needs of pet animals [29]. Children were able to identify the need for farm animals to be free from fear and distress, and along with children's scores on beliefs about animal minds and low acceptance of cruelty to farm animals, this demonstrated that children had some awareness that farm animals are sentient and can feel pain and fear, following the intervention. At pre-test, only 10% of children in the test group were able to identify animal welfare in all farming system images, compared to 36% at post-test, which is still relatively low. However, when looking at the individual animal types, children were better able to identify animal welfare in the cow farming systems, following the intervention (with 67% scoring correctly at post-test, compared to 36% for cows at post-test). These findings reinforce the need to focus on teaching children about specific welfare needs/freedoms, in more depth, in future education programs, focusing on those that children lack knowledge of.

Children scored significantly higher on beliefs about cows' and chickens' minds, following the intervention, compared to the control group who did not. This finding was expected, given previous evaluations showing similar impacts of animal welfare education on children's beliefs about animal minds, for a range of animal types [10]. This finding supports previous hypotheses that such beliefs are easy to change through a simple intervention. These beliefs are a key cognitive factor influencing animal welfare [30–34], with low beliefs in animal minds being predicative of negative interactions child-animal interactions [33,34]. It is, therefore, a positive note that this digital game improved beliefs about farm animal minds, as these beliefs tend to be quite low at baseline (in this study, children scored 39 out of a possible 50, children rated chickens' mental abilities slightly higher than cows), compared to other animal types [29,34]. Furthermore, farm animals can also be a target for animal cruelty and so examining ways of preventing cruelty to farm animals, starting early in childhood, is critical.

Although children scored higher on compassion at post-test, no overall significant difference was found. Compassion scores were relatively high at baseline (mean of 20.71 out of a possible 25), which is consistent with previous studies on children's compassion towards pets and children's compassion towards animals in general [10,29]. No age or gender differences were found for compassion towards farm animals, which was unexpected given research on gender differences in adults' compassion towards animals. Perhaps the sample size was too small to find a difference, or perhaps gender differences do not emerge until later in life, we cannot ascertain these potential developmental trends without an assessment of compassion across a wider age-range. It has been suggested that children might need direct contact and interaction with animals, and a chance to form a bond or attachment to an animal, to develop compassion towards animals. This could be investigated in future research, given such contact and experience might impact children's knowledge of and attitudes towards farm animals.

Farm Animal Welfare did not significantly reduce children's acceptance of cruelty to farm animals, in line with previous evaluations of farm animal education workshops [8]. No age or gender differences were found. A possible reason might be that children were not accepting of cruelty at baseline (scoring a mean of 25.59 out of a possible 90, at pre-test), although these scores did reduce at post-test. Similar scores were found for chickens and cows, when analyzed separately, and children were not more accepting of cruelty to one farm animal type. Farm animals were not exempt from being targets of animal cruelty, and so education interventions should aim to prevent this from occurring, but we need to examine different educational methods. Farm Animal Welfare did not have cruelty content, which might explain why no significant change was found in attitudes towards cruelty. Other game interventions that have focused on preventing animal cruelty through specific content (e.g., accidental cruelty to pets), did find a change in children's acceptance of cruelty to pets [26]. Future interventions that aim to prevent cruelty to farm animals could, therefore, explore how to incorporate the concept of cruelty into the intervention strategies, using child-friendly materials.

5. Conclusions

This study is the first to demonstrate the benefits of utilizing digital gaming technology in animal welfare education and the use of game-based learning, or 'serious' games for promoting children's beliefs about farm animal minds and knowledge of farm animal welfare. Knowledge and beliefs about animal minds are the easiest to change through a digital educational intervention, but compassion seems more resistant to change and needs further investigation. Although this study presents promising findings, longer-term evaluation is required, with a larger population across wider demographics. Long-term follow-up would be useful to examine whether changes are long-lasting. Trialing such computer games, compared with more traditional educational pedagogy with a wider range of ages, could also be beneficial. Children's feedback on the game was extremely positive and children reported wanting more animals and more levels to play and so a more complex game could be developed in the future. This game is a novel small-scale intervention that has shown a positive impact on knowledge of welfare needs, sentience, and production systems. Future games could be developed and evaluated to reveal whether they can have a similar impact on children's welfare knowledge of other types of animals, including pets, wildlife, and zoo animals.

Supplementary Materials: The following are available online at http://www.mdpi.com/2076-2615/9/3/91/s1, Figure S1: Screenshots of Farm Animal Welfare, Table S1: Full content and functionality of Farm Animal Welfare.

Author Contributions: Conceptualization, R.D.H., J.M.W. and G.A.R.M.F.; Supervision of Farm Animal Welfare game development, J.M.W.; data collection and analysis, R.D.H.; writing—original draft preparation, R.D.H.; writing—review and editing, R.D.H. and J.M.W.

Funding: This research received no external funding.

Acknowledgments: We thank the schools, teachers, and children who participated in this research, for their invaluable time and collaboration. We would also like to thank Armand Baboian for developing the Farm Animal Welfare Game as part of his MSc. degree in Computer Science, School of Informatics, University of Edinburgh.

Conflicts of Interest: The authors declare no conflict of interest.

References

1. Clark, B.; Stewart, G.B.; Panzone, L.A.; Kyriazakis, I.; Frewer, L.J. A systematic review of public attitudes, perceptions and behaviours towards production diseases associated with farm animal welfare. *J. Agric. Environ. Ethics* **2016**, *29*, 455–478. [CrossRef]

2. Ventura, B.A.; Von Keyserlingk, M.A.; Wittman, H.; Weary, D.M. What difference does a visit make? Changes in animal welfare perceptions after interested citizens tour a dairy farm. *PLoS ONE* **2016**, *11*, e0154733. [CrossRef] [PubMed]

3. Boissy, A.; Erhard, H.W. How studying interactions between animal emotions, cognition, and personality can contribute to improve farm animal welfare. In *Genetics and the Behavior of Domestic Animals*, 2nd ed.; Elsevier Academic Press: London, UK, 2014; pp. 81–113.

4. Knight, S.; Vrij, A.; Cherryman, J.; Nunkoosing, K. Attitudes towards animal use and belief in animal mind. *Anthrozoös* **2004**, *17*, 43–62. [CrossRef]

5. Janssen, M.; Busch, C.; Rödiger, M.; Hamm, U. Motives of consumers following a vegan diet and their attitudes towards animal agriculture. *Appetite* **2016**, *105*, 643–651. [CrossRef] [PubMed]

6. RSPCA. 'Chickens Trot' Say More than One in 10 Brits, Reveals Freedom Food. 2013. Available online: http://www.freedomfood.co.uk/press/2013/07/%E2%80%98chickens-trot%E2%80%99-say-more-than-one-in-10-brits,-reveals-freedom-food (accessed on 4 August 2015).

7. FAWC. Report on Farm Animal Welfare in Great Britain: Past, Present and Future. 2009. Available online: https://www.gov.uk/government/publications/fawc-report-on-farm-animal-welfare-in-great-britain-past-present-and-future (accessed on 9 February 2019).

8. Jamieson, J.; Reiss, M.J.; Allen, D.; Asher, L.; Wathes, C.M.; Abeyesinghe, S.M. Measuring the success of a farm animal welfare education event. *Anim. Welf. UFAW J.* **2012**, *21*, 65–75. [CrossRef]

9. Animal Welfare Act. 2006. Available online: https://www.legislation.gov.uk/ukpga/2006/45/contents (accessed on 9 February 2019).

10. Hawkins, R.D.; Williams, J.M.; Scottish Society for the Prevention of Cruelty to Animals. Assessing Effectiveness of a Nonhuman Animal Welfare Education Program for Primary School Children. *J. Appl. Anim. Welf. Sci.* **2017**, *20*, 240–256. [CrossRef]

11. Kohlberg, L. The Development of Modes of Thinking and Choices in Years 10 to 16. Ph.D. Thesis, University of Chicago, Chicago, IL, USA, 1958.

12. Muldoon, J.; Williams, J.M.; Lawrence, A.; Lakestani, N.; Currie, C. *Promoting a 'Duty of Care' towards Animals among Children and Young People: A literature Review and Findings from Initial Research to Inform the Development of Interventions (Commissioned by Defra, p. 107)*; Child and Adolescent Health Research Unit, University of Edinburgh: Edinburgh, UK, 2009.

13. Myant, K.A.; Williams, J.M. Children's concepts of health and illness: Understanding of contagious illnesses, non-contagious illnesses and injuries. *J. Health Psychol.* **2005**, *10*, 805–819. [CrossRef]

14. Hawkins, R.D.; Williams, J.M. Children's beliefs about animal minds (Child-BAM): Associations with positive and negative child-animal interactions. *Anthrozoös* **2016**, *29*, 503–519. [CrossRef]

15. Aguirre, V.; Orihuela, A. Assessment of the impact of an animal welfare educational course with first grade children in rural schools in the State of Morelos, Mexico. *Early Child. Educ. J.* **2010**, *38*, 27–31. [CrossRef]

16. Ascione, F.R.; Weber, C.V. Children's attitudes about the humane treatment of animals and empathy: One-year follow up of a school-based intervention. *Anthrozoös* **1996**, *9*, 188–195. [CrossRef]

17. Nicoll, K.; Samuels, W.E.; Trifone, C. An in-class, humane education program can improve young students' attitudes toward animals. *Soc. Anim.* **2008**, *16*, 45–60. [CrossRef]

18. Hawkins, R.D. Psychological Factors Underpinning Child-Animal Relationships and Preventing Animal Cruelty. Ph.D. Thesis, The University of Edinburgh, Edinburgh, UK, 2018.

19. Fletcher, R. Gaming conservation: Nature 2.0 confronts nature-deficit disorder. *Geoforum* **2017**, *79*, 153–162. [CrossRef]

20. Büscher, B. Nature 2.0: Exploring and theorizing the links between new media and nature conservation. *New Media Soc.* **2016**, *18*, 726–743. [CrossRef]

21. Sandbrook, C.; Adams, W.M.; Monteferri, B. Digital games and biodiversity conservation. *Conserv. Lett.* **2015**, *8*, 118–124. [CrossRef]

22. Burke, B. *Gamify: How Gamification Motivates People to Do Extraordinary Things*; Taylor & Francis, Routledge: New York, NY, USA, 2016.

23. Honey, M.A.; Hilton, M. *Learning Science through Computer Games and Simulations*; The National Academies Press: Washington, DC, USA, 2011.

24. Hirsh-Pasek, K.; Zosh, J.M.; Golinkoff, R.M.; Gray, J.H.; Robb, M.B.; Kaufman, J. Putting education in "educational" apps: Lessons from the science of learning. *Psychol. Sci. Public Interest* **2015**, *16*, 3–34. [CrossRef] [PubMed]

25. Annetta, L.A. The "I's" have it: A framework for serious educational game design. *Rev. Gen. Psychol.* **2010**, *14*, 105–113. [CrossRef]

26. Bjerke, T.; Odergårdstuen, T.S.; Kaltenborn, B.P. Attitudes toward animals among Norwegian adolescents. *Anthrozoös* **1998**, *11*, 79–86. [CrossRef]

27. Heleski, C.R.; Zanella, A.J. Animal science student attitudes to farm animal welfare. *Anthrozoös* **2006**, *19*, 3–16. [CrossRef]

28. Wells, D.L.; Hepper, P.G. Attitudes to animal use in children. *Anthrozoös* **1995**, *8*, 159–170. [CrossRef]

29. Hawkins, R.D.; Williams, J.M. The development and pilot evaluation of a 'serious game' to promote positive child-animal interactions. *Hum. Anim. Interact. Bull.* **2019**. under review.

30. Ellingsen, K.; Zanella, A.J.; Bjerkås, E.; Indrebø, A. The relationship between empathy, perception of pain and attitudes toward pets among Norwegian dog owners. *Anthrozoös* **2010**, *23*, 231–243. [CrossRef]

31. Sorabji, R. *Animal Minds and Human Morals: The Origins of the Western Debate*; Cornell University Press: Ithaca, NY, USA, 1995.

32. Burghardt, G.M. Ethics and animal consciousness: How rubber the ethical ruler? *J. Soc. Issues* **2009**, *65*, 499–521. [CrossRef]

33. Baron-Cohen, S.; Tager-Flusberg, H.; Lombardo, M. (Eds.) *Understanding Other Minds: Perspectives from Developmental Social Neuroscience*; Oxford University Press: Oxford, UK, 2013.
34. Menor-Campos, D.J.; Hawkins, R.; Williams, J. Belief in Animal Mind among Spanish Primary School Children. *Anthrozoös* **2008**, *31*, 599–614. [CrossRef]

Article

Attitudes of Young Adults toward Animals—The Case of High School Students in Belgium and The Netherlands

Pim Martens *, Camille Hansart and Bingtao Su

International Centre for Integrated assessment and Sustainable development (ICIS), Maastricht University, P.O. Box 616, 6200 MD Maastricht, The Netherlands; c.hansart@alumni.maastrichtuniversity.nl (C.H.); su_bingtao@126.com (B.S.)

Received: 7 January 2019; Accepted: 27 February 2019; Published: 11 March 2019

Simple Summary: Young adults' attitudes towards animals will be influenced by a number of factors, including: sex, age, nationality/ethnicity, residence area, animal-related activities and hobbies, food habits, culture/religion, education and pet ownership. A case study of Dutch and Belgian high school students shows that levels of concern for animal welfare were distinctly higher among female participants, those who ate little to no meat, Belgian students, pet owners and those who had been to a zoo at least once. In general, students who reported having more contact with animals also had more positive attitudes towards animals.

Abstract: The social context and culture in which individuals grow shapes their perspectives through life. Early on, children learn about animals through storybooks, animated movies, toys, and through interactions with pets and wildlife, and will slowly start to build beliefs around those experiences. Their attitudes towards animals will be influenced by a number of factors, including: sex, age, nationality/ethnicity, residence area, animal-related activities and hobbies, food habits, culture/religion education, and pet ownership. A case study of Dutch and Belgian high school students (aged 12–21) investigated the attitudes of young people towards animals. By using the Animal Attitude Scale (AAS) and the Animal Issue Scale (AIS) questionnaires, our study shows that levels of concern for animal welfare were distinctly higher among: female participants; those who ate little to no meat; Belgian students; pet owners; and those who had been to a zoo at least once. In general, students who reported having more contact with animals also had more positive attitudes towards animals. To understand younger generations and their attitudes toward animals is to understand how future generations will look towards and treat our fellow animals, with which we share the planet Earth.

Keywords: animal welfare; young adult; animal attitudes

---

1. Introduction

Animals have accompanied humans for thousands of years, with a strong bond forged between humans and other species. Our relationships with animals can take different forms. On the one hand, animals can serve instrumental purposes: we currently use animals for clothing, for testing a range of human products, for gaining basic insights into human biology and behavior, and as food. On the other hand, human–animal relations are social. The clearest example is the practice of pet-keeping, with people attributing a special status to their pets [1].

Studies have shown that most children reject the idea of humans being animals [2], although they do have a propensity to anthropomorphise animals [3,4]. Most children have an appreciation for animals on the emotional and recreational levels. They tend to show affection as well as concern for them, in contrast to the more practical and utility-based perspectives of adults [3,5]. Kellert [3] found that children were strongly emotionally attached to individual animals. Hunting was not popular among the children and was only deemed acceptable if the end purpose was to feed oneself as opposed to sheer trophy hunting. Similar results were found by Pagani, et al. [6], who reported the majority of children were against hunting, zoos, animals used in circuses, and their exploitation for leather. Slight preferences were expressed for zoos compared to circuses, perhaps because zoos pursued a greater mission in terms of education and conservation. Earlier, Driscoll [7] outlined the different views adults have on how humans should use animals. Despite considerable opposition, most were in favour of animals being used in medical or scientific research, but did not approve of their use in product testing. Expressing similar ethical concerns, a large proportion of children disapproved of the use of animals in fur farming [3].

Understanding the attitudes that younger generations have toward animals may help us to understand the sustainability of future societies, as our attitude towards animals are central in the sustainability debate. Many factors including gender, age, nationality/ethnicity, residence area, animal-related activities and hobbies, food habits, culture/religion, education, and pet ownership are associated with people's attitudes toward animals [8,9]. The present study was conducted among high school students in Belgium and the Netherlands. Through this study, we aim to find out whether the variables we mentioned above and other variables like household, house type, meat-eating frequency correlate with young adults' attitudes toward animals.

2. Methodology

Research into children's attitudes toward animals in the Netherlands and Belgium was conducted between May and July in 2016. During this period, a paper-based questionnaire was implemented in four different schools, including three schools in the French-speaking province of Walloon, Collège du Christ Roi (N = 54), Athénée Maurice Carême (N = 148), Paul Delvaux (N = 45) and one school in the south of the Netherlands, Rombouts College (N = 120). All participants were high school students aged 12 to 21. Selection of participants in this study was made through simple random sampling; however, only those classes that replied to the invitation to participate in the research are represented here. Schools were contacted via mail and/or telephone prior to visits.

2.1. Questionnaire

The questionnaire consisted of four parts. In the first, we asked respondents to provide their demographic details, including age, gender, nationality, highest level of education, household composition, residence area, type of house, presence of a garden, zoo/aquarium visiting frequency, meat-eating frequency, pet ownership and religious affiliation.

In the second part, we introduced the Animal Attitude Scale (AAS) [10], which was used to assess the participant's attitude toward animals by means of a Likert scale. The questionnaire consisted of 20 questions rated from 1 (strongly agree) to 5 (strongly disagree) for all questions except questions number: 1,3,4,7,10,11,17,19, 20, which were all reverse coded from 1 (strongly disagree) to 5 (strongly agree). Total AAS scores were calculated by adding up all individual questions scores. Higher AAS scores indicated a higher concern and respect for animals. Questions included: "I do not think that there is anything wrong with using animal in medical research" Or "Wild animals, such as mink and raccoons, should not be trapped and their skins made into fur coats".

In the third part of the questionnaire, the Animal Issue Scale (AIS) [11] was introduced and served as a complement to the AAS questionnaire. This scale was much longer, comprised of 43 total questions. These questions were grouped across eight separate sections (use of animals, disrupting animal integrity, killing animals, compromising animal welfare, experimenting on animals, changing animals' genotypes, animals and the environment, and societal attitudes toward animals). Again, respondents had to rate the questions on a five-mark scale ranging from high acceptability to high non-acceptability. Total AIS scores were calculated by adding individual section scores. Akin to the AAS questionnaire, a higher total AIS score indicates a more positive attitude toward animals, and lower acceptability for the issues described [12]. For example, these issues included: "Keeping animals for the education of the public in zoos, wildlife parks, etc." and "Killing young animals that are dependent on their parents".

2.2. Statistical Analysis

Responses from the questionnaires were analysed through IBM SPSS Statistics (version 21, Armonk, NY: IBM Corp). Descriptive statistics were used to analyse the independent variables of age, sex, education level, household composition, residence place, housing type, presence of a garden, zoo and aquarium visiting frequency, meat-eating frequency, pet ownership and religious affiliation. Data was then analysed to check for normality so as to use the best fitted statistical test for analysis. All variables were non-normally distributed. They were thus analysed using a Mann–Whitney U-test and a Kruskal–Wallis test. Furthermore, post hoc analyses on the Kruskal–Wallis tests were conducted to calculate the differences between the average scores of each group. All results are assessed for significance based on the cut-off value of 0.05. Similar shapes of distribution were assumed for the Mann–Whitney U-test for all variables.

3. Results

3.1. Demographics

In total, 367 hard copy questionnaires were distributed in all the four schools. Of these 367 questionnaires, 358 were kept for analysis, and 9 discarded because they were mostly incomplete. 54.5% of the respondents were male; 45.5% female. Considering the relatively small variance in age, participants were divided into only two groups: 12–15 years old (77.7%) and 16–21 years old (22.3%). Mean age and variance across the entire sample was 14.44 ± 1.61years old. 33.4% of students were Dutch and 66.2% were Belgian. All participants had a level of schooling below or equal to the 12th grade. In terms of household composition, 19.2% of the students lived under the guardianship of a single parent and 80.8% under the guardianship of a couple. 44.1 % of the participants lived in urban areas and 55.9% in rural areas. Analysis of the different housing types showed that 7.6% of the children lived in apartments, 42.2% in semi-detached houses, and 50.1% in detached houses. 92.1% reported having a garden, while 7.9% did not. The frequency of visits to zoos and aquariums was predominantly 'once every two years or less' (31.7%), then 'once per year' (27.8%), followed by 'once every six months' (22.1%), 'never' (15.6%), and lastly 'once a month or more' (2.8%). In the sample of students studied, 0.6% were vegetarian, 5.6% ate meat once a week or less, 19% ate meat two or three days a week, 43% four to six days a week, and 31.9% ate meat every day. Regarding pet ownership, 77% replied that they had a pet, while 23% did not.

3.2. The AIS Score

The total AIS score for participants in the study was on average 158.61 out of 215 (see Table 1). Students scored relatively high (high score here means less acceptability) on variables related to: 'deprive animal welfare' (22.82 out of 25), 'harm animals for environment' (23.96 out of 30), 'harm animals for social issues' (24.00 out of 30), and 'killing animals' (18.74 out of 25).

A multiple regression analysis was run to find the factors most predominantly influencing AIS questionnaire scores (significance value $p < 0.05$). Pet ownership was an important determinant, with students who owned any pet scoring 5.41 points higher on average, compared to those who did not. However, only owning a pet dog is significant. AIS scores were 6.30 points higher for students who were dog owners, indicating that owning a dog may strongly influence the AIS scores. Test results also showed that the country of origin influences AIS scores: Belgian students scored 9.81 points higher than Dutch students. Gender was another important determinant in the AIS score; with females scoring 8.54 points higher than males. Lastly, eating meat also influenced total AIS scores (Table 2).

Table 1. Descriptive statistics: total Animal Issue Scale (AIS) and individual variables mean scores.

	Minimum	Maximum	Mean	SD
Total AIS	79	206	158.61	18.76
Use of animals (score: 5–25)	6	24	15.36	3.21
Animal integrity destruction (score: 6–30)	12	30	21.05	3.16
Killing animals (score: 5–25)	9	25	18.74	2.79
Deprive animal welfare (score: 5–25)	5	25	22.82	2.67
Experimentation on animals (score: 5–25)	5	59	16.40	4.38
Changes in animals' genotypes (score: 5–25)	5	25	17.00	4.13
Harm animals for environment (score: 6–30)	0	30	23.96	3.96
Harm animals for social issues (score: 6–30)	0	30	24.00	3.69

Table 2. Correlations between independent questionnaire variables and the total AIS score.

Y: Total AIS Score	Unstandardized Coefficients		Standardized Coefficients	t	p
	B	Std. Error	Beta		
(Constant)	160.80	9.03		17.81	<0.001
X_1: Do you own a pet?	−5.41	3.16	−0.13	−1.71	0.088
X_2: In what country do you live?	9.81	2.38	0.27	4.13	<0.001
X_3: What is your sex?	8.54	2.28	0.24	3.74	<0.001
X_4: How often do you eat meat (including fish) every week?	−3.34	1.35	−0.16	−2.47	0.014
X_5: What pet (s) do you have? Dog (s)	−6.30	2.64	−0.18	−2.39	0.018

3.2.1. Age Groups

There was no significant difference between the two age groups in terms of their total score on the AIS questionnaire (U = 5612.500, N (12–15 years old) = 194, N (16–21 years old) = 63, p = 0.331). This implies that age had no influence on responses. However, the Mann-Whitney U-test did show a significant difference between 12–15 years old participants and 16–21 years old participants in the scoring results for one variable: 'Use of animals' (U = 7823.500, p = 0.007). The median score in this specific section of the questionnaire was significantly higher for 12–15 year olds (Mdn = 16.00) than for 16–21 year olds (Mdn = 15.00).

3.2.2. Gender

Results indicate a significant difference in the total score for the AIS between the two genders (U = 5858.000, N (males) = 150, N (females) = 107, p = 0.000). Females had a higher median total score for the AIS questionnaire (Mdn = 165.00) than males (Mdn = 156.00) (Table 4). This suggests that female students scored higher overall on the AIS questionnaire.

A statistically significant difference between the two genders was observed for all variables, except two: 'Changes in animal's genotypes' (U = 11646.000, p = 0.988) and 'Animal integrity destruction' (U = 13944.000, p = 0.801). Females scored significantly higher than males for the following variables: 'Use of animals' (U = 11680.500, p = 0.017), 'Killing animals' (U = 11704.500, p = 0.009), 'Deprive animal welfare' (U = 11522.000, p = 0.003), 'Experimentation on animals' (U = 9889.500, p = 0.000), 'Harm animals for environment' (U = 7954.500, p = 0.000) and 'Harm animals for social issues' (U = 9102.000, p = 0.000). For all variables, median values were higher for female participants. These results were in accordance with results obtained from the analysis of the total AIS score.

3.2.3. Nationality

Test results for the nationality variable indicate a significant difference between questionnaire scores for variables 'Use of animals' (U = 9561.500, p = 0.000), 'Animal integrity destruction'

(U = 6709.00, p = 0.000), 'Killing animals' (U = 9448.000, p = 0.000), 'Deprive animal welfare' (U = 10113.500, p = 0.000), 'Experimentation on animals' (U = 10346.500, p = 0.023), 'Changes in animal's genotypes' (U = 6544.000, p = 0.000), 'Harm animals for environment' (U = 9807.500, p = 0.036). The only non-significant exception was the following variable: 'Harm animals for social issues' (U = 11475.500, p = 0.392).

This is in alignment with results for the total score of the AIS questionnaire (see below), which indicate a significant difference between Belgian and Dutch respondents (U = 5069.500, N (Belgian) = 157, N (Dutch) = 100, p = 0.000). The total median score was also higher for Belgian participants (Mdn = 164.00) than for Dutch participants (Mdn = 152.50).

3.2.4. Household

The total score of the AIS questionnaire was not found to be significantly different between those students living with a single parent, and those living with a couple (U = 3722.00, p = 0.250). However, there were significant differences in scores for the following separate sections of the questionnaire: 'Use of animals' (U = 5806.000, p = 0.036, Mdn = 17.00) and 'Animal integrity destruction' (U = 5703.500, p = 0.006, Mdn = 22.00), with higher medians attributed to students living with a single parent.

3.2.5. Residence Area

No significant difference was found between the total scores for students who lived in rural areas and those who lived in urban areas (U = 6764.500, p = 0.825), nor between any of the separate sections of the questionnaire ($p < 0.05$ in all cases).

3.2.6. House Type

No significant difference was found between the three groups of participants for residence (either: apartment, semi-detached house, or detached house). All p-values were higher than the threshold value of 0.05, which suggests that the type of house wherein students lived did not influence their AIS questionnaire score.

3.2.7. Owning a Garden

Ownership of, or access to a garden did not influence total scores on the AIS questionnaire (U = 2215.00, p = 0.955). Furthermore, no significant difference in scores was found in any individual section of the questionnaire.

3.2.8. Zoo/Aquarium Visiting Frequency

Equality of variance was assumed for all variables except for: 'Harm animals for social issues'. A significant difference in total scores was found between students, based on how often they visited a zoo or an aquarium ($\chi2$ (2) = 9.624, p = 0.047). Significant differences were also noticed in the following separate variables of the questionnaire: 'Harm animals for environment' ($\chi2$ (2) = 14.823 p = 0.005), 'Deprive animal welfare' ($\chi2$ (2) = 10.936, p = 0.027). A post-hoc analysis of variables showed that students who never visited zoos and/or aquariums had lower total AIS scores than those who did.

3.2.9. Meat-Eating Frequency

Equality of variance was assumed for all variables except: 'killing animals' and 'experimentation on animals'. There was no significant difference in total scores between students who ate meat and those who did not ($\chi2$ (2) = 8.825, p = 0.066). There was however a significant difference for the 'Harm animals for social issues' variable of the questionnaire ($\chi2$ (2) = 3.862, p = 0.425).

3.2.10. Pet Ownership

Data analysis of the AIS questionnaire variables: 'experimentation on animals' (U = 6587.000, p = 0.000), 'Deprive animal welfare' (U = 8299.000, p = 0.021), and 'killing animals' (U = 7792.000, p = 0.003) show pet owners scored significantly higher than non-pet owners. The total AIS score was also significantly higher (U = 3989.000, p = 0.000, N (pet owners) = 198, N (non-pet owners) = 58) for pet owners.

3.2.11. Religious Outlooks

Results indicate that there was no significant difference in the total score in replies to the question: 'is religion important in your life?' (U = 5942.500, N (yes) = 62, N (no) = 195, p = 0.841). Whilst analysis of the total score for the AIS questionnaire did not show any distinction between students who considered religion important and those who did not, there were significant differences in terms of individual sections of the questionnaire. Significant differences only occurred for the following variables: 'Animal integrity destruction' (U = 9626.500, p = 0.008) and 'Killing animals' (U = 9928.500, p = 0.044), with higher median scores for those students who replied religion was important in their lives.

3.3. The AAS Score

The total AAS score for participants in the study was on average 67.94 out of 100 (see Table 3). The highest scores were observed for the following variables: 'it is morally wrong to hunt wild animals just for sport (4.02 out of 5), 'wild animals such as mink and raccoons should not be trapped and their skins made into fur coats' (4.06 out of 5), 'the slaughter of whales and dolphins should be immediately stopped even if it means some people will be put out of work' (4.11 out of 5), and 'breeding animals for their skins is a legitimate use of animals' (4.01 out of 5).

Table 3. Descriptive statistics: total Animal Attitude Scale (AAS) and individual variables mean scores.

	Minimum	Maximum	Mean	SD
Total AAS	37	99	67.94	10.33
1. It is morally wrong to hunt wild animals just for sport.	1	5	4.02	1.18
2. I do not think that there is anything wrong with using animal in medical research.	1	5	3.32	1.66
3. There should be extremely stiff penalties including jail sentences for people who participate in cock-fighting.	1	5	3.56	1.11
4. Wild animals, such as mink and raccoons, should not be trapped and their skins made into fur coats.	1	5	4.06	1.16
5. There is nothing morally wrong with hunting wild animals for food or a better living for poor people.	1	5	2.55	1.07
6. I think people who object to raising animals for meat are too sentimental.	1	5	2.99	1.11
7. Much of the scientific research done with animals is unnecessary and cruel.	1	5	3.21	1.08
8. I think it is perfectly acceptable for cattle and dogs to be raised for human consumption.	1	5	3.08	1.27
9. Basically, humans have the right to use animals as we see fit.	1	5	3.73	1.22
10. The slaughter of whales and dolphins should be immediately stopped even if it means some people will be put out of work.	1	5	4.11	1.13
11. I sometimes get upset when I see wild animals in cages at zoos.	1	5	2.61	1.18
12. In general, I think that human economic gain is more important than setting aside more land for wildlife.	1	5	3.56	1.16
13. Too much fuss is made over the welfare of animals these days when there are many human problems that need to be solved.	1	5	3.16	1.21
14. Breeding animals for their skins is a legitimate use of animals.	1	5	4.01	1.11
15. Some aspects of biology can only be learned through dissecting preserved animals, such as cats.	1	5	3.15	1.22
16. Continued research with animals will be necessary if we are to ever conquer diseases such as cancer, heart disease and AIDS.	1	5	2.64	1.06
17. It is unethical to breed purebred dogs for pets when millions of dogs are killed in animal shelters each year.	1	5	3.48	1.05
18. The production of inexpensive meat, eggs, and dairy products justifies maintaining animals under crowded conditions.	1	5	3.04	1.08
19. The use of animals, such as rabbits, for testing the safety of cosmetics and household products is unnecessary and should be stopped.	1	5	3.82	1.20
20. The use of animals in rodeos and circuses is cruel.	1	5	3.78	1.20

To summarise, results from multiple regression analysis showed that AAS questionnaire scores were mostly influenced by: sex, as females scored 7.85 points higher than males on the questionnaire; meat-eating frequency, as those who ate meat less frequently scored 7.85 points higher than others; pet ownership, as pet owners scored 5.55 points higher than non-pet owners; and nationality, as Belgian students scored 5.13 points higher than Dutch students. Finally, the residence type students lived in influenced AAS total scores, yet as seen with the Kruskal–Wallis test, the religion (Christianity) was not significantly influence respondents' attitudes toward animals (Table 4).

Table 4. Most prevalent correlations between independent questionnaire variables and the total AAS score.

Y: Attitudes towards animals	Unstandardized Coefficients		Standardized Coefficients	t	p
	B	Std. Error	Beta		
(Constant)	74.07	5.77		12.84	<0.001
X$_1$: What is your sex?	7.85	1.12	0.38	7.00	<0.001
X$_2$: How often do you eat meat (including fish) every week?	−2.14	0.67	−0.16	−3.18	0.002
X$_3$: What is your main source of inspiration? Christianity	−2.73	1.46	−0.10	−1.87	0.063
X$_4$: Do you own a pet?	−5.68	1.36	−0.23	−4.18	<0.001
X$_5$: In what country do you live?	4.82	1.25	0.23	3.86	<0.001
X$_6$: In what sort of house do you live?	−2.11	0.97	−0.12	−2.18	0.030

3.3.1. Age

There was no significant difference between the total AAS scores, for students 12–15 years old and those who were 16–21 years old (U = 8759.500, N (10–15 years old) = 245, N (16–21 years old) = 76, $p = 0.436$).

3.3.2. Gender

Female students had significantly higher total AAS scores than males (U = 7522.500, N (females) = 146, N (males) = 175, $p = 0.000$). Higher medians for females were also found in the individual AAS questionnaire questions.

3.3.3. Nationality

The total AAS score was significantly higher for Belgian students than Dutch students (U = 8177.000, N (Dutch students) = 114, N (Belgian students) = 209, $p = 0.000$). Likewise, medians for the individual AAS questions were in general, higher for Belgian students.

3.3.4. Household

Students who lived with a single parent scored significantly higher on the total AAS score than those who lived with a couple (U = 5419.000, N (students living with single parents) = 58, N (students living with parents forming a couple) = 236, $p = 0.036$). Regarding the individual AAS questions, medians were generally higher for students living with a single parent.

3.3.5. Residence Area

Place of residence (either urban or rural area) did not influence total AAS scores in any significant way, as shown in Table 4 (U = 10810.500, N (students living in urban areas) = 130, N (students living in rural areas) = 168, $p = 0.882$).

3.3.6. House Type

Living in either an apartment, a semi-detached house or a detached house did not significantly influence the total AAS score ($\chi2$ (2) = 0.401, N (students living in an apartment) = 22, N (students living in a semi-detached house = 131, N (students living in a detached house) = 157, (p = 0.818) of students.

3.3.7. Owning a Garden

There was no significant difference in the total AAS scores between students who had a garden and those who did not (U = 3455.500, N (students who have a garden) = 297, N (students who do not have a garden) = 24, p = 0.804).

3.3.8. Zoo/Aquarium Visiting Frequency

There was a significant difference in the total AAS score for students depending on how often they visited zoos and/or aquariums ($\chi2$ (2) = 21.989, N (once a month or more) = 9, N (once every six months) = 75, N (once every year) = 88, N (once every two years or less) = 101, N (never) = 47, p = 0.000). A post-hoc analysis revealed that those who never went to a zoo and/or aquarium scored significantly lower than others on the AAS questionnaire.

3.3.9. Meat-Eating Frequency

There was a significant difference in total AAS score ($\chi2$ (2) = 15.420, N (I do not eat meat, I am vegetarian/vegan) = 2, N (once a week or less) = 17, N (2–3 days a week) = 63, N (4–6 days a week) = 134, N (everyday) = 96, p = 0.004). A post hoc analysis revealed higher scores for those who ate meat less frequently (once a week or less) than others.

3.3.10. Pet Ownership

A significant difference was found between students who owned a pet and those who did not, in their total AAS score. Those who had pets scored significantly higher (U = 6359.500, N (pet owners) = 250, N (non-pet owners) = 73, p = 0.000). Overall, medians were also higher for pet owners.

3.3.11. Religious Affiliation

Religious affiliation had no significant influence on total AAS scores from students as shown in Table 4 (U = 10146.000, N (religion/spirituality is important) = 92, N (religion/spirituality is not important) = 232, p = 0.000).

4. Discussion and Conclusion

Results from the study revealed several strong correlates for young adults' attitudes towards animals. The most important factors identified here were: gender, nationality, zoo/aquarium visiting frequency and pet ownership. Similarly to results reported by previous research [5,12,13], students' attitudes towards animals measured by both AIS and AAS were found to be relatively positive.

Female respondents scored higher on both AAS and AIS than their male counterparts. Girls showed more concern for animals specifically in categories where the welfare and life of the animal was compromised (e.g. 'killing animals', 'experimentation on animals', 'Harm animals for environment'). There were no differences between the two genders for items which involved animals being treated to improve their appearance or productivity ('changes in animals' genotypes' and 'animal integrity destruction'). The present results confirm findings of previous studies on gender differences identifying a prevalent female inclination for animal well-being and nurturing [3,6,10,14–17].

Results also showed that Belgian students scored significantly higher in the attitudes questionnaires for most items in contrast to Dutch students. Previous studies have revealed fairly similar attitudes from citizens in both countries, in contrast to the present study. In his study on the welfare of pets in commerce, Dewar found that both Dutch and Belgian respondents were in favour of improving animal welfare in their respective countries. Additionally, in a study investigating the use of animals in society, European students expressed more concern for animal well-being than

Asian students [12,14]. However, more closely aligned to the present study were results from Pifer, Kinya Shimizu and Pifer [15]. These authors found that Belgian respondents expressed more opposition towards animal research than their Dutch counterparts. The two countries do both in fact express concern for animal well-being; however Belgian respondents may be somewhat more passionate about these issues than Dutch students.

The present results also suggest a link between visits to the zoo/aquarium and positive attitudes towards animals. Young adults who reported that they never visited zoos or aquariums had lower AIS and AAS scores than the others. These findings support claims that zoos can fulfil more roles than mere entertainment, by encouraging learning experiences, a sense of connection towards wild animals, and focusing people's attention on conservation issues [18,19]. However, Tunnicliffe, et al. [20] warn that these connections can only lead to a better-educated public if zoos integrate the experience with follow up discussions and leave space for reflection. Special attention should also be paid to less popular animals (e.g. bats, spiders etc.) to increase awareness on the vital role of biodiversity.

Pet ownership was another significant correlate in determining students' attitudes. Analysis showed that students who owned a pet scored higher on the questionnaires and expressed greater concern for animal welfare than students without a pet. These outcomes are consistent with Prokop and Tunnicliffe [21] previous findings on the relation between pet ownership and positive attitudes towards other animals. However research in this area has not yet shown consistent results and theories oscillate between whether pet ownership has a relationship to young adults' attitudes towards animals [22]; for example, how such ownership might correlate with attitudes towards less popular animal species [23].

Regarding diet, significant differences were only observed in response scores for the AAS questionnaire. Those who ate meat once a week or less had higher scores than those who ate meat more frequently. The expression of higher concern for animal welfare from those who report to eat very little to no meat may be explained through the same line of thought found in Amato and Partridge [24] work on vegetarians. Here, the authors reported that a majority of vegetarians had made their dietary choice for ethical beliefs in animal rights. Another study on vegetarian girls revealed that most also made their choice on an ethical basis, and as an effort to reduce animal suffering [25]. Furthermore, these results can also be interpreted in a similar fashion to those of Hagelin, Carlsson and Hau [17] who report that concern for animals killed for food can also be extended to a concern for animal well-being in other domains such as animal research. Finally, a short comment must be made on the relatively small number of self-reported vegetarians in the present study. Most students ate meat at least once a week; however, in those responses a few had added that they wished to be vegetarian, but their parents wouldn't allow it. The results are therefore not entirely reflective of the dietary choices of all students.

Another important variable was household composition. Students living with a single parent demonstrated more concern for animal welfare in the questionnaire than those who lived with two parents. Perhaps these differences can be explained by Albert and Bulcroft [26] work on pet owners, who wrote that, "Attachment to pets is highest among never married, divorced, widowed, and remarried people, childless couples, newlyweds, and empty-nesters. Never married, divorced, and remarried people, and people without children present, are also most likely to anthropomorphize their pets." The young adults' single parents in the present study fall under the category of 'never married' or 'divorced'. As a parents' behaviours influence that of their child [27], it may be possible that these young adults adopted similar attitudes.

No differences were found between young adults who lived in urban areas and those who lived in rural areas, as has also been found in China [8]. This is despite observations that urban and rural citizens have different opportunities to interact with animals, as is reflected in the finding of greater knowledge about animals in rural residents compared to city dwellers [3]. Perhaps one reason that the present study found no differences between urban and rural residents is because the "urban" areas reflected in the present demographic were actually somewhat rural; small towns in close proximity to surrounding rural environments. In further support, neither the type of residence nor garden access correlated with attitudes in the present study.

Religion, more specifically Christianity, showed a weak relationship to young adults' attitudes towards animals but only for a few particular questionnaire items rather than to overall scores. Items on the AIS that asked students for how acceptable they found the killing of animals or the destruction of their integrity correlated with higher values reported for the importance of religion in a respondent's life.

Finally, age did not significantly correlate with attitudes. Because others have reported that significant changes in attitudes towards animals occur throughout childhood [3], this finding was unexpected. However, variability in age in the present sample study group was small, and this may be why no relationship was found between age and attitude ratings.

As shown in the present study, pet ownership is usually associated with positive attitudes towards other animals [21,22]. It is important to note however that pet ownership is not necessarily an end-all contributing factor to more positive attitudes. Although there is a relative correlation between pet ownership and more positive attitudes towards other animals, there is no guarantee that this attitude will extend to all animal species. The popularity, familiarity, biophilia (attraction), and the types of emotions that an animal species triggers can greatly influence the protection and welfare it receives from humans [28]. Likewise, Vining [29] stresses that emotion is at the heart of the actions or inactions of humans in terms of the respect and protection they provide animals. Furthermore, what arguably matters more is the quality of the relationship between young adults and their pets, or other animals in general. A study about animal abuse showed that fear of animals was a considerable determinant of negative attitudes (cruelty, apathy etc.) [6]. This again highlights the importance of engaging in meaningful connections with animals.

Another positive correlated factor to positive attitudes are visits to zoos or aquariums. Young adults gain knowledge and significant appreciation for the environment and its different species, when learning outside the classroom setting, in direct contact with nature and wildlife [30]). Informal educational settings such as zoos and aquariums should work to ensure exposure of their visitors to less popular animal species (e.g., pests, predators), in order to help students to understand the importance of each species in the ecosystem [21,28]. A commitment to education is a common element in the mission statements of contemporary zoos; such institutions can make substantive contributions towards improving public understanding of and appreciation for an animal's specific role in the ecosystem and thus enhance positive attitudes towards that animal [28].

Lastly, the present study found that those who reported eating meat less frequently (once a week or less) also had more positive attitudes towards animals and their welfare as measured in one of the scales used (the AAS). People who opt to eat little or no meat may do so for many different reasons, including reasons having to do with health, economics, and/or an interest in reducing the ecological impact of meat production, as well as for reasons that stem from a moral objection to consuming animals [31]. It should be noted that in the present study, the number of respondents stating that they ate meat only rarely was small (only about 5% of the total number of respondents); nonetheless the significant difference between this group and others in the study suggests that moral convictions that affect dietary choice may also correlate with moral convictions about the humane treatment of animals.

As this paper has shown, a variety of variables correlate with young people's attitudes towards animals and their welfare. A better understanding of the causes of these correlations and the development of these variables over the lifetime of a child may help us to better structure the kinds of experiences that promote empathy and concern for all living things.

Author Contributions: Conceptualization, P.M.; methodology, P.M., C.H., B.S.; formal analysis, C.H., B.S.; investigation, P.M., C.H.; writing, P.M., C.H., B.S.

Funding: This research received no external funding.

Acknowledgments: We thank the anonymous reviewers for their feedback and comments on this paper. Also many thanks to the schools and students that participated in this research.

Conflicts of Interest: The authors declare no conflict of interest.

References

1. Amiot, C.; Bastian, B.; Martens, P. People and companion animals: It takes two to tango. *BioScience* **2016**, *66*, 552–560. [CrossRef]
2. Leddon, E.; Waxman, S.R.; Medin, D.L.; Bang, M.; Washinawatok, K.; Hayes, M.B. One animal among many? Children's understanding of the relation between humans and nonhuman animals. *Psychol. Cult.* **2012**, 105–126.
3. Kellert, S.R. Attitudes toward animals: Age-related development among children. In *Advances in Animal Welfare Science 1984*; Springer: Berlin/Heidelberg, Germany, 1985; pp. 43–60.
4. Archer, J. Why do people love their pets? *Evol. Hum. Behav.* **1997**, *18*, 237–259. [CrossRef]
5. Eagles, P.F.; Muffitt, S. An analysis of children's attitudes toward animals. *J. Environ. Educ.* **1990**, *21*, 41–44. [CrossRef]
6. Pagani, C.; Robustelli, F.; Ascione, F.R. Italian youths' attitudes toward, and concern for, animals. *Anthrozoös* **2007**, *20*, 275–293. [CrossRef]
7. Driscoll, J.W. Attitudes Toward Animal Use. *Anthrozoös* **1992**, *5*, 32–39. [CrossRef]
8. Su, B.; Martens, P. Public attitudes toward animals and the influential factors in contemporary China. *Anim. Welf.* **2017**, *26*, 239–247. [CrossRef]
9. Su, B.; Martens, P. How Ethical Ideologies Relate to Public Attitudes toward Animals: The Dutch Case. *Anthrozoös* **2018**, *31*, 179–194. [CrossRef]
10. Herzog, H.A., Jr.; Betchart, N.S.; Pittman, R.B. Gender, sex role orientation, and attitudes toward animals. *Anthrozoös* **1991**, *4*, 184–191. [CrossRef]
11. Meng, J. Origins of Attitudes towards Animals. Ph.D. Thesis, University of Queensland, Brisbane, Australia, 2009. Unpublished.
12. Phillips, C.; Izmirli, S.; Aldavood, S.; Alonso, M.; Choe, B.; Hanlon, A.; Handziska, A.; Illmann, G.; Keeling, L.; Kennedy, M.; et al. Students' attitudes to animal welfare and rights in Europe and Asia. *Anim. Welf. UFAW J.* **2012**, *21*, 87. [CrossRef]
13. Kidd, A.H.; Kidd, R.M. Factors in children's attitudes toward pets. *Psychol. Rep.* **1990**, *66*, 775–786. [CrossRef] [PubMed]
14. Phillips, C.; McCulloch, S. Student attitudes on animal sentience and use of animals in society. *J. Biol. Educ.* **2005**, *40*, 17–24. [CrossRef]
15. Pifer, L.; Kinya, S.; Pifer, R. Public attitudes toward animal research: Some international comparisons. *Soc. Anim.* **1994**, *2*, 95–113. [CrossRef] [PubMed]
16. Amiot, C.E.; Bastian, B. Toward a psychology of human–animal relations. *Psychol. Bull.* **2015**, *141*, 6. [CrossRef] [PubMed]
17. Hagelin, J.; Carlsson, H.-E.; Hau, J. An overview of surveys on how people view animal experimentation: Some factors that may influence the outcome. *Public Underst. Sci.* **2003**, *12*, 67–81. [CrossRef]
18. Clayton, S.; Fraser, J.; Saunders, C.D. Zoo experiences: Conversations, connections, and concern for animals. *Zoo Biol. Monogr.* **2009**, *28*, 377–397. [CrossRef] [PubMed]
19. Conde, D.A.; Colchero, F.; Gusset, M.; Pearce-Kelly, P.; Byers, O.; Flesness, N.; Browne, R.K.; Jones, O.R. Zoos through the lens of the IUCN Red List: A global metapopulation approach to support conservation breeding programs. *PLoS ONE* **2013**, *8*, e80311. [CrossRef] [PubMed]
20. Tunnicliffe, S.D.; Lucas, A.M.; Osborne, J. School visits to zoos and museums: A missed educational opportunity? *Int. J. Sci. Educ.* **1997**, *19*, 1039–1056. [CrossRef]
21. Prokop, P.; Tunnicliffe, S.D. Effects of having pets at home on children's attitudes toward popular and unpopular animals. *Anthrozoös* **2010**, *23*, 21–35. [CrossRef]
22. Paul, E.; Serpell, A. Childhood pet keeping and humane attitudes in young adulthood. *Anim. Welf.* **1993**, *2*, 321–337.
23. Prokop, P.; Tunnicliffe, S.D. "Disgusting" Animals: Primary School Children's Attitudes and Myths of Bats and Spiders. *Eurasia J. Math. Sci. Technol. Educ.* **2008**, *4*, 87–97. [CrossRef]
24. Amato, P.; Partridge, S. *The New Vegetarians: Promoting Health and Protecting Life*; Springer: Berlin/Heidelberg, Germany, 1989.

25. Kenyon, P.; Barker, M. Attitudes towards meat-eating in vegetarian and non-vegetarian teenage girls in England—An ethnographic approach. *Appetite* **1998**, *30*, 185–198. [CrossRef] [PubMed]

26. Albert, A.; Bulcroft, K. Pets, families, and the life course. *J. Marriage Fam.* **1988**, 543–552. [CrossRef]

27. Maccoby, E.E. Parenting and its effects on children: On reading and misreading behavior genetics. *Annu. Rev. Psychol.* **2000**, *51*, 1–27. [CrossRef] [PubMed]

28. Martín-López, B.; Montes, C.; Benayas, J. The non-economic motives behind the willingness to pay for biodiversity conservation. *Biol. Conserv.* **2007**, *139*, 67–82. [CrossRef]

29. Vining, J. The connection to other animals and caring for nature. *Hum. Ecol. Rev.* **2003**, 87–99.

30. Lindemann-Matthies, P. 'Loveable' mammals and 'lifeless' plants: How children's interest in common local organisms can be enhanced through observation of nature. *Int. J. Sci. Educ.* **2005**, *27*, 655–677. [CrossRef]

31. Steinfeld, H.; Gerber, P.; Wassenaar, T.; Castel, V.; Rosales, M.; Rosales, M.; de Haan, C. *Livestock's Long Shadow: Environmental Issues and Options*; Food & Agriculture Organization: Roma, Italy, 2006.

Article

The Benefits of Improving Animal Welfare from the Perspective of Livestock Stakeholders across Asia

Michelle Sinclair *, Claire Fryer and Clive J. C. Phillips

Centre for Animal Welfare and Ethics, School of Veterinary Sciences, The University of Queensland, Gatton, QLD 4343, Australia; c.fryer1@uq.edu.au (C.F.); c.phillips@uq.edu.au (C.J.C.P.)
* Correspondence: m.sinclair6@uq.edu.au

Received: 7 February 2019; Accepted: 25 March 2019; Published: 28 March 2019

Simple Summary: A previous study into successful international animal welfare management strategy presented the vital need for animal welfare proponents to establish mutual benefits with the livestock industry. What the perceived benefits to addressing farm animal welfare are, is therefore important information not previously researched. This study asked leaders in the livestock industry in regions across six Asian countries what they saw as the key benefits for improving animal welfare, and which of those benefits they found the most compelling. The potentials to increase productivity of the animals and improve meat quality were among the most frequently cited and most highly rated across the countries. Important differences in the focus of other benefits existed by country, with food safety of highest importance in China and Vietnam, and people-focused benefits (such as human health and improved community livelihood) of greater importance in countries with higher rates of poverty such as India and Bangladesh. Animal-based reasons, such as improving animal welfare to the benefit of the animals themselves, were not compelling benefits in any of the investigated countries, other than India. The results of this study could assist in the development of improved animal welfare strategies.

Abstract: In this study, 17 focus group meetings were held with livestock industry leaders in geographically dispersed areas of China, Vietnam, Thailand, Malaysia, India and Bangladesh, regarding animal welfare issues, potential solutions and attitudes. Livestock leaders were asked 'what do you see as the benefits to improving animal welfare' and later to discuss the potential benefits and rank them according to their associated importance. While differences existed by country, the most important perceived benefit area across all countries was financial in nature, primarily focussed on the potential to increase the productive output of the animals and to improve meat and product quality. However, doubt existed around the ability to increase profit against the cost of improving animal welfare, particularly in China. Human health benefits and the tie to human welfare and community livelihood were considered most important in India and Bangladesh, and animal-focussed benefits were not significant in any countries, except India and, to a lesser extent, Bangladesh. Thus, improving animal welfare for the sake of the animals is unlikely to be a compelling argument. The results presented here can be used to create meaningful mutual ground between those that advocate improvement of animal welfare and the stakeholders that have the ability to implement it, i.e., the livestock industry.

Keywords: animal welfare; benefit; profit; human health; Asia; livestock

---

1. Introduction

Farm animal production is arguably the most economically important interface between humans and other animals on this planet. It has the potential to cause suffering in large numbers of animals, over prolonged periods of time, ending in a death that has the conceivable ability to epitomise that suffering. As an industry that has systematically experienced rapid growth and intensification in most regions of the world, methods employed during farming, transport and slaughter are frequently the focus of public concern and advocacy lobby. Ethical arguments exist for addressing farm animal welfare, including it being 'the right thing to do' for the animals themselves; however, these arguments may not be compelling to all important parties engaged in this sector. Depending on the area of the world, other influential factors make more compelling motivators [1], and the literature on this point paints a much more complex landscape than one of basic ethical value, particularly in emerging countries [2,3]. Animal agriculture is not simply a theoretical interface between humans and other species, it is an economic endeavour; it functions foremost as a business, and the stakeholders in the position to have the most power over the welfare of the animals in the sector are those working within the livestock industry.

Key tenets of successful international animal welfare initiatives have been outlined in a recent study, which emphasized the importance of engaging with the industry and establishing mutual benefits as a basis for collaboration [4]. According to the literature, apart from the obvious benefits to the animals themselves, the espoused benefits for improving animal welfare vary. One study focuses on the fact that economic benefits have been historically omitted from consideration and that economists, amongst others, should play an important role when developing animal welfare initiatives [5].

From an economic perspective, improving product quality and reducing animal losses are the potential benefits to improving animal welfare that are found in the scientific animal welfare literature most frequently. Mitigating losses through reduced mortality [6], reducing damage to carcasses through reducing bruising, injuries and the incidence of pale soft meat (PSE) in pigs [7] and dark cutting (dark firm and dry) [8] in beef cattle—both signs of significant stresses caused to the animal before its death—are cited as key benefits of addressing animal welfare concerns [9]. Some studies also cite improved productivity of animals [10], as well as improved reproduction and thrift in livestock [11]. Apart from the product-based economic benefits, there are the strategic business benefits. One benefit that does not appear to have been contested is that improving welfare offers commercial opportunities to market products as being from higher welfare systems, with some studies showing that consumers are willing to pay more (however, not vast amounts more) to purchase meat that makes them feel better about the life the animal had [12]. This, however, is based on studies mostly conducted in western nations, and this may vary in less developed countries where consumers need to buy food as cheaply as possible and do not have the luxury of being discerning. Having noted that, this could still remain a relevant benefit for enterprises in developing countries that are seeking to export, or continue exporting, animal products to western nations.

A major benefit identified in the scientific literature centers on the notion that the public, as evident in many parts of the world, is demanding better treatment of animals [13]. Improving animal welfare offers the business benefit of mitigation of risk to the brand through bad publicity, loss of purchase partnerships and even the jeopardy of a whole industry. In some parts of the world, this is a concern, fuelled by advocacy lobby efforts that have seen reformation of farming practices, such as the abolition of veal crates, cage eggs and sow stalls in the European Union [14], and it has likewise caused periodic market collapses, such as that experienced by the live export industry in Australia [15]. In addition to avoiding poor publicity, improving animal welfare also provides positive marketing, which has the opportunity to improve the public's perception of the livestock industries as a whole [13]. Finally, in terms of business benefits, the scientific literature identifies employment benefits. By going through 10 years of industry data, it was discovered that improved animal welfare makes the animals safer and easier to handle, which results in a need for fewer staff, who are more satisfied, likely to have substantially less time off and have less medical expenses [9,16].

Apart from the benefits received by the animals directly from improving their animal welfare and the business and economic benefits, some wider community-based social benefits are also reported in the scientific literature [17,18]. This includes mitigating environmental despoliation and mitigating the non-therapeutic use of antibiotics driving the emerging anti-microbial resistance crisis. A close connection between animal and human welfare has been advanced under the umbrella of the One Welfare concept [19]. As a demonstration of that, attention has been drawn to the risks to human health of operating in environments that are poor for animal welfare, including the incubation of pathogens found in high-confinement situations, along with respiratory problems and low-level antibiotic resistance [20]. One study has even found that where animal welfare has become a priority, it has contributed to positive competition within communities to improve the health and strength of the animals they care for [21].

Despite the potential benefits of improving the animal welfare of farmed animals, introducing these changes can be expensive, and some of the scientific literature is cautious of overstating the economic benefits awarded [9]. This is particularly true when it comes to space allowances and stocking density, where profits may be increased by maintaining more animals to a smaller space, however detrimental this is to the welfare of the individual animal and even if it requires animals to be pushed beyond their biological limits through selective breeding or husbandry practices such as introducing medication and chemical supplementation [9].

It has been argued that, while animal welfare improvements are often perceived to conflict with economic gain, which causes hesitation within the industry, modelling financial benefits may provide compelling motivation to overcome that perception [6]. Considering the costs of implementing higher welfare systems, cost benefit analyses have found that the total income potential was still increased [12,22,23].

Because most of the literature is western-based, it is not clear whether the livestock stakeholders' perception of benefits vary in other parts of the world, in particular in developing countries. Asia is home to the biggest livestock-producing country in the world, the People's Republic of China (henceforth referred to as China) [24], and no animal welfare legislation exists. Globalisation of society requires that we assess stakeholder perceptions and understand their priorities in major livestock-producing nations in order to provide incentives that make solutions realistically attainable [5]. For those involved in governance, domestic enterprise, export/import business enterprise, or animal welfare advocacy, understanding the potential benefits of addressing animal welfare, as perceived (or not perceived) by livestock stakeholders, provides an important step in identifying mutual benefits to create partnerships, improve initiatives and/or enact policy reform across borders. This study begins addressing this gap by reporting the outcomes of a series of focus groups held across six culturally diverse countries in Asia that addressed the issue of what benefits might derive from improvements in animal welfare.

2. Methods and Materials

To gather data for this project, 17 focus groups were held in geographically dispersed locations across Vietnam, Malaysia, Thailand, China, India and Bangladesh (see Table 1). Locations were chosen in different areas of each country (i.e., south, north, central, capital and regional) in an effort to capture potential varied sentiments between domestic regions. Industry leaders were invited as representatives for the livestock industry to discuss the state of animal welfare in their country, in the context of major welfare issues, challenges, solutions, opportunities and perceived benefits to improving animal welfare, with the benefits being the focus of this paper.

Table 1. Location of focus groups and abbreviation codes used in quote citations.

Country	Abbreviated Code	City/Town	Participant N	
China	CH	Guangzhou	7	23
		Zhengzhou	7	
		Beijing	9	
Vietnam	VN	Hanoi	7	20
		Ban Me Thout	5	
		Ho Chi Minh City	8	
Thailand	TL	Bangkok	10	19
		Khon Kaen	3	
		Chiang Mai	6	
Malaysia	MAL	Negeri Sembilan	6	19
		Kuala Lumpur Selangor	13	
India	IN	Banglaore	6	15
		Kolkata	5	
		Trivandrum	4	
Bangladesh	BA	Dhaka	13	43
		Savar	13	
		Mymensingh	17	

Participants were invited through country-based collaborators based on the following selection criteria: that they were leaders in the animal production sector, working for an organisation with a maximum of two government vets and with the ability to implement change into private businesses (see Table 2). The majority were private industry leaders (e.g., pig or poultry slaughterhouse, or production managers or owners). In some groups, some participants were known as professional colleagues. Although plans were made for 5 to 7 participants in each session, the actual number of participants present on the day varied from 3 to 14, with some participants cancelling and others indicating their desire to attend just before the event.

Table 2. Breakdown of stakeholder participant roles within the livestock industry, by country.

Country	Stakeholder Role			
	Private Industry Leaders	Private Industry Veterinarians	Government Representatives	Agricultural Academics
China	15	0	1	9
Vietnam	4	3	13	1
Thailand	11	4	2	2
Malaysia	9	5	5	1
India	3	5	1	6
Bangladesh	4	2	17	21

In this instance, focus groups were selected as the method of data collection in preference of surveys due to their scope to collect broader qualitative data, which, once identified, can then be measured quantitatively in future studies. Likewise, focus groups were chosen in preference of individual interviews to enable the collection of data from a wider sample size, to encourage cross-participant discussion that may lead to more in-depth and honest data collection and to allow for frequent consensus checks where the facilitator can ascertain if sentiments are shared, or contested, by the participant group. Within the focus groups, participants were asked an open-ended question at the onset of the focus group by the lead researcher (MS), i.e., 'what do you see as the benefits to addressing animal welfare'. Their collective round table responses were recorded and presented back to the participants later during an activity that invited a group discussion on ranking the benefits from most important to least important, as a group. Participants that felt strongly towards certain benefits chose to advocate for a higher ranking of those benefits, in discussion form, and the groups ultimately voted democratically by raising their hands to vote for the 'most important benefit' from top to bottom.

The discussions surrounding the benefits of improving animal welfare and the final rankings delivered through the activity were documented and form the basis of this study.

The remainder of focus group content considered specific animal welfare issues, solutions, meanings and motivations, which will be presented in separate manuscripts.

All contributions were voice-recorded during the sessions, and additional field notes were taken by a research assistant (CF). Both data sets were used to create abridged transcripts of each session. As participation was subject to translation by a third-party translator and then presented in English to the researchers, word for word transcripts were not possible.

The average time of sessions was 3.5 h, with an average of eight participants. Sessions with higher numbers of participants often ran approximately 30–45 min past the scheduled 3.5 h to offer all participants adequate opportunity to contribute. Transcripts were uploaded into NVivo software for Mac 11.4.3 for analysis.

Analysis

Benefits were identified and coded as a primary node/theme in Nvivo. The benefits were then classified into broader categories depending on who or what they fundamentally benefitted (human benefit, business benefit, animal benefit or community benefits). Data were then divided into relevant logical sub-themes, where present, identified by careful inspection and familiarization of the data, and within each benefit, key quotes that demonstrate the sentiments towards that benefit were manually selected for inclusion in this paper. At the completion of analysis and coding of themes and sub-themes, no new benefits emerged from the data, suggesting data saturation. The same lead researcher (MS) that conducted the focus groups also coded all themes/nodes and conducted the analysis.

To avoid presenting misleading data, linguistics and tone are not reported, as all data were translated, abbreviated, and summarised through six translators, from six different languages to English. For this reason, rather than focussing on word usage, more attention was paid to the careful analysis on the key themes (benefits), the frequency of their appearance across countries, the general context and meanings that were applied to them by the participants and how they relate to one another. However, word frequency functions in NVivo were utilised in the identification of sub themes and were reported infrequently in the results. Direct quotes are presented in the results according to the country in which they were collected, with the abbreviated codes presented in Table 1.

This study was granted human ethics approval by the University of Queensland Ethics Committee, approval number: 2017000628.

3. Results

In general, stakeholder leaders were positive about animal welfare and forthcoming with potential benefits; however, it is important to note that, in some regions, the existence of the reported benefits was met with scepticism. In some groups (listed individually below), while benefits were raised as worthwhile, some participants were dubious about the ability to obtain the benefits by addressing animal welfare. This was particularly the case for benefits tied to economics and productivity. In some sessions, participants struggled to identify benefits to addressing animal welfare at all. On one rare occasion, addressing animal welfare was openly associated with liability and cost, rather than benefit.

In Zhengzhou (CH), participants were not confident in listing any benefits, with one participant stating that 'why some people don't improve welfare is because of the limitation of economic (factors), not because of their consciousness'.

It is impossible in this experimental setting to quantify how many individuals had this sentiment of doubt, as constructs such as conformity and groupthink play a role in the data that are shared; however, this sentiment could be valuably followed up with an individual-level study. For that reason, the following report represents general sentiments of scepticism only and is not a reflection of opinions shared across the country (or even across the entire group in some instances).

In Chiang Mai (TL), one participant raised the point that they are doing well with animal welfare in their business; however, in the end, the price they receive for their product is the same as those who have practices that are bad for welfare. 'It's not very fair because when you do (improve) animal welfare you have to put in more effort, but you gain back just about the same' <TL>. While some agreed with this comment, another participant stated that 'if we take care of the animals well, you don't need to spend much on the medical expense and so on, and then the costs will be reduced' <TL>. Similar sentiments were also expressed regarding the existence of financial benefit in Guangzhou (China), with one participant stating 'we want to know if there's any specific data to prove a positive connection between animal welfare and economic benefits to company ... if we have such data (it) will become much easier to promote (the) concept' <CH>. On the other hand, in response, it was also stated that 'if we don't have an improvement in production rate or, even worse, production potential but we do have better flavour or meat quality, that is also acceptable' <CH>.

These sentiments were also present in Negeri Sembilan (Malaysia), where a 'conflict between making money and (animal) welfare' was expressed. 'They know if they take good care it will benefit them, but in terms of making profit with limited space and budget, the issue is that it is hard to make a profit' <MAL>. 'It needs a lot of investment not only in new facilities but also to improve old facilities to use better technology and housing. For us businessmen to improve animal welfare, or make any changes, we need money ... at the same time when we improve animal welfare, we want to improve the output. Businessmen will rarely see the benefit' <MAL>. By way of further example, one farm manager stated that '(if) the handling of chicken during harvest is more gentle, we can expect reduced processing time', then added 'damage (can) be reduced, that's the benefit, but stakeholders might not understand this clearly yet ... benefits are not clearly understood' <MAL>.

Lastly, it was also acknowledged by one participant in Malaysia that the benefits that had been raised by the group along with their ranking would be 'seen differently by NGOs, that it would be upside down', and then commented (with agreement from the group) that NGOs don't seem to understand the livestock industry, due to different goals and priorities. 'The industry wants to make a profit, but the NGOs don't, this issue will be questioned again and again' <MAL>.

4. Nature of Benefit

In terms of species context, leaders in China and Vietnam mostly gave examples relating to poultry and pig production, in Malaysia and Thailand, the focus was primarily on poultry production, in Bangladesh, on cattle, goat and sheep, and in India, comments were less species-specific, with only rare comments regarding cattle (likely due to the beliefs of the Hindu population) and pigs (likely due to the beliefs of the Muslim community responsible for slaughter in India).

Table 3 quantifies and outlines all benefits identified by the livestock leaders, according to country and region, and the subsections below aim to provide further contextual information by presenting illustrative quotes around these benefits. Table 4 presents the outcome of an activity within the focus groups where leaders were asked to rank the benefits they presented at the onset of the session, in order of importance. Figure 1 presents the most frequently identified benefits by respondents in each country, while Figure 2 presents benefit categories, and providing country comparisons. Within Figure 2, all benefits were placed into categories based on their intended beneficiary (i.e., business, human, community, animal). Where the suitable category for a benefit was not clear, confirmation was sought from the original data.

Animals **2019**, *9*, 123

Table 3. Benefits identified by participants in each region, in each country, presented in order from the most frequently identified benefit to the least cited (top to bottom).

	China			Vietnam			Thailand			Malaysia		India			Bangladesh		
	Beijing	Guangzhou	Zhengzhou	Hanoi	Ho Chi Minh City	Ban Me Thout	Bang-kok	Chiang Mai	Khon Kaen	Kuala Lumpur	Negeri Sembil-an	Kolkata	Bangal-ore	Trivan-drum	Dhaka	Myme-nsingh	Savar
Productivity of animals	X	X	X	X	X	X		X	X	X	X	X	X			X	X
Improve quality of meat or animal product	X	X	X	X	X	X	X	X	X	X	X	X	X	X		X	X
Reduce disease and injury and treatment costs		X		X	X		X	X						X		X	
Increased revenue/profit		X									X			X		X	
Avoid cruelty and reduce animal suffering	X	X	X		X	X		X	X	X	X	X	X	X	X	X	X
Human health/zoonosis						X		X		X		X	X			X	
Protection of natural resources/ecosystem development	X		X			X				X		X			X		
Food safety/biosecurity	X	X				X		X		X				X	X	X	
International trade opportunities		X															X
Stronger/healthier animals	X				X				X							X	
People feel better for the animals	X	X	X					X	X			X	X	X			X
Improve human/animal relationship			X	X	X				X			X	X	X	X		
Addressing the animals' rights/sanctity of life	X												X	X			
Improved community livelihood			X									X	X	X			
Public concern/consumer confidence						X	X										
Relationship between way humans and animals are treated, tie to human welfare	X			X	X	X											
Improved taste of animal product		X	X								X						
International recognition (not being left behind)	X	X									X						
Allowing natural behaviour of animals							X									X	X
Compliant with international regulation					X												
Human responsibility to give a good life					X		X	X						X	X		
Lower mortality							X	X									
Ease of handling calmer animals																	
Improved commercial promotion	X																

Note: 'X' signifies the presence of the theme in the focus group session in that region.

Key:

Profit- or business-driven benefits
Animal-focussed benefits
Human-focussed benefits
Bigger picture community-focussed benefits

Table 4. Ranking of importance of benefits, by country.

Rank	China	Vietnam	Thailand	Malaysia	India	Bangladesh
1	Improve quality of meat or animal product	Improve quality of meat or animal product	Improve quality of meat or animal product	Productivity of animals	Avoid cruelty and reduce animal suffering	Productivity of animals
2	Stronger/healthier animals	Productivity of animals	Reduce disease and injury and treatment costs	Increased revenue/profit	Improved community livelihood	Reduce disease and injury and treatment costs
3	Protection of natural resources/sustainable development	Reduce disease and injury and treatment costs	Stronger/healthier animals	Food safety/biosecurity	Human health/zoonosis	Food safety/biosecurity
4	Productivity of animals	Human health/zoonosis	Human responsibility to give a good life	Improve quality of meat or animal product	Reduce disease and injury and treatment costs	Increased revenue/profit
5	People feel better for the animals	Improve human/animal relationship	Increased revenue/profit	Reduce disease and injury and treatment costs	Addressing the animals' rights/sanctity of life	Stronger/healthier animals
6	Public concern/consumer confidence	Relationship between way humans and animals are treated, tie to human welfare	Avoid cruelty and reduce animal suffering	Human health/zoonosis	Productivity of animals	Avoid cruelty and reduce animal suffering
7	Increased revenue/profit	Protection of natural resources/sustainable development	Food safety/biosecurity	Avoid cruelty and reduce animal suffering	People feel better for the animals	Addressing the animals' rights/sanctity of life
8	Improved taste of animal product	Food safety/biosecurity	Ease of handling calmer animals	Improved taste of animal product	Increased revenue/profit	Improve quality of meat or animal product
9	Avoid cruelty and reduce animal suffering	Increased revenue/profit	International trade opportunities	Protection of natural resources/sustainable development	Food safety/biosecurity	Protection of natural resources/ecosystem development
10	Reduce disease and injury and treatment costs	Avoid cruelty and reduce animal suffering	People feel better for the animals	International recognition (not being left behind)	Improve human/animal relationship	Allowing natural behaviour of animals
11	International trade opportunities	International trade opportunities	Public concern/consumer confidence	Allowing natural behaviour of animals	Human responsibility to give a good life	Compliant with international regulation
12	Improved commercial promotion	People feel better for the animals	Productivity of animals		Improve quality of meat or animal product	Human health/zoonosis
13	International recognition (not being left behind)	Compliant with international regulation	Lower mortality		Protection of natural resources/ecosystem development	Improve human/animal relationship
14		Lower mortality				International trade opportunities

Notes: Shaded cells indicate that the benefits were ranked equally.

Key benefits by country

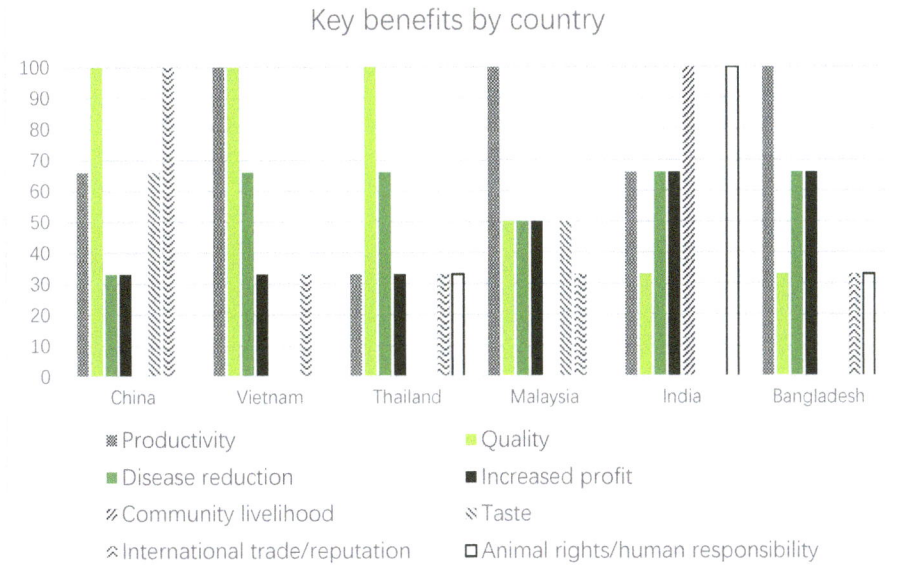

Figure 1. Comparison within countries regarding the appearance of certain perceived benefits for addressing animal welfare. Note: the values represent the % of focus groups that indicated the selected benefit within that country.

Country comparison of benefit categories

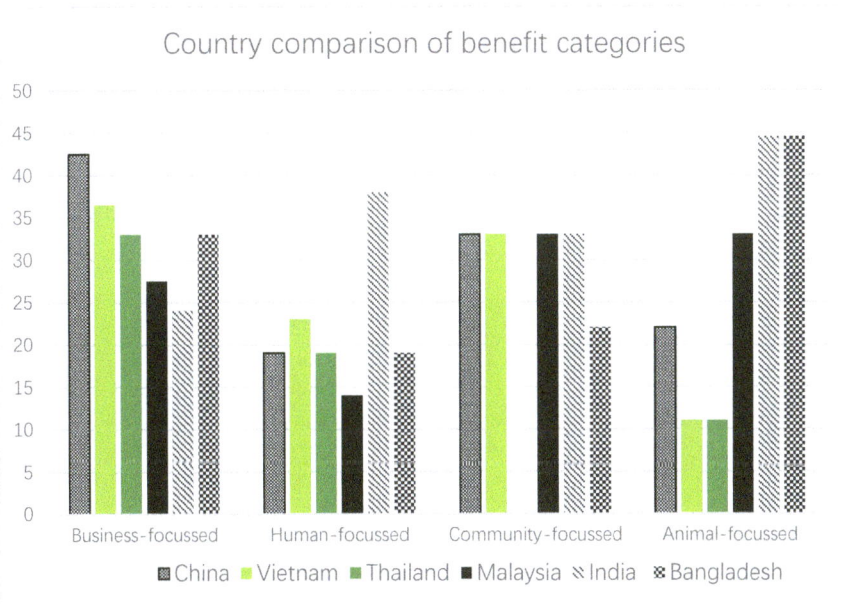

Figure 2. Comparison between countries of benefit categories, based on frequency of appearance. Note: Amount of times a benefit falling under this category appeared, presented in a percentage according to the amount of opportunities to raise it as a benefit per country. Individual benefits associated to categories as per the colour key in Table 1.

4.1. Financial Benefits: Improved Animal Productivity

'For businessmen, most of us we see that to improve animal welfare, or any changes, we need money' <MAL>. 'I think man is the most selfish creature, so revenue profit is number 1' <IN>.

Although some aforementioned doubt existed as to the actual financial benefits (particularly when considering increased profit or return from investment in higher welfare systems), leaders were, however, mostly positive towards the existence of potential financial benefits. This was particularly the case with reducing economic losses (reduction of treatments and antibiotic usage) and with animal-based profit measures such as increasing the productivity of the animals themselves and improving the quality of the meat/animal product.

Improved animal productivity along with improved meat quality were the most important benefits identified. In regard to productivity, leaders made statements such as 'when body condition is good, production is also high, so profit and productivity is increased <BA>'. '(I am) a farmer, (I) produce chicken and laying hens and observe (that) if chickens are given a good climate, good environment, ventilation, and space it improves (their) productivity' <VN>. '(The birds) need to perform optimally in terms of productivity, that is why we make sure they are not too hot . . . happy birds make more money, that's what we understand about welfare' <MAL>. 'When the pigs are very depressed or under stress they will grow slowly' <CH>. After describing the situation of most stockholders in Bangladesh, in that animals frequently share houses with families, one participant stated (with general agreement amongst the group), that where the animals are given love and affection, when they come when their name is called and when psychological welfare is high 'in that environment, the meat production is very high . . . the reproduction and meat production is great . . . it is the most important thing' <BA>.

4.2. Financial Benefits: Improved Meat and Product Quality

Along with improved productivity, meat quality was the most important benefit described. 'I am a farmer of pork . . . feed, housing, water quality and slaughter . . . all these things improve meat quality, economic efficiency and value' <VN>. 'All species can be eaten, and most people treat them inhumanely especially in slaughterhouse, and to improve animal welfare can improve meat quality' <VN>. 'At slaughterhouse, (I) observed and realise if you improve handling, with stunning, it improves meat quality' <VN>.

In Thailand; 'I come from a slaughterhouse and I think that if they have good animal welfare, it reduces defects, the animals are more convenient and easier to handle, and it's a good product' <TL>. 'The benefit for animal welfare is you will get a good quality product, reduced PSE, and also if you give (the animal) good welfare you will get a good yield' <TL>. 'To improve animal welfare means we will get good quality of products and customer will be happy with that' <TL>. In China; 'with the development of ecological agriculture, the importance of animal welfare has been emerging, so we need to improve both the management and give (the) animals some humane treatment to improve the quality of the products; that way we make the whole chain happy' <CH>. 'In (the) slaughterhouse, the brokers who buy and transports pigs have realised they need to rest the pigs for some days before slaughter, and the meat quality will become much better and taste good' <CH>. 'If we can have better animal welfare for the chicken, we can also have better benefits for our economy and our livelihood <CH>'.

4.3. Financial Benefits: Risk Avoidance And Business Loss Mitigation and Opportunity

In addition to the animal and product-based financial benefits, leaders also shared financial benefits in the form of risk and loss mitigation, through reducing the costs of medicine and treatments and lowering mortality. 'Animal welfare requirements in standards should be satisfied from farm to slaughterhouse, even in the lairage and on the truck, feed, water, handling, all steps . . . (this will result in the) best quality and also improvement in health, it will reduce economic loss' <VN>. 'Improving animal welfare will improve animal health, less disease, lower mortality and improve growth rate (to) improve economic efficiency' <VN>.

The business risk for the domestic brand and product sales of not addressing animal welfare was also raised. 'In recent years, the people in China value animal welfare more, for example, during slaughter, they (try to) use the knife to bleed quickly and they use stunning first to lower stress before slaughter' <CH>. 'You can see some dogs or cats are abused by people (online); someone will put the videos up, and people involved will be cursed by the public' <CH>. Likewise, the risk of losing international export clientele was also raised; '(improving animal welfare offers) benefits to the business owner in terms of if they supply products for export, the customer is concerned about animal welfare, and, at same time, if they provide good animal welfare our products will be good quality' <TL>.

Other than the benefit of risk mitigation, maintenance of current markets (domestic and international) and reducing costs through treatments and stock loss, 5/17 of the groups suggested that improving animal welfare standards would open up new markets, particularly those in export 'in the time of globalisation and industrialisation, improved animal welfare gives better opportunity to export products' <VN>.

Finally, in regard to financial benefits, one leader briefly touched on the possibility of increased product promotion ability in China, and another on the procedural benefit of handling calmer, less stressed animals in Thailand.

4.4. Human Benefits: Physical Health

In most countries, the benefits were almost entirely business- and financially focussed, except in India and to some extent Bangladesh, where more emphasis was given to protecting human health, feeding the community and community livelihood. This tied into the perceived benefit of animal welfare in the form of food safety and biosecurity. '(Regarding) food animals, I think if you can ensure animal welfare, the food will be safe' <BA>.

The ties between animal welfare and human welfare, in the shape of the One Welfare initiative, was well perceived in India, with prevalent comments such as 'animals and human welfare is the same' <IN>. 'Indian people live so close to animals so there's a lot of mixing . . . not like in Western countries where animals are not living in close proximity . . . even when we go to work we meet so many animals on the road including dogs, cats, buffaloes, so there is a lot of interaction between humans and animals. Because of that, there is a lot of linking between human and AW' <IN>. 'If animals are healthy and happy, humans will benefit . . . promoting animal welfare means humans also gain welfare' <IN>. 'Human and animal welfare is tied . . . . (for example), rabies is transferred from animal to human . . . but still street dogs are not vaccinated' <IN>. 'One health; the health of animals and health of humans are interlinked . . . we need to improve animal health to improve human health as many diseases are zoonotic' <IN>. 'Also, saving money from (human) diseases . . . improving (the) health of animals reduces the cost of treating ourselves' <IN>.

4.5. Human-Based Benefits: Psychological Wellbeing

In relation to benefits for human health, the benefits were not restricted to physical health benefits but also included benefits for human mental health. Unlike the emphasis on physical health, the inclusion of mental health benefits was not restricted to India (where it took the form of satisfying religious duty to the animals), rather, some version of mental health benefits appeared in every country. The first seemed to be the satisfaction of empathy and vindication from a perceived guilt for involvement in killing animals for both the consumers and the livestock workers themselves. 'I am involved in animal production and I think in terms of the consumer, that if the production section is managed with good welfare then the consumer will be happy and feel good that product has come from good management, the customer will feel good about the product; hence, we can eat animals don't feel bad' <TL>. 'When we do the production line and then kill the animals you feel bad about that, but at the same time that kind of animal is food for people; so we should be kind to the animals to take care of them well before they become our food' <TL>. 'If you look after them well, the benefit you get is good animals with less disease, increased efficiency and productivity . . . also, workers, livestock men, feel good doing that' <TL>. 'I have a small farm with free-range chickens . . . what I see is that the taste is very good . . . I enjoy watching the chickens get to live as chickens, run around flapping wings and fighting' <MAL>. 'I work in a layer company . . . most direct benefit for me personally is good feeling when I see my layers well taken care of' <CH>.

The link between poor treatment of animals and poor treatment of other people was also recognised in this context; 'when you provide humane treatment to animals, there is a relationship with providing humane treatment to humans' <VN>.

4.6. Human-Based Benefits: Community Livelihood

How the health and wellbeing of the animals directly affects community livelihood in India and Bangladesh were benefits that were emphasised frequently in those countries, although not in any others. 'Through history, the animal has assisted human population for livelihood development' <IN>'. 'In our country, we use animals for working, if welfare is ok, this will help us use animal more for ploughing' <BA>. 'Unlike in other countries, you can't see animals as a separate entity, animals form an important and integral part of livelihood . . . If you want the benefits of improving the welfare, the prime thing is if welfare of animal is improved automatically human beings are improved' <IN>. 'All human beings only survive because of their livelihood, they're dependent on this animal . . . if the welfare is ensured that will ensure the people's livelihood' <IN>. The beliefs of Hinduism were also tied to community livelihood in some instances. 'Indians are a people who worship animals . . . so taking care of their welfare is equal to taking care of God, but most people are not aware of that . . . so you can direct things into that angle' <IN>. 'You cannot see the issue as two sides—cannot see animal and man as separate . . . consider gods of Hindus, every god is related to some animal, and actually we have a great culture of worshipping these animals; welfare is taking care of needs, people will see that as caring for gods' <IN>.

4.7. Societally Based Benefits

To a lesser extent, benefits that appear to be societally focussed, bigger picture and holistic were also presented by the leaders. In some instances, particularly in China and Vietnam under the local concept of 'ecological agriculture', animal welfare fitted into the objectives of protecting the environment in general. This may potentially suggest that animals are seen as a part of, rather than separate from, the natural environment, from which humans are still separate. 'if we ensure their rights, they are part of the environment, so if we improve their lifestyle and provide basic needs, they will help sustain our environment' <VN>.

The importance of keeping up with international progress was also raised, with improving international standing as the perceived benefit to improving animal welfare. This was specifically pertinent to government representatives in the groups. 'From my department's point of view, we need to achieve international recognition . . . without this, we will be left behind . . . this was evident when we had an issue regarding slaughter of animals through ESCAS (Exporter Supply Chain Assurance System) and we had to develop a whole protocol' <ML>. Lastly, 5/17 of the groups raised the benefit of improving the human/animal relationship in general. 'There are no definite rules and regulations in our country, no compliance with international standard . . . without any know-how, some people relate animal welfare to their affection, to love animals' <BA>.

4.8. Animal-Based Benefits

Leaders spent the least amount of time discussing improving animal welfare for the benefit of the animals themselves; however, this was more discussed in India. Where it was discussed, it was in the context of a desire to avoid cruelty and reduce suffering. 'We want to improve welfare for the sake of the animals' <IN>. Other than the context of animal welfare for the animals, the other context was that of animal rights. 'Every living being has its basic rights to ensure his life is lived in proper way, and we have to ensure all basic needs and reduce the physical and psychological suffering' <BA>.

5. Discussion

According to this study, the most important perceived benefit for improving animal welfare amongst livestock stakeholders is financial, primarily through increased productivity and yield from the animals in question, and improved quality of the end product (including taste in China, and not elsewhere). Other business benefits that directly or indirectly impact on the profit were ranked with high importance, including meeting customer demands and expectations, particularly with export customers, creating new markets through offering higher welfare products (again, particularly for export markets) and reducing expenses, such as treatment for disease and injury, and stock losses. Throughout the study, leaders presented benefits as if they 'could' be important benefits for them. So, although they were raised as important, the benefits were not necessarily without skepticism that they were necessarily achievable. Doubt around the actual existence of the perceived benefits was entirely in regard to one benefit category, increased profit, and it was particularly present in China. This suggests a need to conduct economic evaluations of financial gains that may be possible with improving individual aspects of animal welfare, ensuring reliable information is available to leaders within the livestock industry. Some well-cited studies outlined the relationship between animal welfare and economic productivity and found the attitudes of the public are intrinsically tied to this relationship [25,26] and necessarily offset by cost [12].

Similar work has been conducted in the field of environmental conservation and protection, with studies highlighting that financial benefits do exist and must be isolated and understood [27–29].

Along with environmental protection, animal protection has been hypothesized to exist in a 'nature trifecta' of importance to the general public [30], making it a social issue that is highly valued across borders. Examples of attitudes to animals and their perceived welfare impacting profits can be seen in select case studies. One example of this, specifically on mitigating losses rather than increasing profit, is seen with the live export industry in Australia. Media exposés highlighting animals in conditions that were distasteful to the public resulted in lobbying and temporary shut-down of the Australian live export industry in its entirety, equating to reported agricultural losses in excess of the millions [31]. On the same animal welfare issue, economic modelling found that more profit could be accessed by processing the animals in Australia, rather than sending them overseas, a solution to animal welfare concerns and an opportunity for eventual increased profit [32]. In another study, the transition from battery cages for layers hens to alternatives that increase the welfare of the birds were assessed to be economically favorable in conditions that need to be carefully measured and implemented [33]. In another case, economic modelling of the relationship between milk production and dairy cow welfare found that a herd of 100 head could increase profit margins by £10,000 (over $13,000 at the time of publishing) by implementing attainable welfare-related target rates, which is likely to have increased at the present day [34]. Likewise, profit was again related to the welfare of the dairy heifer in financial models that measured the cost of production diseases [35]. Despite the literature that argues that financial benefits exist in addressing animal welfare, making changes requires financial outlay. It is also important to note here that this paper is not affirming that financial benefits are present in all animal welfare improvement, but that, in line with the data collected, where economic modelling can be completed and financial benefit demonstrated, it could provide a largely compelling benefit that is likely to result in increased motivation to address the animal welfare change modelled.

While financial benefits were raised as important in all countries, it was particularly the case in China and South East Asia (Malaysia, Vietnam and Thailand). All of these countries, except Malaysia, have large agriculture industries that are exporting internationally and seek to increase their export markets. However, the focus of benefits changes when looking at India and, to a lesser extent, Bangladesh, where human-focussed benefits are prioritised. Benefits such as improved human physical health through reduction of zoonotic disease risks and the relationship between animal welfare and the livelihood of the community may be the result of a culture that lives in close proximity to farm animals. As a direct result of their reverence as holy animals and their legislative protection from slaughter, cows join the ranks of commonly straying animals alongside cats and dogs in India, and interaction is frequent [36].

Likewise, in Bangladesh, a majority of farming is by subsistence farming, where it is commonplace that animals may be sharing a domestic environment and sometimes a home with their carers [37]. Cognitive dissonance theory refers to the pressure felt to convince oneself that immoral activities are in fact moral, to avoid uncomfortable inconsistency between attitudes and behavior [38,39]. This often unconscious human practice frequently relies on avoidance or disconnect of information or situations that result in this feeling of uncomfortable inconsistency [39]. Therefore, in the situations where avoidance or disconnection from farm-based animals is less possible, given the proximity between people and animals, cognitive dissonance from any suffering the animal may be presented with may become more difficult, and vicarious suffering may be increased. This may further explain the higher associated importance of the human–animal relationship in Bangladesh and India and the importance of benefits to the animals themselves in India.

This is also consistent with an increased concern for the 'psychological' wellbeing of humans and the perceived benefit that improving animal welfare will 'make humans feel better' in these countries, as shown by statements in this region such as 'happy animals happy people', and 'if animals are healthy and happy, humans will benefit, promoting animal welfare means humans also have better welfare'. In Bangladesh, a country where cattle and buffalo are still used for work by small-farm holders, the link between the health and strength of the animal, underpinned by their welfare, and the livelihood of the community is clear. According to a comprehensive economic data analysis of 189 countries conducted by the Human Development Program at the United Nations in 2018, Malaysia (ranking 57th), Thailand (ranking 83rd) and China (ranking 86th) were considered high to very high in development, Vietnam (ranking 116) was medium, and India (ranked 130th) and Bangladesh (ranked 136th) were in the bottom of the medium development category [40]. India and Bangladesh are considered in earlier stages of development. This is consistent with the findings of this study, in which countries placed in earlier stages of development have presented the importance of human-based benefits. In this case, presenting initiatives centered on a positively impact for both human welfare and prospects for the wider community may be more likely to succeed. This is in contrast with the financially focussed profit-driven countries investigated, which are placed higher in the human development scale.

In general, benefits that are received by the animals themselves were not often presented or ranked with any great importance in most countries, despite animals being the most logical beneficiary of their welfare improvement. This could be that livestock leaders considered these benefits to be too obvious to raise; however, considering they were, on occasion, indicated along with other benefit categories, it is more likely that animal-focussed benefits were just not considered that important. Improving animal welfare for the sake of the animals is rarely a compelling argument to livestock industry leaders. One exception to this existed, i.e., again, India and Bangladesh. Improving animal welfare for the purpose of reducing suffering, respecting the animal's rights and fulfilling the duty to appropriately consider the care of other living species could be partially attributable to the pervasion of Hindu religious beliefs in India and again due to physical proximity to the animals resulting in more developed empathy in Bangladesh. The notion of Ahimsa, non-violence, in Hinduism includes all life, and appropriate treatment of animals is tied to the tenet of Karma, where causing ill to another will result in ill to oneself. Inherently tied to rebirth, Hindus believe they may be reborn as an animal, and an animal may be reborn as a human, the specifications of which depend on their state of karma [41]. Lastly, the Vedas describe the code of sarva–bhuta–hita (devotion to the good of all creatures) [42].

Surprisingly, benefits to addressing animal welfare with the purpose of managing branded images and avoiding negative media and even market collapse, as has been seen with the livestock export industry in Australia [15,43,44], did not appear to be mentioned with any significance in this study.

This may be due to a reduced concern in Asian countries that citizens may lobby and protest in an attempt to challenge an industry. This is potentially underpinned by the cultural dimension of 'power distance' (the degree to which a hierarchy and the directions provided by it are accepted without question), which is often higher in Asian countries [45]. This may also be attributable to a greater concern for maintaining economic stability as compared to western countries that enjoy a higher development ranking.

Previous research has demonstrated the importance of valuating potential changes to practices on animal welfare grounds by estimating and understanding the benefits and the costs in doing so [12,26]. Understanding the value citizens place on animal welfare benefits is deemed worth of exploration, as it directly relates to the indirect loss of profit for the industry [25]. Likewise, a better understanding of the strength of animal welfare benefits according to the livestock industry itself could have great utility.

The findings of this study suggest potential grounds for presenting more compelling mutual benefits to livestock industry leaders when seeking to improve farm animal welfare internationally. By applying this information and creating education and awareness initiatives in line with benefits that are more likely to appeal to the livestock community, it is likely that an increased engagement with animal welfare initiatives will be seen. For countries within this study, that includes the creation and presentation of reliable data sets that demonstrate profit opportunities where they exist, and an increased effort to reach business owners and senior managers in production companies with this information.

6. Conclusions

This study explored benefits of improving animal welfare as perceived by livestock stakeholders in China, Vietnam, Thailand, Malaysia, India and Bangladesh. Although the overarching importance of benefits that yield financial gain was shared across all countries, mostly through improved productivity of the animals and improved product quality, regional differences were present. This was most noticeably the case with India and Bangladesh being more concerned with human- and community-focussed benefits of improving animal welfare and the tie between human and animal welfare by reducing zoonotic risks, but also in regard to the potential increase in human psychological welfare by observing the animals in more positive states. Animal-focussed benefits were not presented with any significance by the livestock leaders included in this study, with the exception of India and Bangladesh, suggesting that improving animal welfare for the sake of the animals is unlikely to be a compelling reason to act in most cases of livestock enterprise, unless it is directly related to the productive output of the animal or another financial indicator such as reducing the risk of stock losses. This study does not investigate the presence of any benefits of improving animal welfare, rather, it investigates which benefits, should they be present, would be most valued by stakeholders. If applied to international animal welfare initiatives with the purpose of finding mutual benefits to initiate collaborations, this founding information could be useful. In addition, if the more compelling benefits presented in this study can be investigated and demonstrated, it is suggested that stakeholders will be more likely to engage in change to improve animal welfare.

Author Contributions: M.S. conceptualised the study, developed the methodology, conducted the investigation, collected the data, conducted data analysis, and wrote the paper. C.F. assisted in data collection, and editing the paper. C.J.C.P. edited the paper and supervised the study. M.S. and C.J.C.P. acquired the funding.

Funding: This research was funded by Open Philanthropy and the Good Ventures Foundation.

Conflicts of Interest: The authors declare no conflict of interest. The funders had no role in the design of the study; in the collection, analyses, or interpretation of data; in the writing of the manuscript, or in the decision to publish the results.

References

1. Sinclair, M.; Phillips, C. Key motivators and meanings: A follow up study with Asian livestock stakeholders on improving animal welfare. *Anim. Welf.* **2019**, submitted.

2. Keyserlingk, M.; Hötzel, A. The Ticking Clock: Addressing Farm Animal Welfare in Emerging Countries. *J. Agric. Environ. Ethics* **2015**, *28*, 179–195. [CrossRef]

3. Tao, B. A stitch in time: Addressing the environmental, health, and animal welfare effects of China's expanding meat industry. *Georget. Int. Environ. Law Rev.* **2003**, *15*, 321–357.

4. Sinclair, M.; Phillips, C.J.C. Key Tenets of Operational Success in International Animal Welfare Initiatives. *Animals* **2018**, *8*, 92. [CrossRef] [PubMed]

5. Christensen, T.; Lawrence, A.; Lund, M.; Stott, A.; Sandøe, P. How can economists help to improve animal welfare? *Anim. Welf.* **2012**, *21*, 1–10. [CrossRef]

6. Dawkins, M.S. Animal welfare and efficient farming: Is conflict inevitable? *Anim. Prod. Sci.* **2017**, *57*, 201–208. [CrossRef]

7. Hambrecht, E.; Eissen, J.J.; Newman, D.J.; Smits, C.H.M.; Verstegen, M.W.A.; den Hartog, L.A. Preslaughter handling effects on pork quality and glycolytic potential in two muscles differing in fiber type composition. *J. Anim. Sci.* **2005**, *83*, 900–907. [CrossRef] [PubMed]

8. Gruber, S.L.; Tatum, J.D.; Grandin, T.; Scanga, J.A.; Belk, K.E.; Smith, G.C. *Is the Difference in Tenderness Commonly Observed between Heifers and Steers Attributable to Differences in Temperament and Reaction to Preharvest Stress*; National Cattlemen's Beef Association: San Antonio, TX, USA, 2006.

9. Grandin, T. The Effect of Economic Factors on the Welfare of Livestock and Poultry. In *Improving Animal Welfare: A Practical Approach*; Grandin, T., Ed.; Cabi: Oxfordshire, UK, 2015.

10. Aguayo-Ulloa, L.A.; Miranda-de La Lama, G.C.; Pascual-Alonso, M.; Olleta, J.L.; Villarroel, M.; Sañudo, C.; María, G.A. Effect of enriched housing on welfare, production performance and meat quality in finishing lambs: The use of feeder ramps. *Meat Sci.* **2014**, *97*, 42–48. [CrossRef]

11. Green, L.; Kaier, J.; Wassink, G.; King, E.; Grogono, T. Impact of rapid treatment of sheep lame with footroot on welfare and economics and farmer attitudes to lameness in sheep. *Anim. Welf.* **2012**, *21*, 65–71. [CrossRef]

12. Bennett, R.; Kehlbacher, A.; Balcombe, K. A method for the economic valuation of animal welfare benefits using a single welfare score. *Anim. Welf.* **2012**, *21*, 125–130. [CrossRef]

13. Grandin, T. The economic benefits of proper animal welfare. In *Reciprocal Meat Conference Proceedings*; American Meat Science Association: Savoy, IL, USA, 1995.

14. Compassion in World Farming. *Strategic Plan 2013–2017*; Compassion in World Farming: Surrey, UK, 2013.

15. Future Beef. Live Export. 2018. Available online: https://futurebeef.com.au/knowledge-centre/live-export/ (accessed on 1 June 2018).

16. Douphrate, D.L.; Rosecrance, J.C.; Stallone, L.; Reynolds, S.J.; Gilkes, D.P. Livestock handling injuries in agriculture: An analysis of workers compensation data. *Am. J. Ind. Med.* **2009**, *52*, 391–407. [CrossRef] [PubMed]

17. McGlone, J.J. Farm animal welfare in the context of other society issues: Toward sustainable systems. *Livest. Prod. Sci.* **2001**, *72*, 75–81. [CrossRef]

18. De Passillé, A.M.; Rushen, J. Food safety and environmental issues in animal welfare. *Revue Scientifique et Technique-Office International des Épizooties* **2005**, *24*, 757.

19. Pinillos, R.G.; Appleby, M.C.; Manteca, X.; Scott-Park, F.; Smith, C.; Velarde, A. One welfare–a platform for improving human and animal welfare. *Vet. Rec.* **2016**, *179*, 412–413. [CrossRef]

20. Rollin, B. Why is Agricultural Animal Welfare Important? The Social and Ethical Context. In *Improving Animal Welfare: A Practical Approach*; Grandin, T., Ed.; Cabi: Oxfordshire, UK, 2015.

21. Pritchard, J.C.; van Dijk, L.; Ali, M.; Pradhan, S.K. Non-economic incentives to improve animal welfare: Positive competition as a driver for change among owners of draught and pack animals in India. *Anim. Welf.* **2012**, *21*, 25–32. [CrossRef]

22. Burgess, D.; Hutchinson, W.G. Do people value the welfare of farm animals? *EuroChoices* **2005**, *4*, 36–43. [CrossRef]

23. Vetter, S.; Vasa, L.; Ózsvári, L. Economic aspects of animal welfare. *Acta Polytech. Hung.* **2014**, *11*, 119–134.

24. FAOSTAT. Data: China. 2017. Available online: http://www.fao.org (accessed on 1 March 2019).

25. Bennett, R. The value of farm animal welfare. *J. Agric. Econ.* **1995**, *46*, 46–60. [CrossRef]

26. McInerney, J.P. *Animal Welfare: An Economic Perspective*; Valuing Farm Animal Welfare: Oxford, UK, 1993.

27. Pimentel, D.; Wilson, C.; McCullum, C.; Huang, R.; Dwen, P.; Flack, J.; Tran, Q.; Saltman, T.; Cliff, B. Economic and environmental benefits of biodiversity. *Bioscience* **1997**, *47*, 747–757. [CrossRef]

28. Ackerman, F.; Heinzerling, L. Pricing the priceless: Cost-benefit analysis of environmental protection. *Univ. Pa. Legal Rev.* **2001**, 1553–1584. [CrossRef]

29. Schaltegger, S.; Synnestvedt, T. The link between 'green'and economic success: Environmental management as the crucial trigger between environmental and economic performance. *J. Environ. Manag.* **2002**, *65*, 339–346.

30. Sinclair, M.; Phillips, C. The Cross-Cultural Importance of Animal Protection and Other World Social Issues. *J. Argric. Environ. Ethics* **2017**, *30*, 439–455. [CrossRef]

31. Sinclair, M.; Derkley, T.; Fryer, C.; Phillips, C. Australian public opinions regarding the live export trade before and after an animal welfare media expose. *Animals* **2018**, *8*, 106. [CrossRef] [PubMed]

32. Davey, A.; Fisher, R. *Economic Issues Associated with the West Australian Live Sheep Export Trade*; Pegasus Economics: Canberra, Australia, 2018.

33. Foelsch, D.; Huber, H.; Boelter, U.; Gozzoli, L. Research on alternatives to the battery system for laying hens. *Appl. Anim. Behav. Sci.* **1988**, *1*, 29–45. [CrossRef]

34. Esslemont, R.J.; Peeler, E.J. The scope for raising margins in dairy herds by improving fertility and health. *Br. Vet. J.* **1993**, *149*, 537–547. [CrossRef]

35. Kossaibati, M.A.; Esslemont, R.J. The costs of production diseases in dairy herds in England. *Vet. J.* **1997**, *154*, 41–51. [CrossRef]

36. Kennedy, U.; Sharma, A.; Phillips, C. The Sheltering of Unwanted Cattle, Experiences in India and Implications for Cattle Industries Elsewhere. *Animals* **2018**, *8*, 64. [CrossRef] [PubMed]

37. Baul, T.K.; Moniruzzaman, M.M.R.; Nandi, R. Status, utilization, and conservation of agrobiodiversity in farms: A case study in the northwestern region of Bangladesh. *Int. J. Biodivers. Sci. Ecosyst. Serv. Manag.* **2015**, *11*, 318–329. [CrossRef]

38. Rabin, M. Cognitive dissonance and social change. *J. Econ. Behav. Organ.* **1994**, *23*, 177–194. [CrossRef]

39. Festinger, L. *A Theory of Cognitive Dissonance*; Stanford University Press: Palo Alto, CA, USA, 1957.

40. United Nations Development Programme. *Human Development Report*; United Nations: New York, NY, USA, 2018.

41. Krishna, N. *Sacred Animals of India*; Penguin Books: New Dehli, India, 2010.

42. Szucs, E.; Geers, R.; Jezierski, T.; Sossidou, E.N.; Broom, D.M. Animal welfare in different human cultures, traditions and religious faiths. *Asian—Australas J. Anim. Sci.* **2012**, *25*, 1499–1507. [CrossRef]

43. Everingham, S.; O'Brien, K. *Cattle Industry Launches Class Action against Federal Government, Seeking Compensation over Live Export Ban*; ABC News: Melbourne, Australia, 2014.

44. Petrie, C. Live Export: A Chronology. 2016. Available online: https://www.aph.gov.au/About_Parliament/Parliamentary_Departments/Parliamentary_Library/pubs/rp/rp1617/Chronology/LiveExport (accessed on 1 June 2018).

45. Hofstede Insights. Country Comparison. 2018. Available online: https://www.hofstede-insights.com/country-comparison/china,malaysia,thailand,vietnam/ (accessed on 1 March 2019).

Article

Knowledge of Stakeholders in the Livestock Industries of East and Southeast Asia about Welfare during Transport and Slaughter and Its Relation to Their Attitudes to Improving Animal Welfare

Ihab Erian, Michelle Sinclair and Clive J. C. Phillips *

Centre for Animal Welfare and Ethics, School of Veterinary Science, University of Queensland,
Gatton, QLD 4343, Australia; ihab.erian@gmail.com (I.E.); m.sinclair6@uq.edu.au (M.S.)
* Correspondence: c.phillips@uq.edu.au

Received: 12 March 2019; Accepted: 18 March 2019; Published: 19 March 2019

Simple Summary: The potential to improve stakeholders' knowledge of animal welfare in the livestock industries through training programs and its influence on their attitudes to livestock welfare is unclear. Stakeholders in East and Southeast Asia responded to a questionnaire on their knowledge of animal welfare considerations during livestock transport and slaughter, as well as indicating their attitudes towards the welfare of livestock at these times. They then received training, after which their knowledge scores increased. Knowledge scores had few connections to attitudes, but whether the respondents were certain or not about their attitudes to livestock welfare was most likely to have the strongest correlation to knowledge. Regional differences were evident and suggested that these differences should be considered in future training provisions.

Abstract: The World Organisation for Animal Health (OIE) sets standards and guidelines for international animal welfare for the international livestock trade. The growing economic advancement in the East and Southeast Asian region suggested the potential benefit of a research study to examine stakeholders' understanding of animal welfare during the transport and slaughter of livestock. A survey of stakeholders' knowledge of livestock welfare in the transport and slaughter industries was conducted in four Southeast Asian countries, Malaysia, China, Vietnam and Thailand, in association with trainer and stakeholder workshops conducted in each country. The attitudes of participants towards animal welfare during slaughter and transport were also identified. Knowledge scores were in accordance with the respondents' assessment of their own knowledge level. The biggest knowledge improvement was among Thai respondents, who tended to be younger and less experienced than in other countries. The respondents with the biggest improvement in knowledge scores were most likely to be involved in the dairy industry and least likely to be involved in the sheep and goat industries, with meat processors and those involved in pig or poultry production intermediate. The respondents who obtained their knowledge from multiple sources had most knowledge, but it increased the least after training. Connections between attitudes to improving animal welfare and knowledge were limited, being mainly confined to ambivalent responses about their attitudes. The study suggests that knowledge can be improved in animal welfare training programs focused on livestock welfare around transport and slaughter, but that local cultural backgrounds must be considered in designing the program.

Keywords: animal welfare; Asia; knowledge; slaughter; transport; training

1. Introduction

Animal welfare issues are recognized as an important concern associated with animal production in many countries, particularly those with existing animal welfare policies, legislation and public awareness [1]. It is important to also recognize that there are significant differences in attitudes to animal welfare issues between regions [2]. With the current trend for the expansion of animal production in developing countries, livestock legislation is beginning to be promoted internationally through animal welfare codes of practice and minimal animal welfare standards. In 2002, the World Organisation for Animal Health (OIE) began the process of creating animal welfare standards, which are largely derived from scientific and technical knowledge. Such knowledge has been developed through informal and formal processes [3], in particular the acquisition of facts, theories and ideas, through education, reading from reliable sources, peers, consultation and the media [4].

The theory of Planned Behavior acknowledges that knowledge and attitudes are important elements of changing behavior towards ethical animal welfare practices [5]. The extent of knowledge, the role and relevance of such knowledge and the complexity of the knowledge are all acknowledged to have an effect on attitudes and behavior [6]. The theory has been widely used, for example to predict alcohol consumption among school students [7], to change the behavior of employees on construction sites so that waste was reduced [8], to predict intentions to care for patients with alcohol dependence by nursing students [9] and to assess the importance of self-belief for developing ways to rectify alcohol problems [10]. It is acknowledged that knowledge alone is insufficient to change beliefs, and positive behavior strategies and regular checks are needed to increase compliance with the procedures (e.g., of hand hygiene in nursing students [11]). Other research has demonstrated the necessity of understanding the relationships between planned behavior and attitudes (e.g., in perceptions of workplace health and safety) [12].

The OIE Regional Animal Welfare Strategies (RAWS) aim to facilitate the regional implementation of the Terrestrial Animal Code, (Section 7 Animal Welfare), which presents animal welfare guidelines for safe international trade in terrestrial animals and their products, that has been adopted by the OIE member states. A regional strategy was developed for Asia, the Far East and Oceania in 2008 to provide a vision for this region in which the welfare of animals is respected, promoted and incrementally advanced, concurrent with the pursuit of progress and socioeconomic development [13,14]. The achievement of this goal in the region is expected to require education in the form of training stakeholders in key aspects of welfare identified by the OIE standards.

Asia accounts for 39% of global meat production, with China producing almost twice as much meat globally as the second highest producer, the United States of America [15]. In 2016, China produced 76.4 million metric tons of meat (beef, pork and chicken), the second highest year on record [16] and driven by the fact that meat consumption in this region is rapidly increasing. Annual animal slaughter increased from 10.2 to 13.5 billion animals between 1996 and 2016 [17]. Domestic beef consumption increased by 111% from 3.5 million tons in 1996 to 7.3 million tons in 2015 [18]. The improvement of animal welfare practices in developing countries can also lead to improved product quality and increased export trade opportunities. There is also an important social element, since approximately 70% of the world's poorest economies are tied to livestock industries, including many Asian countries [19], which have the advantage of cheap labor and land [20].

The recent animal welfare guidelines of the OIE have advocated a number of ways of influencing regional and global approaches to animal welfare standards, but particularly suggest the implementation of the OIE guidelines in legislation and education [13]. Apart from seeking to identify knowledge in the livestock industry stakeholders, there have been few attempts to understand public knowledge of animal production systems [21] and to discover their sources of information. One qualitative study used the concept of Planned Behavior theory to identify the most noticeable consumer beliefs regarding dairy products in the food markets [22]. These studies can be used to guide implementation of the OIE standards, as knowledge gaps can be identified and rectified. The recommendation of the OIE is for each country to implement animal welfare standards or introduce a "Code of Practice" for animal welfare conditions. However, there remain instances where developing Asian countries have accepted lower animal welfare standards of practice than would be acceptable in developed countries, e.g., [23]. Stakeholders in the livestock industries now have roles and responsibilities to effect changes in the way society treats and views animals [24] and in providing the care and conditions to make the animals contented, not just avoiding pain and stress [25].

This paper focused on the importance of improving local husbandry knowledge of, and attitude towards, animal welfare during transport and slaughter in four diverse countries in southeast Asia (the Federation of Malaya, hereafter referred to as Malaysia, the People's Republic of China, hereafter China, the Socialist Republic of Vietnam, hereafter Vietnam, and the Kingdom of Thailand, hereafter Thailand). Attitudes included towards the effect of local laws, the effect of personal and religious beliefs and the importance of any improvement to workplace, community and peers. It was hypothesized that knowledge of the standards would improve attitudes towards animal welfare. The study also examined the cross-cultural differences between these countries and the way that the knowledge had been acquired. Previous work with these stakeholders has examined their intention and ability to enhance animal welfare [26], as well as the differences between the different countries and stakeholders in their attitudes to livestock welfare during transport and slaughter [27,28].

2. Materials and Methods

Human ethical clearance was obtained from the University of Queensland Ethics Committee (Reference Number 2015000059). The study used a quantitative questionnaire that was administered at training sessions in Malaysia, China, Vietnam and Thailand (Supplementary Materials Section 1), which addressed stakeholders' knowledge of livestock production systems, with a focus on livestock welfare during transport and slaughter and their attitude towards livestock welfare. These countries were selected due to their future export potential for international livestock products within the next few decades and the diverse cultures they represented.

Four two-day 'train the trainer' workshops were conducted under the auspices of an Animal Welfare Standards Project (http://www.animalwelfarestandards.net/), one in each country. These were led by four international livestock experts in animal slaughter and livestock transport. The participants comprised 118 trainers (30 from Malaysia, 46 from China, 20 from Vietnam and 22 from Thailand) (Table 1), who were given access to a resource package with presentations in the local language (http://www.animalwelfarestandards.net/). The trainers were chosen by country coordinators on the basis of involvement in the livestock industry, slaughter or transport.

Table 1. Number of participants in the initial 'train the trainer' and livestock stakeholders' workshops.

	Malaysia	China	Vietnam	Thailand	Total Participants
Trainers	30	46	20	22	118
Stakeholders	94	338	196	268	896
Total	124	384	216	290	1014

The questionnaire was also answered by 896 stakeholders (94 in Malaysia, 338 in China, 196 in Vietnam and 268 in Thailand), who were trained by the trainers in a series of forty-four one-day regional workshops (Table 2). All stakeholders received travel allowances, free lunch and per diem expenses. The project coordinators in each country were responsible for selecting participants for the workshops based on their involvement in the livestock slaughter or transport industries in the different capacities identified in the questionnaire. The workshop invitees included delegates from local OIE veterinary services, local animal welfare focal point personnel, and personnel working directly with livestock in the transport and or slaughter industries. Farmers, team leaders who supervise people who work directly with the animals, business owners in the livestock industries, business managers in the industry, veterinarians who treat animals or work for the government as advisors and university researchers were also invited. After completion of the training on the OIE standards in each country, the same knowledge questionnaire was reissued and answered by the majority of the participants.

Table 2. Number of participants in the livestock stakeholders' workshops in each country.

Country	Locations of the Workshops
Malaysia	Zon Selatan, Tengah, Utara, Sabah, Sarawak, Pantai Timiur and Kuala Lumpur
China	Guandong, Hain, Hubei, Hun, Shandong, Zhejiang and Jiangxi
Vietnam	Hanoi, Halphong, Vinh, Dang, Vungtau, Binhduong and Cantho
Thailand	Khon Ratchasima, Udon Thani, Champon, Khon Kaen, Sakon Khon, Petchaburi and Bangkok

The questions (Supplementary Materials Questionnaire Section 1) were based on the OIE standards for international farm animal welfare in addition to the animal welfare material contained in the presentations to participants by the international experts in each field at the 'train the trainer' workshops. A pilot survey with 10 respondents from all the countries was conducted through industry stakeholders.

The initial questionnaires were developed in English in collaboration with researchers, academics, international experts in the animal welfare domain and literature. Adjustments were made to avoid leading questions or a possible bias. The final questionnaire was translated to Bahasa, Mandarin, Vietnamese and Thai and then back-translated to English to ensure that the translated version in the local language was consistent with the original questionnaire. The questionnaire was delivered as a hard copy to all the trainers, in their local language. Participants were given a unique code to facilitate identification before and after the workshops.

The first section of the questionnaire comprised 18 questions that focused on the trainers' knowledge of the OIE animal welfare standards and common local livestock practices in each country, with specific reference to livestock welfare during transport and slaughter (Supplementary Materials).

The second section examined the participants' attitude to animal welfare during transport and slaughter, how satisfactory they believed animal welfare to be, whether any improvements at their workplace could be initiated by new legislation or by international monetary gains for their local products [26–28]. The attitude questions were grouped into four sections, personal assessment, community assessment, ability to make improvements, all separately identified for slaughter and transport, thereby covering the central components of the theory of Planned Behavior. The response to each question was ranked on a five-point Likert scale, from strongly disagree to strongly agree. The first eight questions focused on general attitudes to animal welfare. The second set of questions investigated the key factors influencing the stakeholders' assessments of animal welfare during slaughter and transport, which included: personal beliefs, relevant laws, the importance of animal welfare within the workplace, community and peers, the benefits of improving animal welfare within the community, workplace and industry, in general and in terms of monetary gain. The thirteen questions also investigated the factors influencing the stakeholders' evaluation, capacity to improve and sources of improvement of animal welfare practices during slaughter and transport [26–28]. The final section contained demographic questions: the participants' country, their age, sex, religion and residential region, their role within their industry, how long they have been involved in the industry and how their knowledge was gained (formal, employment or otherwise).

Statistical Analysis

The questionnaire response data was initially tabled in a Microsoft Excel 2013 spreadsheet for each country and then transferred to Minitab Version 17 for analysis. A total of 330 stakeholders (Malaysia 7, China 294, Vietnam 16 and Thailand 13) were excluded from the analysis because their data was incomplete, mostly due to a failure to redo the knowledge test after the training. Only invited attendees' data were included in the analysis. A husbandry knowledge score (K score) was determined for each respondent from the total number that were correct, out of eighteen knowledge questions.

The change in numerical distribution of K scores was examined after the completion of the workshops to determine the usefulness and benefits of the training sessions.

A principal component analysis of the respondents' answers was undertaken first but demonstrated little evidence of clustering. A stepwise general linear model regression analysis assessed the significance of the relationships between the respondents' demographic data, as the independent variables, and the distribution of the Likert scale responses for the total K score, as the dependent variable, using a Logit link function. Least square means are quoted, and Alpha values for parameters to enter or leave the equation were set at 0.1. Plots of residuals were examined to ensure that the correlated factors between attitude and knowledge questions and demographics approximated a normal distribution and all probability values were considered significant at $p \leq 0.05$. Post hoc comparisons of individual means was by Tukey's test.

3. Results

The mean response time to answer the questionnaire was 33 min. A total of 683 male respondents (67.4%), 310 female respondents (30.6%) and 21 respondents who did not indicate their gender (2.0%) completed the workshops and questionnaires (Table 3). The gender balance appeared different in Thailand, being almost equal, whereas in the other three countries the majority were male. The most common age group among all the countries was 26–45 (n = 634; 62.5%), only 45 respondents (4.4%) were in the category 56–65 years of age. Thai respondents were also more likely to be under 25 and to have less than one year of experience than in other countries. The most common source of gaining knowledge was from formal qualifications in the livestock industry, either through a relevant degree or training course (n = 464; 45.8%), with 14.0% gaining their knowledge from hands-on experience through employment in the relevant livestock industry. Vietnamese respondents were more likely to indicate that they gained their knowledge from all possible sources. Slightly more respondents, 575 (56.7%), indicated that they lived most of their lives in metropolitan or urban areas than those who indicated they lived in rural areas, 416 (41.0%). Chinese were more likely to be from an urban zone than those from other countries. Of 998 respondents (98.4%) who identified a religious affiliation, 411 respondents (40.5%) were Buddhist, 393 (38.8%) did not follow a religion, 77 (7.6%) were Muslim and 38 (3.7%) were Christian. In Malaysia and Thailand, most indicated that they were moderately or very religious, but in China and Vietnam most said that they were not religious.

Table 3. Demographic characteristics of respondents (n = 1014) in Malaysia, China, Vietnam and Thailand.

Demographic	Respondents, n (% of Total Responses within Country)			
	Malaysia	China	Vietnam	Thailand
Total	124 (12)	384 (38)	216 (21)	290 (29)
Gender				
Male	90 (73)	294 (77)	157 (73)	142 (49)
Female	30 (24)	87 (23)	53 (25)	140 (48)
No answer	4 (3)	3 (1)	6 (2)	8 (3)
Residential zone				
Rural	60 (48)	121 (32)	87 (40)	148 (51)
Urban/metropolitan	59 (48)	258 (67)	123 (57)	135 (47)
No answer	5 (4)	6 (1)	6 (3)	7 (2)
Age				
Under 25	7 (6)	50 (13)	13 (6)	81 (28)
26–35	47 (38)	157 (41)	99 (46)	52 (18)
36–45	26 (21)	117 (30)	64 (30)	72 (25)
46–55	29 (23)	50 (13)	29 (13)	41 (14)
56–65	11 (9)	2 (0.5)	3 (1)	29 (10)
Over 65	1 (1)	5 (1.3)	1 (1)	7 (2)
No answer	3 (2)	3 (0.8)	7 (3)	8 (3)

Table 3. *Cont.*

Demographic	Respondents, *n* (% of Total Responses within Country)			
	Malaysia	China	Vietnam	Thailand
Religion				
Buddhist	15 (12)	47 (12)	69 (32)	280 (97)
Atheist/don't follow religion	0 (0)	259 (67)	134 (62)	0 (0)
Muslim	74 (60)	1 (0)	0 (0)	2 (1)
Christian	23 (18)	10 (3)	3 (1)	2 (1)
Other	12 (10)	71 (18)	10 (5)	6 (2)
Religiosity				
Not religious at all	2 (2)	193 (51)	106 (49)	31 (11)
Not very religious	5 (4)	63 (16)	64 (30)	66 (23)
Moderately religious	70 (57)	109 (28)	21 (10)	176 (61)
Very religious	40 (32)	16 (4)	5 (2)	12 (4)
No answer	7 (5)	3 (1)	20 (9)	5 (1)
Job role				
Work with animals	33 (27)	150 (39)	45(21)	76 (26)
Team Leader	24 (27)	61 (16)	11 (5)	30 (10)
Business owner	2 (1)	11 (3)	3 (1)	20 (7)
Business manager	11 (9)	41 (11)	5 (2)	9 (3)
Farmer	7 (6)	6 (2)	14 (7)	130 (45)
Practicing veterinarian	11 (9)	94 (25)	28 (13)	13 (5)
Veterinary advisor	19 (15)	17 (4)	101 (47)	7 (2)
No answer	7 (6)	4 (1)	9 (4)	5 (2)
Level of industry understanding				
Expert	2 (2)	56 (15)	1 (1)	2 (0.7)
Good knowledge	24 (20)	121 (32)	56 (26)	51 (18)
Some knowledge	55 (44)	151 (39)	106 (49)	136 (47)
Little knowledge	36 (29)	30 (8)	32 (15)	80 (28)
No knowledge	3 (2)	21 (5)	6 (2)	14 (5)
No answer	4 (3)	5 (1)	15 (7)	7 (2)
Knowledge acquisition				
Formal qualifications	43 (35)	206 (54)	104 (48)	111 (38)
Farm Employment	35 (28)	63 (16)	1 (1)	43 (15)
Personal interest	13 (10)	37 (10)	7 (3)	35 (12)
Friends	5 (4)	9 (2)	2 (1)	54 (19)
All of the above	22 (18)	60 (16)	83 (38)	35 (12)
No answer	6 (5)	9 (2)	19 (9)	12 (4)
Type of livestock involvement				
Beef/buffalo production	27 (22)	22 (6)	15 (7)	66 (7)
Dairy industry	8 (6)	12 (3)	2 (1)	30 (10)
Abattoirs/meatworks	26 (21)	151 (39)	87 (40)	27 (9)
Sheep/goat meat production	13 (11)	10 (3)	0 (0)	29 (10)
Wool/hair production	0 (0)	4 (1)	0 (0)	2 (1)
Poultry industry	25 (20)	59 (15)	15 (7)	62 (22)
Meat processing	2 (2)	14 (4)	6 (3)	10 (3)
Pig production	9 (7)	108 (28)	24 (11)	52 (18)
No answer	14 (11)	4 (1)	67 (31)	12 (4)
Years of industry involvement				
Up to 1 year	9 (7)	62 (16)	6 (3)	74 (26)
2–3 years	20 (16)	59 (15)	28 (13)	51 (18)
3–5 years	20 (16)	76 (20)	29 (13)	44 (15)
5–9 years	28 (23)	71 (19)	57 (26)	28 (10)
10–15 years	11 (9)	59 (15)	41 (19)	40 (14)
Over 15 years	27 (22)	53 (14)	39 (18)	30 (10)
No answer	9 (7)	4 (1)	16 (8)	23 (8)

The respondents' roles in the livestock transport and slaughter industry were varied: 304 (30.0%) worked directly with animals, 238 (23.5%) were supervisors, team leaders, business owners or managers within the industry and 290 (28.6%) were veterinarians working "hands-on" in the field or in a government advisory role. Thais were more likely to be farmers and the Vietnamese were more likely to be vets. Three hundred and thirteen respondents (30.9%) regarded themselves as being experts or having good knowledge of their relevant livestock industry, but 222 respondents (21.9%) regarded themselves as having little or no knowledge in the livestock production systems regarding animal welfare standards during transport or slaughter.

The most common involvement of the respondents was in abattoirs or in meat processing facilities (*n* = 291, 28.7%), followed by the respondents involved in the pig (*n* = 193, 19%) and poultry (*n* = 161, 15.9%) production industries. The most common length of involvement in the relative industry was 2–5 years (*n* = 327, 32.2%). One hundred and eighty four (18.1%) had had 5–9 years involvement in their respective industries, and 149 respondents (14.7%) had had over 15 years.

Respondents' Husbandry Knowledge

The respondents' self-assessed level of knowledge of livestock production systems was most commonly reported as some, little or no knowledge (*n* = 670, 66.1%) with fewer than 6.0% (*n* = 61) of respondents regarding their knowledge as expert. Very few Chinese or Vietnamese claimed little knowledge, but a significant proportion of Malaysian and Thai respondents did. Most gained their knowledge from training courses or relevant degrees (*n* = 464, 45.8%), with a significant number gaining their knowledge from personal interest such as the internet, television programs, journals and newspaper articles (*n* = 92, 9.1%) (Table 3).

Overall, in the pre-workshop knowledge test, Thai respondents had a lower proportion correct, compared with Malaysian and Chinese respondents (Malaysia, 3.80 [a]; China, 3.44 [a], Vietnam 3.23 [ab], Thailand 2.80 [b], out of maximum 15, Standard Error of the Difference between two means, SED 0.221, F-value, 4.56, *p* = 0.004). However, the improvement in the number correct by the end of the workshop was greater in Thai respondents than in Chinese or Vietnamese respondents (Malaysia, +5.80 [ab]; China, +5.50 [b], Vietnam +4.46 [c], Thailand +6.24 [a], SED 0.291, F-value, 5.05, *p* = 0.002).

4. Influencing Factors

4.1. Attitude Effects on Knowledge Score

There were six attitude questions that significantly influenced the respondents' K scores (Table 4). Mostly, the difference related to greater or lesser K scores for those who neither agreed nor disagreed, compared with the other responses. For the following four attitude questions, the K score was lower: I intend to make improvements to welfare of animals in my care; the laws on animal slaughter and transport influence my assessment of their animal welfare (AW) at this time; my knowledge about animal slaughter and transport limits my ability to improve AW during transport; and changes prescribed by my company encourage me to change practices. For the following two attitude questions, the score was greater: vehicle design makes improvement to AW during transport difficult; and changes prescribed by my supervisor encourage me to change practices.

4.2. Attitude Effects on Change in Knowledge Score Post-Training

There were nine attitude questions that significantly influenced the respondents' change in K scores (Table 5). Mostly, the differences again related to a greater or lesser score improvement for those who neither agreed nor disagreed, compared with the other responses. For the following attitude questions, the improvement was greater: welfare of transported animals is satisfactory; importance to my peers of factors influencing welfare of animals; and encouraged to change if prescribed by government. For the following two attitude questions, the improvement was less: encouraged to change if prescribed by law; and monetary gain influences my personal assessment of welfare. When the respondents strongly disagreed with the statement "Importance of welfare to peers influences ability to make improvement during slaughter", the improvement in their knowledge score was very low. Those strongly agreeing that their personal beliefs influence their ability to make improvement during transport also had reduced.

Table 4. Significant ($p < 0.05$) effects of attitudes to the welfare of animals during transport and slaughter on Knowledge Scores (SED = 0.221).

Attitude Question	Strongly Disagree	Disagree	Neither Agree nor Disagree	Agree	Strongly Agree	F-Value	p-Value
I intend to make improvements to welfare of animals in my care	4.62 [ab]	2.89 [ab]	2.54 [b]	3.17 [a]	3.38 [a]	2.61	0.03
The laws on animal slaughter and transport influence my assessment of their AW at this time	4.02 [a]	2.99 [ab]	2.63 [b]	3.33 [a]	3.62 [a]	4.30	0.002
My knowledge about animal slaughter and transport limits my ability to improve AW during transport	3.46 [ab]	2.56 [b]	2.84 [b]	3.14 [b]	3.91 [a]	2.34	0.04
Vehicle design makes improvement to AW during transport hard	2.34 [c]	3.34 [bc]	4.07 [a]	3.74 [ab]	3.10 [bc]	3.68	0.006
Changes prescribed by my company encourage me to change practices	5.52 [a]	3.20 [b]	2.27 [c]	2.56 [bc]	3.03 [b]	4.40	0.002
Changes prescribed by my supervisor encourage me to change practices	2.43 [b]	3.56 [ab]	3.90 [a]	3.95 [a]	2.75 [b]	3.87	0.004

AW = animal welfare. Means with different superscripts (a, b or c) within rows are significantly different ($p < 0.05$) by Tukey's test.

Table 5. Significant ($p < 0.05$) effects of attitudes to the welfare of animals during transport and slaughter on improvement in Knowledge Scores after training (SED = 0.300).

Attitude Question	Strongly disagree	Disagree	Neither Agree nor Disagree	Agree	Strongly Agree	F-value	p-Value
Welfare of transported animals is satisfactory	5.90 [ab]	5.07 [b]	6.04 [a]	5.30 [b]	5.20 [ab]	2.38	0.05
Importance to my peers of factors influencing welfare of animals	5.52 [abc]	4.88 [bc]	6.53 [a]	5.78 [b]	4.79 [c]	4.81	0.001
Company approval towards improving the welfare of animals	5.94 [ab]	5.74 [a]	4.50 [b]	5.28 [a]	6.03 [a]	2.92	0.02
Vehicles design influences ability for improvement	7.03 [a]	5.75 [a]	4.34 [b]	4.65 [b]	5.73 [a]	5.46	0.001
Encouraged to change if prescribed by government	3.95 [b]	5.62 [b]	6.74 [a]	5.66 [b]	5.54 [b]	3.85	0.004
Encouraged to change if prescribed by law	7.31 [a]	4.59 [bc]	4.90 [c]	4.87 [c]	5.83 [ab]	2.83	0.02
Monetary gain influences my personal assessment of welfare	6.61 [a]	5.27 [bc]	4.79 [c]	5.48 [ab]	5.36 [abc]	2.95	0.02
Importance of welfare to peers influences ability to make improvement during slaughter	1.04 [c]	7.39 [a]	6.49 [ab]	6.69 [ab]	5.89 [b]	3.84	0.004
My personal beliefs influence my ability to make improvement during transport	5.41 [ab]	6.07 [a]	5.87 [a]	5.86 [a]	4.29 [b]	2.61	0.03

Means with different superscripts (a, b or c) within rows are significantly different ($p < 0.05$) by Tukey's test.

4.3. Demographic Effects on Knowledge Score

Apart from their country, there were two demographic questions that significantly influenced K scores. The respondents who considered that they had good knowledge of livestock production systems actually had higher K scores compared with those who said that they had just some knowledge (expert, $n = 54$, 3.36 [ab], good, $n = 195$, 3.78 [a], some, $n = 296$, 3.18 [b], little, $n = 105$, 3.29 [ab], none, $n = 21$, 2.97 [ab], SED 0.221, F-value 2.65, $p = 0.03$). In relation to where they got their knowledge, those who indicated that they got it from all the possible sources had a higher K score than those who indicated any one particular source (formal qualifications, $n = 326$, 3.27 [b], farm employment $n = 102$, 3.25 [b], personal interest (internet, journals, newspapers, TV), $n = 67$, 2.95 [b], friends and acquaintances, $n= 47$, 3.21 [ab], and all of these $n = 129$, 3.91 [a]; SED 0.221, F-value 3.00, $p = 0.02$).

4.4. Demographic Effects on Change in Knowledge Score Post-Training

Apart from their country, there were three demographic questions that significantly influenced the respondents' increase in K scores. Age had a significant effect on K Score improvement, with this being highest in those over 65 and lowest in those 56–65 (18–25, $n = 101$, 5.60 [b], 26–35, $n = 232$, 5.44 [b], 36–45, $n = 197$, 4.97 [bc], 46–55, $n = 95$, 5.54 [b], 56–65, $n = 23$, 4.10 [c], >65, $n = 13$, 7.35 [a], SED 0.221, F-value 3.00, $p = 0.01$). In relation to where they got their knowledge, those who indicated that they got it from all the possible sources had a lower K score increase than those who indicated that they got their knowledge from formal qualifications or friends (formal qualifications, $n = 321$, 5.59 [ab], farm employment $n = 103$, 5.09 [bc], personal interest (internet, journals, newspapers, television), $n = 65$, 5.53 [abc], friends and acquaintances, $n = 47$, 6.34 [a], and all of these $n = 125$, 4.94 [c]; SED 0.300, F-value 2.48, $p = 0.04$). In relation to the type of livestock industry, the respondents involved with the dairy industry had greater improvement than meat processors or those involved in the pig industry, which along with those involved in cattle, abattoirs and meat processors, in turn had greater improvement than those in the sheep and goat industries (cattle, $n = 98$, 6.12 [ab], dairy, $n = 35$, 6.71 [a], abattoir, $n = 221$, 5.77 [ab], sheep/goat, $n = 30$, 3.69 [c], poultry $n = 101$, 5.58 [b], meat processors, $n = 25$, 5.28 [ab], pigs, $n = 151$, 5.36 [b], SED 0.300, F-value 3.89, $p = 0.001$).

5. Discussion

The measurement of knowledge in this study appears to have been successful, since scores broadly agreed with stakeholders' self-rated knowledge assessment. Also, those citing multiple knowledge sources had greater knowledge scores but the lowest increase after training, as expected. The questions were chosen to be at a mixture of knowledge levels and to be generic to the four countries.

Demographics

The recruitment and selection criteria for the trainers had some potential bias when the local participants were selected for workshops. The trainers were nominated by the local authority in each country, who were under instruction to choose trainers with extensive involvement in livestock slaughter or transport. Although the respondents at the workshops were selected by the prospective authorities in their countries, some selection bias may have occurred by not inviting or including legislative personnel and animal welfare regulators in the workshops.

The respondents from the four countries identified that the main source of gained knowledge was through training courses and relevant formal qualifications ($n = 464$, 45.8%) from professional institutions such as universities, agricultural and veterinary colleges, livestock processing companies and livestock transport bodies (Table 3). Training could be tailored to the local needs by exploring the gaps in husbandry knowledge of the OIE animal welfare recommendations by implementing a mechanism, in the form of questionnaires, to identify these needs and to ensure the success of the training to deliver the desired outcome and give local people an understanding of the OIE animal welfare recommendations [29]. The study also identified that the least likely source of gaining knowledge was through television programs, internet, academic journals, newspaper articles, friends and acquaintances (162, 16%), which could be attributed to the limited access of some respondents to multimedia means of dissemination of information (Table 3). There were just 70 respondents (*n* = 1014, 6.9%) who gained their animal welfare knowledge through their friends and acquaintances, and 92 respondents (9.1%) gained it through personal interest via internet, journals, newspaper articles and television programs. However, those who did gain knowledge from friends had a greater increase in K score. This supports the contention that the peer effect in the East and Southeast Asian region is a strong influence on disseminating knowledge [30]. Previous research regarding the relationship between knowledge and animal welfare issues found that knowledge supplied by animal protection organizations was the most credible source of knowledge; however, this is likely to be because the respondents were from these organizations [31]. Further research on the effect of peers on the increase of knowledge in similar cultural backgrounds could be beneficial for the success of animal welfare programs and industry engagement. Research of animal welfare knowledge of advocates [31] has also identified that original scientific literature was highly regarded, which accords with the recognition in our study that gaining animal welfare knowledge through formal education is likely to lead to increased receptivity to new knowledge.

6. Respondents' Husbandry Knowledge

The data collected from the respondents identified that the oldest respondents (>65 years) improved their knowledge most, even though the sample size was small. Human morality is recognized as increasing with age in Kohlberg's progression of moral reasoning [32]. Also the respondents who are involved in the dairy cattle industry had a higher K score increase post training than those involved in other industries, particularly sheep and goats. Dairy farming is a labor-intensive, capitally-intensive industry, offering considerable benefits to those who do it well, and milk production closely related to cow welfare, whereas sheep farming is extensive and offers less financial benefit to welfare improvement. Hence, it is possible that those drawn to these industries have different levels of motivation for knowledge improvement, with dairy farmers likely to benefit most.

The study highlighted the strong differences in religion between the different countries, with a contrast between two countries with high religiosity, Malaysia and Thailand, and two with low levels, China and Vietnam. In addition, Malaysia and Thailand had different dominant faiths, Islamism and Buddhism, respectively. Despite these strong contrasts, religion did not influence knowledge or the acquisition of knowledge in the training sessions. It could be hypothesized that the religious beliefs of Malaysian participants may have hindered them from, for example, improving their knowledge of the pig welfare standards, since 60% of respondents were Muslims, similar to 61% of Malaysians overall [33]. However, there was no evidence of this. Islamic doctrine advocates prevention of unnecessary suffering of any animal before and during slaughter [34]. Malaysia is among the most religious countries in the world [35], whereas China and Vietnam are amongst the least, and this was reflected in the survey respondents' beliefs—67% of respondents claimed not to follow any religion, the same proportion as has been recorded nationally [36]. Agriculture is very important to both of the less religious countries, China and Vietnam (44% of the working population in Vietnam are employed in agriculture according to 2015 statistics [37]).

Thai respondents showed the lowest level of knowledge and were more likely to indicate that they thought they had little knowledge, but this is likely to reflect the fact that the stakeholders were younger and more likely to have less than one year of experience than in other countries. They showed the most significant improvement, with a mean difference of +33.9% after the workshops. Thailand's agriculture sector is also very important, employing over 70% of the working population [38] and it has a particularly strong dairy industry, which is used for social support [39,40]. As the respondents associated with the dairy industry improved their knowledge more than those in other industries, as did those from Thailand, compared with the other countries, further training sessions to livestock stakeholders regarding the OIE animal welfare standards would probably be of substantial benefit to Thai stakeholders involved in the slaughter and transport of dairy cattle. The Thai government may be supportive; in 2016, it introduced "Thailand 4.0", an economic model which aims to achieve a 7-fold increase in the average annual income of farmers from 56,450 baht (5,470 USD) to 390,000 baht (15,000 USD) by 2037 [41].

7. Attitude Effects on Knowledge Score

People that were uncertain about improving animal welfare generally, the influence of the law and their company, and the limitations of their knowledge actually had less knowledge. This suggests a common approach in some respondents of not considering their beliefs and not attempting to acquire knowledge. This could reflect a lack of incentive or capacity to make change [26]. Conversely, uncertainty about the importance of vehicle design, a much more tangible topic, was associated with higher K scores. This may be because the respondents were not livestock transporters and a recognition of this uncertainty appears to have been more common amongst more knowledgeable respondents. Understanding the impact of vehicle design would be limited in those in other professions. Similarly, uncertainty about the importance of encouragement by their supervisor appears to have been more highly associated with a high level of knowledge. Thus the personal influence of the supervisor appears to be differently perceived to that of the company's influence, and appears potentially more influential on knowledge, whereas the company's influence is generally antagonistic to knowledge. Those who were uncertain or agreed that the supervisor was influential had more knowledge. For knowledge improvement, there was evidence that those with uncertainty about the importance of peers and knowledge on their ability to make improvements, or uncertainty about whether welfare of transported animals was satisfactory had greater improvement. This is to be expected because, at least for knowledge and law importance, the scores were lower in the first assessment; therefore, they were likely to rise more in the second. Similarly, those uncertain about the importance of vehicle design had greater scores in the first assessment and they increased less than those with firm views in the second.

Overall, there was little impact of knowledge on attitudes to animal welfare, in agreement with a recent study on the effects of knowledge of meat production systems on attitudes of people in Brisbane, Australia, towards chicken welfare and consumption, which concluded that increasing knowledge of the industry does not necessarily increase empathy towards animal welfare [21].

8. Demographic Effects on Knowledge Score and Its Improvement

Gender was not an apparent barrier to knowledge. Wambui et al. 2018 [42] reported that female stockpeople, aged over 50, and with livestock experience greater than 10 years had a significantly higher level of animal welfare knowledge, which is reflected in their attitudes and livestock practices.

The workshops highlighted the fact that significant improvement in animal welfare understanding can be achieved, which may lead to improved behavior in interactions with animals. Coleman and Hemsworth (2014) [43] found that training had the capacity to improve stockpersons' beliefs and behaviors towards enhancing animal welfare. Our study also suggested that ethical and cultural backgrounds, rather than necessarily people's religion, must be considered in designing training programs for the region. They should highlight the commercial advantages for each country to adopt animal welfare practices as well as the forecast economic future of a progressive advancement that is anticipated for the region. The major importance of country in the study suggests benefit from the integration of local communities to develop specific tailored training to deliver successful programs to improve the understanding of the OIE animal welfare standards in relation to livestock transport and slaughter practices. The tailored training programs should take into consideration the cultural and socioeconomic measures and encourage local relevant bodies to take responsibility for monitoring the agreed program.

Formal training was an important predictor of knowledge score improvement, indicating that this provided an improved ability to learn. It would be, therefore, desirable to the students in current university agriculture, veterinary studies and other animal related fields in these countries to receive a compulsory animal welfare and ethics syllabus to ensure that they have a good understanding of contemporary welfare issues.

The feedback received from participants in the workshops identified that improved knowledge of the OIE animal welfare standards during transport and slaughter by training is achievable, effective, well received by locals in the four East and Southeast Asian countries and forms a future opportunity for the OIE in spreading animal welfare standards in the region. Although the animal welfare concept is relatively new to the region [44], there are encouraging steps which have been taken by the four countries. Thailand has passed a Prevention of Cruelty and Animal Welfare Provision Act 2014 [45] and Malaysia has developed an Animal Welfare Bill that has recently been enacted [46]. Currently, there is no animal welfare legislation in Vietnam, or national animal welfare legislation in China [47,48]; however, some animal protection control was introduced in China in September 2009 [49].

Improving knowledge of the OIE recommendations during slaughter and transport has become a pre-requisite for any future improvement of animal welfare practices in these four east and southeast Asian countries, which may facilitate a change of attitude towards animal welfare [27,28]. Our research shows that improved knowledge will help people to define their attitudes to welfare issues. Other research has suggested behavior improvements as well; a study by Hemsworth (2003) [50] suggested that cognitive-behavioral intervention training programs designed to specifically target livestock stakeholders would have a direct effect on animals' level of fear, welfare and productivity.

9. Conclusions

Animal welfare improvement in the four Southeast Asian countries should focus on:

(a) Improving the OIE animal welfare standard and husbandry knowledge of livestock industry stakeholders.

(b) Local research and training programs based on moral and ethical concepts about animal welfare during slaughter and transport. The education programs should be aimed at all age groups.

(c) Adopting local public animal welfare awareness campaigns aimed at students in education, multimedia platforms and social organizations, which will bring about improvements in knowledge about animal welfare.

Supplementary Materials: The following are available online at http://www.mdpi.com/2076-2615/9/3/99/s1, Questionnaire Section 1: Descriptive statistics are provided for each husbandry knowledge question in the stakeholders workshops, Questionnaire Section 2: Attitudes towards livestock management, with special emphasis on transport and slaughter, Questionnaire Section 3: Demographic Background.

Author Contributions: Conceptualization, C.J.C.P.; Data curation, I.E. and M.S.; Formal analysis, I.E. and C.J.C.P.; Funding acquisition, C.J.C.P.; Methodology, M.S.; Project administration, M.S. and C.J.C.P.; Supervision, C.J.C.P.; Writing—original draft, I.E.; Writing—review and editing, I.E. and C.J.C.P.

Acknowledgments: The authors acknowledge the financial support of the governments of New Zealand, Australia, Malaysia, and the EU, as well as Universiti Putra Malaysia, World Animal Protection, and the Humane Slaughter Association for a Dorothy Sidley Memorial Award provided by the Humane Slaughter Association to Ihab Erian. We also acknowledge the support of the World Animal Health Organisation, OIE.

Conflicts of Interest: The authors declare no conflict of interest.

References

1. Veissier, I.; Butterworth, A.; Bock, B.; Roe, E. European approaches to ensure good animal welfare. *Appl. Anim. Behav. Sci.* **2008**, *113*, 279–297. [CrossRef]

2. Phillips, C.; Izmirli, S.; Aldavood, S.; Alonso, M.; Choe, B.; Hanlon, A.; Handziska, A.; Illmann, G.; Keeling, L.; Kennedy, M.; et al. Students' attitudes to animal welfare and rights in Europe and Asia. *Anim. Welf.* **2012**, *21*, 87–100. [CrossRef]

3. Kumar, R. Evaluation of two instrumental methods of comparing writing paper. *J. Forensic Sci.* **2011**, *56*, 514–517. [CrossRef]

4. Jurcoane, A.; Draghici, M.; Popa, M.; Niculita, P. Consumer choice and food policy. A literature review. *J. Environ. Prot. Ecol.* **2011**, *12*, 708–717.

5. Ajzen, I. The theory of planned behaviour. *Organ. Behav. Hum. Decis. Process.* **1991**, *50*, 179–211. [CrossRef]

6. Fabrigar, L.; Petty, R.; Smith, S.; Crites, S. Understanding knowledge effects on attitude-behavior consistency: The role of relevance, complexity, and amount of knowledge. *J. Pers. Soc. Psychol.* **2006**, *90*, 556–577. [CrossRef]

7. Cooke, R.; Dahdoh, M.; Norman, P.; French, D. How well does the theory of planned behaviour predict alcohol consumption? A systematic review and meta-analysis. *Health Psychol. Rev.* **2014**, *10*, 148–167. [CrossRef] [PubMed]

8. Li, J.; Zuo, J.; Cai, H.; Zillante, G. Construction waste reduction behaviour of contractor employees: An extended theory of planned behaviour approach. *J. Clean. Prod.* **2017**, *172*, 1399–1408. [CrossRef]

9. Talbot, A.; Dorrian, J.; Chapman, J. Using the Theory of Planned Behaviour to examine enrolled nursing students' intention to care for patients with alcohol dependence: A survey study. *Nurse Educ. Today* **2015**, *35*, 1054–1061. [CrossRef]

10. Finch, T.; Chae, S.; Shafaee, M.; Siegel, K.; Ali, M.; Tomei, R.; Panjabi, R.; Kishore, S. Role of students in global health delivery. *Mt. Sinai J. Med.* **2011**, *78*, 373–381. [CrossRef] [PubMed]

11. Jeong, S.; Kim, K. Influencing factors on hand hygiene behaviour of nursing students based on theory of planned behaviour: A descriptive survey study. *Nurse Educ. Today* **2016**, *36*, 59–164. [CrossRef]

12. Guerin, B.; Serano, P.; Iacono, M.; Herrington, T.; Widge, A.; Dougherty, D.; Bonmassar, G.; Angelone, L.; Wald, L. Realistic modelling of deep brain stimulation implants for electromagnetic MRI safety studies. *Phys. Med. Biol.* **2018**, *63*, 18. [CrossRef]

13. OIE Global Animal Welfare Strategy. Available online: http://www.oie.int/fileadmin/Home/eng/Animal_Welfare/docs/pdf/Others/EN_OIE_AW_Strategy.pdf (accessed on 7 January 2019).

14. OIE—PVS Gap Analysis Report in Nigeria 2010. Available online: http://www.oie.int/fileadmin/Home/eng/Support_to_OIE_Members/pdf/PVS_GapAnalysisReport-Nigeria.pdf (accessed on 3 February 2019).

15. Food and Agriculture Organization of the United Nations 2016. FAOSTAT. Available online: http://www.fao.org/faostat/en/#home (accessed on 16 November 2017).

16. Cook, R. China Meat Production by Year. Available online: http://beef2live.com/story-china-meat-production-year-0-113958 (accessed on 11 June 2018).

17. Food and Agriculture Organization of the United Nations 2018. The State of Food and Agriculture 2018. Available online: http://www.fao.org/3/I9549EN/i9549en.pdf (accessed on 13 November 2018.).

18. Xiang, Z.; Chang, G.; Lin, S. Current situation and future prospects for beef production in China—A review. *Asian-Australas. J. Anim. Sci.* **2018**, *31*, 984–991. [CrossRef]

19. World Bank 2016. World Bank Annual Report 2016. Available online: https://www.google.com.au/search?q=world+bank+annual+report&oq=World+Bank+Annual&aqs=chrome.2.0j69i57j0l4.23334j0j8&sourceid=chrome&ie=UTF-8 (accessed on 16 November 2017).

20. World Bank 2017. World Bank Annual Report 2017. Available online: https://www.google.com.au/search?q=world+bank+annual+report&oq=World+Bank+Annual&aqs=chrome.2.0j69i57j0l4.23334j0j8&sourceid=chrome&ie=UTF-8 (accessed on 16 November 2017).

21. Erian, I.; Phillips, C. Public understanding and attitudes towards meat chicken production and relations to consumption. *Animals* **2017**, *7*, 20. [CrossRef] [PubMed]

22. Nolan-Clark, D.; Neale, E.; Probst, Y.; Charlton, K.; Tapsell, L. Consumers' salient beliefs regarding dairy products in the functional food era: A qualitative study using concepts from the theory of planned behaviour. *BMC Public Health.* **2011**, *11*, 843. [CrossRef] [PubMed]

23. Tiplady, C.; Walsh, D.; Phillips, C. Cruelty to Australian cattle in Indonesian abattoirs-how the public responded to media coverage. *J. Agric. Environ. Ethics* **2012**, *26*, 869–885. [CrossRef]

24. Haynes, R.P. *Animal Welfare: Competing Conceptions and Their Ethical Implications*; Springer Science: Berlin, Germany, 2008; p. 82. Available online: https://www.springer.com/gp/book/9781402086182 (accessed on 19 March 2019).

25. Rollin, B.E. *Farm Animal Welfare: Social, Bioethical and Research Issues*; Wiley International: Hoboken, NJ, USA, 1995.

26. Sinclair, M.; Morton, J.; Phillips, C. Turning intentions into animal welfare improvements in the Asian livestock sector. *J. Anim. Welf. Sci.* **2018**, *26*, 1–15. [CrossRef] [PubMed]

27. Sinclair, M.; Zito, S.; Idrus, Z.; Yan, W.; van Nhiem, D.; Lampang, P.; Phillips, C. Attitudes of stakeholders to animal welfare during slaughter and transport in SE and E Asia. *Anim. Welf.* **2017**, *26*, 417–425. [CrossRef]

28. Sinclair, M.; Zito, S.; Phillips, C. The impact of stakeholders' roles within the livestock industry on their attitudes to livestock welfare in Southeast and East Asia. *Animals* **2017**, *7*, 6. [CrossRef]

29. Hodge, B.; Wright, B.; Bennett, P. The role of grit in determining engagement and academic outcomes for university students. *Res. High. Educ.* **2018**, *59*, 448–460. [CrossRef]

30. Green, A. *Proceedings of the 10th International Conference on Intellectual Capital Knowledge Management & Organisational Learning*; The George Washington University: Washington, DC, USA, March 2018. Available online: https://books.google.com.au/books?id=mZ4TBAAAQBAJ&pg=PA326& dq=influencing+factors+for+disseminating+knowledge&hl=en&sa=X&ved=0ahUKEwjjraHquc_XAhWKTbwKHc6NAfkQ6AEIKDAA#v=onepage&q=influencing%20factors%20for%20disseminating%20knowledge&f=false (accessed on 19 March 2019).

31. Ross, T.; Phillips, C. Relationships between knowledge of chicken production systems and advocacy by animal protection workers. *Soc. Anim.* **2018**, *26*, 73–92. [CrossRef]

32. Kohlberg, L. The claim to moral adequacy of a highest stage of moral judgment. *J. Phil.* **1973**, *70*, 630–646. [CrossRef]

33. Caumont, A. Pew Research Centre. Available online: http://www.pewresearch.org/fact-tank/2016/12/29/pew-research-centers-most-read-research-of-2016/ (accessed on 14 February 2019).

34. Halal Food Authority. 2016. Available online: https://www.halalfoodauthority.com/animal-welfare (accessed on 10 June 2018).

35. Mooney, S. Global Index of Religion and Atheism. Available online: https://sidmennt.is/wp-content/uploads/Gallup-International-um-tr%C3%BA-og-tr%C3%BAleysi-2012.pdf (accessed on 11 June 2018).

36. Religion Prevails in the World. Available online: http://gallup-international.bg/en/Publications/2017/373-Religion-prevails-in-the-world (accessed on 11 June 2018).

37. Trading Economics—The World Bank Collection of Development Indicators. Available online: https://tradingeconomics.com/vietnam/employment-in-agriculture-percent-of-total-employment-wb-data.html (accessed on 14 February 2019).

38. Leturque, H.; Wiggins, S. Thailand's Progress in Agriculture: Transition and Sustained Productivity Growth. Available online: https://www.odi.org/publications/5108-thailand-agriculture-growth-development-progress#downloads (accessed on 13 November 2017).

39. Thongnoi, J. Milking the System. Available online: https://www.bangkokpost.com/archive/milking-the-system/733380 (accessed on 13 November 2017).

40. Suwanabol, I. School Milk Programme in Thailand. Available online: http://www.fao.org/fileadmin/templates/est/COMM_MARKETS_MONITORING/Dairy/Documents/School_Milk_Programme_in_Thailand.pdf (accessed on 18 October 2015).

41. Poapongsakorn, N.; Chokesomritpol, P. Agriculture 4.0: Obstacles and How to Break Through. Available online: https://www.bangkokpost.com/opinion/opinion/1278271/agriculture-4-0-obstacles-and-how-to-break-through (accessed on 13 November 2017).

42. Wambui, J.; Lamuka, P.; Karuri, E.; Matofari, J. Animal welfare knowledge, attitudes, and practices of stockpersons in Kenya. *Anthrozoos* **2018**, 397–410. [CrossRef]

43. Coleman, G.; Hemsworth, P. Training to improve stockperson beliefs and behaviour towards livestock enhances welfare and productivity. *Rev. Sci. Tech.* **2014**, *33*, 131–137. [CrossRef] [PubMed]

44. Phillips, C.; Sinclair, M. Livestock welfare in China and surrounding countries. In *International Cooperation Committee of Animal Welfare (ICCAW) conference proceedings of the World Conference of Animal Welfare*; Bao, J., Ed.; ICCAW: Hangzhou, China, 2017; pp. 200–211.

45. Cruelty Prevention and Welfare of Animal Act, B.E. 2557. Department of Livestock Development, Ministry of Agriculture and Cooperatives. Available online: http://www.dld.go.th/th/images/stories/law/english/en_cruelty_prevention_act2014.pdf (accessed on 14 February 2019).

46. World Animal Protection 2016. World Animal Protection Index. Available online: https://www.worldanimalprotection.org.au/search?query=Vietnam (accessed on 14 November 2017).

47. Wei, S. China Animal Welfare Legislation: Current Situation and Trends—From Analysis of Three Cases in Recent Years. Available online: http://animallawconference.org/wp-content/uploads/sites/7/2015/09/China-Animal-Welfare-Legislation.pdf (accessed on 9 June 2018).

48. Wang, C.; Liu, W.; Li, B.; Xin, H. Overview of 2015 International Symposium on animal environment and welfare held in Chongqing, China. *Int. J. Agric. Biosyst. Eng.* **2015**, *8*, 179–180.

49. ESDAW, European Society of Dog and Animal Welfare, Animal Welfare and Rights in China. Available online: http://www.esdaw.eu/animal-welfare-and-rights-in-china.html (accessed on 9 June 2018).

50. Hemsworth, P. Human-animal interactions in livestock production. *Appl. Anim. Behav. Sci.* **2003**, *81*, 185–198. [CrossRef]

![animals logo] *animals*

MDPI

Article

Sheep Farmers' Perception of Welfare and Pain Associated with Routine Husbandry Practices in Chile

Cristian Larrondo [1,*], Hedie Bustamante [2] and Carmen Gallo [3]

[1] Escuela de Graduados, Facultad de Ciencias Veterinarias, Universidad Austral de Chile, Valdivia 5090000, Chile

[2] Instituto de Ciencias Clínicas Veterinarias, Facultad de Ciencias Veterinarias, Universidad Austral de Chile, Valdivia 5090000, Chile; hbustamante@uach.cl

[3] Instituto de Ciencia Animal, Facultad de Ciencias Veterinarias, OIE Collaborating Centre for Animal Welfare and Livestock Production Systems—Chile, Universidad Austral de Chile, Valdivia 5090000, Chile; cgallo@uach.cl

* Correspondence: cristian.larrondoc@gmail.com; Tel.: +569-82223104

Received: 16 October 2018; Accepted: 24 November 2018; Published: 28 November 2018

Simple Summary: Lambs are simultaneously subjected to several routine husbandry practices that cause pain. One of the main factors that limit the use of analgesics in lambs is the difficulty in pain recognition by sheep farmers. This study aimed to determine how husbandry practices are carried out in Chilean farms, the sheep farmers' perception of animal welfare and pain, and the factors that affect them, as well as the level of agreement among farmers in the recognition of pain associated with these practices. Farmers were invited to participate in a workshop and they were asked through a survey about their sociodemographic information, how husbandry practices are being performed in their farms, and were asked to score the intensity of pain associated to seven of these practices. Castration and tail docking were perceived as the most painful practices and farmers agreed among them that these routine husbandry practices cause severe pain to animals. Several factors were associated with the farmers' pain perception, such as the method used for the specific husbandry practices and the farmers' educational level. In general, routine husbandry practices were carried out without using analgesics and with painful methods despite the agreement among farmers regarding the recognition of pain associated with these procedures.

Abstract: Considering the public concern about the welfare of farm animals during routine husbandry practices, this study aimed to determine how husbandry practices are carried out in Chilean farms, sheep farmers' perceptions of animal welfare and pain, and factors that affect them, as well as the level of agreement among farmers in the recognition of pain associated with these practices. Using a self-administered survey, participants were asked about their sociodemographic information, how husbandry practices are carried out in their farms, and their pain perception for seven of these common husbandry procedures using a numerical rating scale (0 to 10). A total of 165 farmers completed the survey and perceived castration and tail docking as the most painful practices in lambs (median pain score 10 vs. 8, $p < 0.05$). Pain perception was associated with the method used for the specific husbandry practices, the farmers' educational level, the farm size, and flock size ($p < 0.05$). There was a fair to good level of agreement beyond chance ($p < 0.05$) in the recognition of pain associated with the most painful practices. In general, husbandry practices are not carried out in young animals, use painful methods, without using analgesics, which may have a negative impact on animal welfare.

Keywords: Animal welfare; husbandry practices; lambs; pain; sheep farmers; perception; agreement

1. Introduction

Pain is defined as a complex and subjective experience associated with actual or potential tissue damage [1], which not only depends on the severity of tissue damage, but also on the length of exposure of the animals to a painful stimulus [2]. Therefore, pain may potentially induce physical and behavioral changes that are the evidence of suboptimal animal welfare [2,3].

In production systems, sheep may experience pain due to different diseases, e.g., mastitis and lameness [4]. Also, lambs in extensive production systems, the predominant husbandry system in Chile [5], are subjected simultaneously to several routine husbandry practices that cause pain and distress (ear tagging, tail docking, castration, vaccination), which may have a negative impact on their welfare [6]. Different researchers [7,8] and international recommendations [9,10] challenge the use of some of these procedures in a routine way, such as castration and tail docking, because there is no consensus regarding the productive and welfare impact of these procedures on the animals [8,11,12]. Furthermore, society has become increasingly concerned about the welfare of farming animals, especially when they are subjected to painful husbandry practices [4,13]. Painful husbandry procedures in lambs induce an increase in locomotor activity, including abnormal postures, jumping, rolling, tail wagging, and repetitive standing and lying, regardless of the age and method with which they are performed [3,6,14,15].

Scientific evidence has shown that analgesics reduce pain during husbandry procedures [6,16,17]. The World Organisation for Animal Health recommends that when painful practices cannot be avoided, pain should be managed accordingly [18]. Similarly, the Chilean legislation [19] demands mitigation of animal suffering during painful husbandry practices and proposes four possible ways to reduce pain and improve animal welfare: (1) Replacing the current husbandry practice by another non-surgical procedure that has been demonstrated to improve animal welfare; (2) carrying out the husbandry practices at the earliest possible age; (3) use of analgesics during these procedures; or (4) genetic selection, selecting for animals that do not present the feature that requires the husbandry procedure. However, so far, there is still a low or lack of use of pain relief drugs in sheep farms when these procedures are performed [2,20]. Several reasons have been elicited for not using analgesics, including drug costs, availability, and practical difficulties associated with time intervals, herds' sizes, and farm employees' availability [4,21]. In addition to these limitations, both veterinarians and sheep farmers have reported difficulties in the recognition and evaluation of pain in lambs [4,22], a problem that further limits the operators' motivation to reduce pain.

Pain evaluation and recognition in lambs is a complex process, mainly due to the fact that active pain behaviors in these animals are not as intense as in other species [4]. Moreover, pain recognition may be influenced by the incorrect assumption that younger animals have an immature central nervous system [23,24]. As prey animals, ruminants tend to be more stoic and not express pain, added to the impossibility in its verbalization [23]. Therefore, it is necessary that veterinarians and sheep farmers know how to correctly recognize pain in these species, and implement pain management strategies when husbandry practices are carried out. The aims of this study were to determine how husbandry practices are carried out in Chilean farms, to describe sheep farmers' perception of animal welfare and pain, the factors that affect them, as well as the level of agreement among farmers in the recognition of pain associated with these practices.

2. Materials and Methods

This study was approved by the Committee of Bioethics and Use of Animals in Research of the Universidad Austral de Chile (N° 288/217). The study was conducted between November 2016 and May 2018. A self-administered survey was applied to sheep farmers based in three Chilean regions: Magallanes and the Chilean Antartic (sheep population: 1,571,056; 77.1% of the national population), Aysén (177,972; 8.7%), and Libertador Bernardo O'Higgins (123,715; 6.07%). These regions were selected because they have the largest number of sheep and flocks of the country [25]. Through government and private agencies, sheep farmers in each region were invited to participate in a workshop entitled, "Improving handling practices on sheep farms", where veterinarian experts spoke about sheep production, livestock practices, and animal welfare. Before each workshop started, a self-administered, confidential, and anonymous survey was given to participants to be completed. Of a total of 747 existing farmers in the mentioned regions [25], 180 attended the workshops and took the survey, and 165 completed all questions. However, in the case of the numerical rating scale for pain perception of routine husbandry practices, only 125 farmers rated all husbandry practices. The data obtained represent a sample size with a confidence level of 95%, with an error equal to 7% (R Core Team, Vienna, Austria) [26], and correspond to 22% of the sheep farmers in the above-mentioned regions.

2.1. General Sociodemographic and Farm Information of Sheep Farmers

The survey consisted of three sections. In the first section, open-ended questions about the farmers' sociodemographic information were obtained, including gender, age, region, educational level, as well as information related to their time experience as a sheep farmer (years), farm size (hectares), and flock size (all sheep). Also, farmers were asked to select one or more options regarding if their farm had technical advice in the following areas: Nutritional, reproductive, health, or none of them.

2.2. Routine Husbandry Practices

The second section of the survey contained open-ended questions of how sheep farmers carry out the following husbandry practices in lambs: Animal identification, tail docking, and castration. For each procedure, they were asked by open questions if they carry out each practice, and if they do, "at what age they carry it out", the method used, the reasons for performing the practice, person responsible for performing the procedure, and if they use analgesics or not. Additionally, for tail docking, a diagram was used to mark the length at which the tail was docked and to justify the answer. The diagram was adapted from Fisher et al. [8], and farmers had to mark one of three options for tail length: Short = docked leaving little tail, not enough to cover the tip of the vulva in ewe lambs and at similar length in male lambs; Medium = tail stump covers the tip of the vulva in ewe lambs and at similar length in male lambs; and Long = docked longer than Medium.

2.3. General Animal Welfare Perception

The third section aimed to establish the farmer's knowledge concerning animal welfare, including individual perception and importance. Farmers were asked if they had previously heard about animal welfare (yes/no), and to choose what animal welfare means to them by selecting one or more of the following options: Animals do not suffer from hunger and thirst; animals do not suffer from pain, injury, and diseases; animals do not suffer from discomfort; animals do not suffer from fear and distress; animals can express normal behavior. Participants were also asked to select one or more options from a pre-defined list regarding the importance of having good animal welfare in their farms, considering the following alternatives: It affects animal health and production; it influences meat quality; it is a consumer requirement; or animal welfare is indifferent to me. Also, they were asked to select and rank in order of importance the three main animal welfare issues in their farms from a pre-defined list: Predation, lamb mortality, rounding up/transportation, husbandry practices, slaughter, facilities, nutrition/water availability, and diseases.

2.4. Farmers' Perceptions of Pain during Routine Husbandry Procedures

Finally, sheep farmers were asked to describe how they perceive the intensity of pain associated with each of the following husbandry procedures: Animal identification, tail docking, castration, hoof trimming, deworming, vaccinating, and shearing. For this, a table with an eleven-point (0 to 10) numerical rating scale (NRS) was used, considering 0 as "no pain" and 10 as the maximum pain [26].

2.5. Data Management and Statistical Analyses

Data of a total of 165 completed surveys were analyzed using the statistical program, R (R Core Team, Vienna, Austria) [27]. Descriptive analysis of sociodemographic information, husbandry procedures, and the animal welfare section were performed. Sociodemographic information and open-ended responses from the second and third sections (routine husbandry procedures and general animal welfare perception) were coded as frequencies and percentages. The scores obtained in the NRS for each husbandry practice were summarized with their mode, median, and range. Also, a radar chart was made with the median pain scores of each husbandry practice. Spearman correlations were made between pain scores (median pain scores) that sheep farmers perceived as being associated with husbandry practices. Mann-Whitney non-parametric tests ($p < 0.05$) were used to determine the association of the method used for each husbandry practice with the sheep farmers' pain perception. Linear regression models with Poisson distribution were fitted to assess the effect of the continuous variables (age, time experience as a sheep farmer) and factors (gender, educational level, flock size, farm size) on the pain scores associated to each husbandry practice.

To estimate the degree of agreement among sheep farmers in the recognition of pain associated to husbandry practices, pain scores of NRS were categorized as follows: No pain = 0; Mild to moderate pain = 1 to 5; and Severe pain = 6 to 10. The degree of agreement beyond chance was estimated by the Fleiss' kappa coefficient (κ) for multiple raters using Epidat 4.1 software, where κ values: <0.40 indicated poor agreement; between 0.40 and 0.75 fair to good agreement; >0.75 indicated excellent agreement [28].

3. Results

3.1. General Sociodemographic and Farm Information of Sheep Farmers

The majority of sheep farmers were males ($n = 145$; 87.9%), 18 (10.9%) were females, and only two persons did not answer this question (1.2%). The mean age of the participants was 47.3 ± 14.5 years, the mean time experience as sheep farmer was 23.5 ± 19.9 years. More than half of the farms had technical advice, either from the government (63/165; 38.2%) or private (41/165; 24.8%). In addition, most of the farms (115/165; 69.7%) were visited by a veterinarian, at least once a year. Table 1 shows further information about the age of farmers, their educational level, number of ewes per farmer (flock size), and farm size (ha).

Table 1. Sociodemographic data of sheep farmers ($n = 165$) and farm information.

Variable	Frequency	Percentage
Age (years)		
18–32	25	15.8
33–47	64	38.8
48–62	48	29.1
≥ 63	26	15.8
Educational level		
Elementary [1]	57	34.5
Secondary [2]	34	20.6
Technical	21	12.7
Professional	37	22.4
No schooling	7	4.2
Flock size		
≤ 50	60	36.4
51–200	33	20
201–500	10	6.1
501–1000	6	3.6
≥ 1001	44	26.7
Farm size (ha)		
≤ 50	57	34.5
51–1000	40	24.2
1001–5000	13	7.9
5001–10,000	12	7.3
≥10,001	13	7.9

[1] Eight years of elementary education; [2] four years of secondary education.

3.2. Routine Husbandry Practices

3.2.1. Identification

Animal identification was carried out (85.1%) on lambs at a mean age of 3.8 ± 3.1 months (range= 0.5–18). The most common method for identification was ear notching (52.1%), followed by the use of an ear tag (39.4%), and both methods (8.5%). The totality of farmers (100%) considered identifying their animals as a sign of property and because the procedure allowed them to maintain accurate records. No farmers reported the use of analgesics (100%) during this procedure, which is mainly performed by themselves (33.9%) or by farm personnel (26.3%). Nonetheless, 33% of farmers did not answer the question of who performed the procedure.

3.2.2. Tail Docking

Of the sheep farmers, 91.2% answered that they dock lamb tails at a mean age of 3.4 ± 1.9 months (range = 1–12). The most common method reported was the use of a knife (67.3%), followed by a rubber ring (18.7%), and hot iron (11.2%). Reasons for tail docking included: To improve mating (29.1%), to improve animal sanitary conditions (22.3%), for easier handling (21.4%), esthetic reasons (10.7%), and to improve reproduction (6.8%); 3.9% of farmers answered that they do not know why they do it. Other less frequent answers (< 2%) included: Tradition, to improve animal productivity, and as a method of animal identification. The majority of sheep farmers (85.9%) do not use analgesics during this husbandry practice. Similar to animal identification, tail docking is performed by sheep farmers (40.6%) and by farm personnel (29.7%).

The majority of sheep farmers (49.7%) performed tail docking only in animals that will remain in the farm for breeding purposes, 25.9% only dock ewe lambs and 10.9% perform it on male and ewe lambs. Most farmers (55.9%) leave a short tail stump that does not cover the tip of the vulva in ewe lambs and at a similar length in male lambs; 33.6% dock lambs' tails at a medium length and 0.8% at a long length. The main reasons given by farmers in relation to why they dock lambs' tails at these lengths were: To improve mating (33.8%), to improve animal sanitary conditions (18.4%), tradition (12.2%), esthetic reasons (10.2%), and to protect the perineum (5.4%).

3.2.3. Castration

Regarding castration, 52.3% of the sheep farmers castrate their male lambs at a mean age of 7.6 ± 15 months (range = 1–72). The most common method used was a rubber ring (75.5%), 14% use a knife, and 10.5% use other methods. The main reasons for castrating their lambs included: Handling (48%), commercial requirements (20%), improvement of animal productivity (17%), and to avoid breeding (15%). Most of the farmers (92.6%) indicated that they do not use analgesics during castration with the procedure mainly being performed by farm personnel (40.8%) or by themselves (32.4%).

3.3. General Animal Welfare Perception

The majority of sheep farmers (83.6%; 138/165) had previously heard about animal welfare. Regarding animal welfare [29], 83.6% of participants selected all the options given, as explained in Section 2.3, meaning that animals do not suffer from thirst, hunger and malnutrition, pain, injuries, diseases, discomfort, fear, and distress, and the possibility to express normal behaviors. Five percent of farmers perceive and associate animal welfare only with good nutrition and 3.6% associate it only with the fact that animals may express their natural behavior.

The importance that sheep farmers give to animal welfare was mainly associated with the facts that it affects animal health and production (53.55%), it influences meat quality (28.9%), and it is a consumer requirement (12%). A 5.6% of participants answered that animal welfare was something indifferent to them. The most important animal welfare concerns reported by sheep farmers on their farms, in order of importance, included: Nutrition and water availability > diseases > predators.

In general, sheep farmers perceived that animal welfare in their farms was good (65.7%) and regular (32.8%), with a minority perceiving that it was bad (1.5%).

3.4. Sheep Farmers' Perceptions of Pain during Routine Husbandry Procedures

A total of 125 answers were obtained for the pain NRS, which represents a response rate of 75.8% (125/165). Only data from farmers that completed all the NRS (seven husbandry practices) was included. Husbandry practices that farmers scored as the most painful procedures included both castration and tail docking (Table 2). Castration was perceived to be more painful than tail docking (median pain score 10 vs. 8; $p < 0.05$). The most frequent pain scores for castration and tail docking were 10 (mode 10), with 55.2% and 35.2% of sheep farmers scoring both procedures with a score of 10, respectively (Table 2). As shown in Figure 1, animal identification was perceived as the third most painful husbandry practice followed by vaccination (median pain score 4 vs. 2; $p < 0.05$). Hoof trimming, shearing, and deworming were scored with the lowest pain scores in the NRS (Figure 1).

Table 2. Summary statistics of sheep farmers' ($n = 125$) perceptions of pain intensity associated to seven husbandry practices in lambs using a numerical rating scale, from 0 (no pain) to 10 (maximum pain).

Husbandry Practice	Mode	Median	Range
Castration	10	10	2–10
Hoof trimming	0	2	0–10
Deworming	0	1	0–7
Vaccinating	2	2	0–10
Identification	3	4	0–10
Tail docking	10	8	2–10
Shearing	0	1	0–6

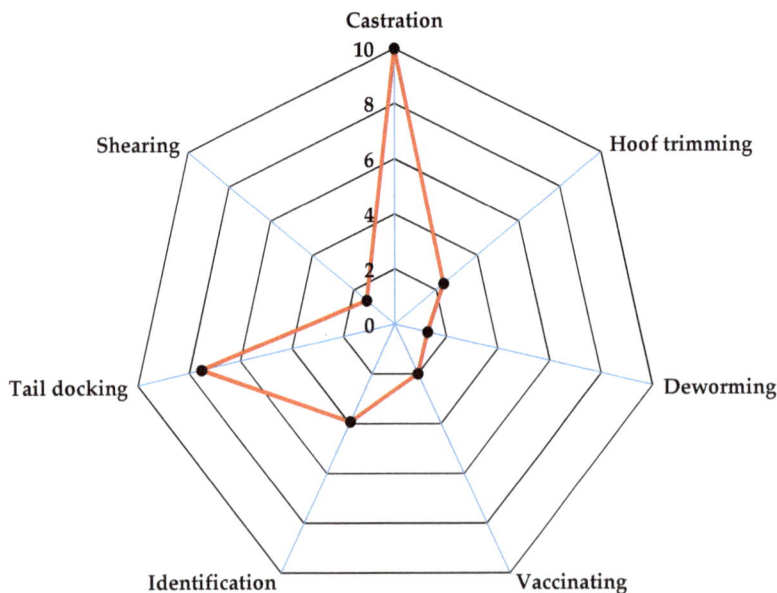

Figure 1. Median pain scores associated to husbandry practices in lambs using a numerical rating scale, from 0 (no pain) to 10 (maximum pain).

A positive and significant correlation was obtained (rho = 0.48; $p < 0.05$) between castration and tail docking pain scores and between tail docking and identification pain scores (rho = 0.21; $p < 0.05$), regardless of the method used for these procedures. In contrast, castration and identification pain scores were not correlated (rho = 0.07; $p > 0.05$). Tail docking and animal identification using a knife were scored with higher scores ($p < 0.05$) by sheep farmers, compared to when these procedures were performed using a rubber ring and ear tagging, respectively (Table 3). The farmers did not perceive differences in pain ($p > 0.05$) between tail docking with a knife and hot iron, nor between castration methods ($p > 0.05$).

Table 3. Median numerical rating scale pain scores perceived by sheep farmers according to husbandry practice and method used.

Husbandry Practice	Method			
	Knife	Ear Tagging	Rubber Ring	*p*-Value
Identification	5	3	-	$p < 0.05$
Tail docking	9	-	5–6	$p < 0.05$
Castration	10	-	8–9	$p > 0.05$

Numerical rating scale, from 0 (no pain) to 10 (maximum pain).

The age of farmers affected ($p < 0.05$) their pain perception only for deworming and vaccinating (Table 4). There was a significant effect ($p < 0.05$) of farmers' educational level on pain perception associated with the majority of husbandry practices (Table 4). Pain scores were higher in farmers who did not complete a formal education curriculum and who had elementary education compared to those with professional education ($p < 0.05$). The time experience as sheep farmers ($p < 0.05$) only affected the farmers' pain perception for vaccinating. Flock size and farm size were factors that affected ($p < 0.05$) farmers' pain perception associated with hoof trimming, deworming, and vaccinating. Farmers who had larger flocks and farms gave lower pain scores ($p < 0.05$) for these husbandry practices than farmers who had smaller flocks and farms (Table 4).

Table 4. Effect of sociodemographic factors on pain perception associated with routine husbandry practices in lambs. Results of the linear regression models testing pain scores for each husbandry practice.

Main Effect	Husbandry Practice						
	Castration	Hoof Trimming	Deworming	Vaccinating	Identification	Tail Docking	Shearing
Gender	ns	ns	ns	ns	ns	ns	ns
Age (years)	ns	ns	$p < 0.05$	$p < 0.05$	ns	ns	ns
Education	ns	$p < 0.05$	$p < 0.05$	$p < 0.05$	$p < 0.05$	$p < 0.05$	ns
Experience	ns	ns	ns	$p < 0.05$	ns	ns	ns
Flock size	ns	$p < 0.05$	$p < 0.05$	$p < 0.05$	ns	ns	ns
Farm size	ns	$p < 0.05$	$p < 0.05$	$p < 0.05$	ns	ns	ns

* ns: Not significant effect ($p > 0.05$).

Overall, agreement beyond chance among sheep farmers for pain recognition associated to husbandry procedures was poor (Fleiss' κ = 0.3353, $p < 0.05$; Table 5). Similarly, agreement for pain recognition associated with the "No pain" (κ = 0.1896) and "Mild to moderate pain" (κ = 0.2311) categories was also poor ($p < 0.05$). Nonetheless, for the "Severe pain" category, the agreement among farmers was fair to good (κ = 0.5603, $p < 0.05$; Table 5).

Table 5. Level of agreement among sheep farmers (*n* = 125) in the recognition of pain associated with husbandry practices in lambs. Fleiss' kappa coefficient values for pain categories: No pain (0), Mild to moderate pain (1–5), Severe pain (6–10), according to pain intensity scores used in a numerical rating scale.

Category	Kappa	Confidence Interval (95%)		Z-Value	*p*-Value
No pain	0.1896	0.0572	0.3163	44.1683	*p* < 0.05
Mild to moderate pain	0.2311	−0.0384	0.4894	53.8356	*p* < 0.05
Severe pain	0.5603	0.2204	0.8776	130.4994	*p* < 0.05
Global kappa	0.3353	0.0507	0.6049	105.6649	*p* < 0.05

4. Discussion

4.1. Sheep Farmers and Routine Husbandry Practices' Description

Sociodemographic features of sheep farmers (Table 1) are similar to the current national situation, in which the majority of farmers are males [25]. The average age of sheep farmers is in agreement with the increased aging tendency of Chilean farmers and the general population [25,30] and also corresponds with the experience as sheep farmers, which in this study exceeded 20 years. The educational level here reported for farmers is in agreement with another Chilean study [30], in which an average of 6.2 years of education with incomplete secondary education is described. Additionally, in the present study, the majority of sheep farmers (38.7%) had completed elementary education and more than half (51%) had completed secondary education (Table 1).

A marked heterogeneity in flock size is described, characterized by a similar percentage of sheep farmers with less than 50 (36.4%) and more than 500 ewes (30.3%). Sheep farmers in the Chilean Patagonia (Regions of Magallanes and Aysén) have large flocks (more than 500 ewes) in extensive production systems and large farms (more than 1,000 hectares). In contrast, sheep farmers of central Chile (Region of Libertador Bernardo O'Higgins) are characterized by small flocks (less than 50 ewes) and less than 50 hectares. These differences between regions in terms of flock and farm size are in agreement with Gallo et al. [5], and further relate to differences in transport and preslaughter conditions of the lambs produced.

Current European and Chilean legislation [9,10,19] mandate performing painful husbandry practices in young animals. However, the results of the present study indicate that painful husbandry practices in lambs, such as animal identification, tail docking, and castration, are not performed accordingly. The age at which sheep farmers perform these procedures could be associated with the predominant sheep production conditions in Chile, where lambs are reared in large herds (thousands of animals) and extensive grass pastures [5]. These factors make the rounding up and gathering of animals in their early weeks of life quite difficult. Consequently, the vast majority of routine husbandry practices are performed simultaneously [31], usually at ages older than three months, which may have negative implications on animal welfare [20,21,23,32]. Moreover, tail docking at 45 days may induce long-term consequences, such as primary hyperalgesia and chronic pain [33].

The most frequent method used by sheep farmers to carry out identification and tail docking in lambs was the knife. Several authors [16,34,35] have identified the knife as the most painful method for both ear notching and tail docking, resulting in greater behavioral and physiological changes than ear tagging and using a hot iron, respectively [35]. Therefore, current scientific evidence recommends avoiding the use of a knife, selecting instead methods that are less painful [16,34–36]. Similarly, rubber ring castration was preferred by Chilean sheep farmers, probably due to the lower costs implied, easiness to perform, and relative quickness. However, some studies suggest that rubber ring castration causes greater pain when associated with behavioral changes compared to other castration methods [6,23,37].

Animal identification and castration are performed mainly for handling and commercial reasons, while there are several reasons why sheep farmers carry out tail docking, including improving both mating and farm sanitary conditions. According to the scientific evidence, these arguments are questionable because we are not aware of any studies reporting that tail docking improves mating [38,39]. Moreover, according to Orihuela et al. [39], rams prefer to court and mate with intact tail ewes over tail docked ewes. Also, there is no consensus in the scientific literature about a potential relationship between dags/flystrike and tail presence. Fecal soiling and dags (fecal material around the anus) could be mainly associated with fecal consistency [12,40], so it is essential that farms have proper management practices, such as deworming, shearing, and crutching. Furthermore, in countries where flystrike is not a welfare and health issue, like Chile, tail docking may be more difficult to justify as a routine husbandry practice [10].

Sheep farmers mostly performed tail docking to ewe lambs and to those ewe and male lambs that remain in the farms for breeding purposes. This fact could also be associated with the main reason why they perform this practice: To improve mating. Similarly, the length at which farmers tail docked their animals may be also influenced by this factor, mainly due to the fact that the vast majority of sheep farmers carry out this procedure, leaving a short tail that does not cover the tip of the vulva. However, it has been demonstrated that short-tail docked animals have a greater incidence of rectal prolapses and even a higher risk of flystrike [8,41]. Therefore, several authors [8,41,42] have recommended that when tail docking is performed, a tail stump that covers the tip of the vulva in ewe lambs with a similar length in male lambs should be maintained.

When painful husbandry practices cannot be performed at an early age, legislation mandates for the use of analgesics [19,20]. This is supported by the fact that an increased chronic inflammatory reaction has been demonstrated in lambs that have been castrated at later ages in contrast to when these practices are performed in younger lambs [43]. These differences could be associated with an increased tissue trauma due to the procedure in older animals, rather than differences in pain sensitivity related to age or an underdeveloped central nervous system [4,20,23]. According to the present study, nearly all farmers carry out husbandry practices in lambs without the use of pain relief drugs. These results are in agreement with other international results, where analgesic drugs are not routinely administered during husbandry practices, despite the fact that international legislation from some countries requires or promote their use [9,10]. There are many reasons why farmers do not administer analgesics to lambs, including management and economical arguments [44]. Nevertheless, analgesics' use during and after various husbandry practices have shown to significantly improve animal welfare [16,17]. Accordingly, Small et al. [17] administered meloxicam before tail docking and reported a seven-fold reduction in abnormal pain associated behaviors. Furthermore, Phillips et al. [45] mentioned that Australian sheep farmers considered analgesic administration more important than the method used to carry out castration and tail docking.

4.2. Sheep Farmers' Perceptions of Animal Welfare and Pain

The majority of sheep farmers have previously heard about animal welfare and they associate it to physical, mental, and behavioral conditions of animals. Also, more than half of the surveyed farmers agreed that animal welfare is important to ensure animal health and production. These findings are in agreement with those reported by Australian [46] and Brazilian farmers [47]. The most important animal welfare issues identified by sheep farmers were associated with nutrition and water availability. Interestingly, almost all surveyed farmers perceived that animal welfare on their farms was "regular to good", which, according to them, would be associated to the extensive production conditions, in which animals are able to express their natural behavior and human handling is infrequent [21,48,49]. However, animal welfare in extensive production systems is affected by several environmental challenges and conditions, such as drought, snow, food and water scarcity, and predation. Moreover, the close observation of animals by a stock person or farmers is difficult [21,48,49]. Consequently, animal welfare perception in extensive production systems may be overestimated in contrast to intensive production systems [48].

Castration and tail docking were the only husbandry practices in which the totality of sheep farmers agreed as inducing pain (Table 2). These results are contrary to those reported by Dwyer [21] and Tamioso et al. [47] in the United Kingdom and Brazil, respectively. In the study of Dwyer [21], 15.8% of farmers indicated that tail docking was a painless procedure and of the remaining 84.2% only associated this practice with mild to moderate pain (range 2–6, in a 1 to 10 scale). The results of Dwyer [21] and Tamioso et al. [47] reported that castration was associated with mild to severe pain. However, Dwyer [21] found a positive correlation between tail docking and castration scores, similar to the findings obtained in the present study.

Castration was perceived by sheep farmers as the most painful husbandry practice for lambs, followed by tail docking and animal identification (Table 2), which is in agreement with some scientific evidence [6,26,47]. Using a similar pain scale, Scott et al. [26] also found that veterinarians gave higher pain scores to rubber ring castration than tail docking (median pain score 6 vs. 4). These results are different from those obtained in the present study, where the sheep farmers' pain perception was higher for castration and tail docking regardless of the method used (median pain score 10 vs. 8).

The pain perception differences found between sheep farmers in relation to the method used to perform certain husbandry practices (Table 3) are in agreement with those reported in other pain perception studies [26,45] and animal behavior research [6,16,50]. Routine husbandry practices, such as animal identification, tail docking, and castration carried out using a knife, were perceived by sheep farmers as more painful. However, the use of a knife was the preferred method used by sheep farmers to perform the vast majority of painful husbandry practices in lambs, which does not agree with their pain perception. It is important to highlight that sheep farmers' perception could be based only on acute pain experienced by animals, and chronic pain associated with some husbandry practices, e.g., tail docking, is invisible to sheep farmers [33].

Flock size and farm size influenced the perception of pain (Table 4). Sheep farmers with large flock and farm sizes perceived less pain associated to husbandry practices than those farmers with smaller flock and farm sizes, presumably because animals in extensive production conditions are infrequently gathered and observed, and are subjected at the same time to the vast majority of painful practices [31]. When flock size is large, husbandry practices must be performed faster, thus limiting the observation of any pain related behaviors, which are mostly acute and occur during the first 30 to 60 minutes after painful procedures [37,50]. Additionally, one of the actual trends in extensive sheep production systems is to decrease the stockperson:sheep ratio [21,51]. Therefore, sheep farmers' perceptions of animal welfare and pain associated to husbandry practices in large flocks may be deficient, mainly to a decreased human:animal interaction, a situation that could be different for farmers with small flocks. Managers of larger flocks may be distanced from participating in painful routine procedures.

The majority of sheep farmers perceived hoof trimming, deworming, vaccinating, and shearing as painless practices, although a high variability in their pain perception was observed (Table 2). This variability could be explained by the effect of sociodemographic factors on pain perception associated with these husbandry practices (Table 4). According to the results of our study, farmers who had a higher educational level had a lower pain perception; these farmers also had larger flock sizes. Having larger flocks may reduce the human:animal interaction and also pain perception because they do not see and handle their animals frequently; on the contrary, farmers with smaller flocks may have a closer human-animal bond and hence a higher perception of pain in their animals because they round them up every day and have greater chances of watching them closer. Additionally, this variability could be associated to the fact that these husbandry procedures have the potential to induce pain when they are carried out improperly, e.g., when the shearer cuts the sheep's skin [52]. These husbandry practices could be more directly related to handling procedures, such as the previous rounding up, confinement, food and water deprivation, and isolation from the flock [53,54]. Therefore, it is important to train stockmen and sheep farmers to handle animals carefully and carry out these routine practices properly to improve animal welfare.

The poor degree of agreement observed among sheep farmers for pain recognition associated to husbandry practices (Table 5) could be also explained by difficulties in animal pain assessment [4,22,48,55] rather than farmers being insensible to the pain animals may experience [46]. Pain assessment difficulties in ruminants have been described as one of the most important reasons why veterinarians and farmers do not administer analgesics when they carry out painful husbandry practices in sheep [4,22,55]. For these reasons, several researchers have developed guidelines and methodologies to more accurately assess pain in sheep [22,50]. Another possible explanation of the poor agreement among sheep farmers is the variability of the scoring obtained mainly for routine husbandry practices that should not cause pain to animals, such as deworming, vaccination, shearing, and hoof trimming (Table 2). However, for the "severe pain" category, the level of agreement among sheep farmers was fair to good, indicating that there is an agreement in the recognition of intense pain associated to husbandry practices causing mild to severe pain to animals, such as castration and tail docking.

This study may be limited by the selection of farmers that attended an industry-government continuing education event and a self-administered survey format. This process may have selected from a cohort of more generally literate farmers than the population targeted. This is the first study in Chile that approaches how sheep husbandry practices are performed and the farmers' perception of pain and animal welfare; further studies on the subject should follow.

5. Conclusions

Animal identification, tail docking, and castration are painful husbandry practices that are carried out by Chilean sheep farmers at later ages than recommended by the international literature, using methods that may have a negative impact on animal welfare and analgesia is rarely used. Sheep farmers perceived castration and tail docking as the most painful husbandry practices, however, they carried out these procedures with methods that they perceived as the most painful for animals. The vast majority of sheep farmers perceived a regular to good level of animal welfare in their farms, possibly due to the fact that they considered other animal welfare issues as more important than husbandry practices, such as animal nutrition and water availability. Although there was a poor global agreement among sheep farmers in the recognition of pain associated to routine husbandry practices, there was a good level of agreement for the most painful husbandry practices, such as castration and tail docking, notwithstanding that no analgesia is used for both procedures. It is inferred that the lack of use of analgesics would not be explained by sheep farmers' pain perceptions nor their time experience as sheep farmers, but other important factors that influence this decision would exist, such as practical and economic reasons, as well as the fact that in extensive pasture production systems, sheep farmers do not have a close human:animal interaction and there are difficulties to actually observe animals.

Author Contributions: C.L.: conceptualization and methodology, validation of survey, project administration, statistical analysis, original draft preparation, writing, and editing. H.B.: conceptualization and methodology, statistical analysis, writing review, and editing. C.G.: conceptualization and methodology, resources, supervision, original draft preparation, review and editing.

Funding: This research was funded by CONICYT Doctoral National Fellowship N° 21150880 and by Escuela de Graduados, Facultad de Ciencias Veterinarias, Universidad Austral de Chile, Valdivia, Chile.

Acknowledgments: The authors would like to thank to Tamara Tadich, Valeska Covacich and Cristian Aguila for their collaboration during the application of surveys.

Conflicts of Interest: The authors declare no conflict of interest.

References

1.   International Association for the Study of Pain. Part III: Pain Terms, A Current List with Definitions and Notes on Usage. In *IASP Task Force on Taxonomy*; Merskey, H., Bogduk, N., Eds.; IASP Press: Seattle, WA, USA, 2012; pp. 209–214.

2. Fitzpatrick, J.; Scott, M.; Nolan, A. Assessment of pain and welfare of sheep. *Small Rum. Res.* **2006**, *62*, 55–61. [CrossRef]

3. Molony, V.; Kent, J. Assessment of acute pain in farm animals using behavioral and physiological measurements. *J. Anim. Sci.* **1997**, *75*, 266–272. [CrossRef] [PubMed]

4. McLennam, K. Why pain is still a welfare issue for farm animals, and how facial expression could be the answer. *Agriculture* **2018**, *8*, 127. [CrossRef]

5. Gallo, C.; Tarumán, J.; Larrondo, C. Main factors affecting animal welfare and meat quality in lambs for slaughter in Chile. *Animals* **2018**, *8*, 165. [CrossRef] [PubMed]

6. Grant, C. Behavioural responses of lambs to common painful husbandry procedures. *Appl. Anim. Behav. Sci.* **2004**, *87*, 255–273. [CrossRef]

7. French, N.P.; Wall, R.; Morgan, K.L. Lamb tail docking: A controlled field study of the effects of tail amputation on health and productivity. *Vet. Rec.* **1994**, *134*, 463–467. [CrossRef] [PubMed]

8. Fisher, M.W.; Gregory, N.G.; Kent, J.E.; Scobie, D.R.; Mellor, D.J.; Pollard, J.C. Justifying the appropriate length for docking lambs' tails-a review of the literature. *Proc. New Zeal. Soc. Anim. Prod.* **2004**, *64*, 293–296.

9. Department for Environment, Food and Rural Affairs (DEFRA). Code of Recommendations for the Welfare of Sheep. 2003. Available online: https://assets.publishing.service.gov.uk/government/uploads/system/uploads/attachment_data/file/69365/pb5162-sheep-041028.pdf (accessed on 2 August 2018).

10. Farm Animal Welfare Council (FAWC). Report on the implications of castration and tail docking for the welfare of lambs. 2008. Available online: https://www.gov.uk/government/publications/fawc-report-on-the-implications-of-castration-and-tail-docking-for-the-welfare-of-lambs (accessed on 2 August 2018).

11. Webb Ware, J.K.; Vizard, A.L.; Lean, G.R. Effects of tail amputation and treatment with an albendazole controlled-release capsule on the health and productivity of prime lambs. *Aust. Vet. J.* **2000**, *78*, 838–842. [CrossRef]

12. Sutherland, M.; Tucker, C. The long and short of it: A review of tail docking in farm animals. *Appl. Anim. Behav. Sci.* **2011**, *135*, 179–191. [CrossRef]

13. Ferguson, D.M.; Fisher, A.; Colditz, I.G.; Lee, C. Future challenges and opportunities in sheep welfare. *Adv. Sheep Welf.* **2017**, *7*, 285–293. [CrossRef]

14. Thornton, P.D.; Waterman-Pearson, A.E. Quantification of the pain and distress responses to castration in young lambs. *Res. Vet. Sci.* **1999**, *66*, 107–118. [CrossRef] [PubMed]

15. Guesgen, M.J.; Beausoleil, N.J.; Minot, E.O.; Stewart, M.; Stafford, K.J. Social context and other factors influence the behavioural expression of pain by lambs. *Appl. Anim. Behav. Sci.* **2014**, *159*, 41–49. [CrossRef]

16. Mellor, D.J.; Stafford, K.J. Acute castration and/or tailing distress and its alleviation in lambs. *New Zeal. Vet. J.* **2000**, *48*, 33–43. [CrossRef] [PubMed]

17. Small, A.H.; Belson, S.; Holm, M.; Colditz, I.G. Efficacy of a buccal meloxicam formulation for pain relief in Merino lambs undergoing knife castration and tail docking in a randomised field trial. *Aust. Vet. J.* **2014**, *92*, 381–388. [CrossRef] [PubMed]

18. World Organisation for Animal Health (OIE). Animal Welfare. In *Terrestrial Animal Health Code*; OIE, World Organisation for Animal Health: Paris, France, 2018; Chapter 7.1.

19. Ministerio de Agricultura. *Aprueba Reglamento sobre Protección de los Animales Durante su Producción Industrial, su Comercialización y en otros Recintos de Mantención de Animales*; Decreto N° 29; Publicado en el Diario Oficial: Santiago, Chile, 2013. (In Spanish)

20. Dwyer, C.M. The welfare of the neonatal lamb. *Small Rum. Res.* **2008**, *76*, 31–41. [CrossRef]

21. Dwyer, C.M. Welfare of sheep: Providing for welfare in an extensive environment. *Small Rum. Res.* **2009**, *86*, 14–21. [CrossRef]

22. Manteca, X.; Temple, D.; Mainau, E.; Llonch, P. Evaluación del dolor en el ganado ovino. Farm Animal Welfare Education Centre (FAWEC). 2017. Available online: https://www.fawec.org/es/fichas-tecnicas/49-ganado-ovino/237-evaluacion-dolor-ovino (accessed on 20 September 2018). (In Spanish)

23. Impey, S. The effect of early post-natal castration on subsequent electroencephalogram response to tail docking in lambs. Master's Thesis, Massey University, Palmerston North, New Zealand, 2015.

24. Windsor, P.A.; Lomax, S.; White, P. Progress in pain management for livestock husbandry procedures. In Proceedings of the Combined ACV/ASV Annual Conference, Hobart, Australia, 11–13 February 2015; pp. 312–319.

25. Instituto Nacional de Estadísticas (INE). Encuesta de ganado ovino 2017. Número de ovejerías y existencia de ganado ovino por categoría, según región y provincia. 2017. Available online: https://www.odepa.gob.cl/wp-content/uploads/2017/01/Encuesta-de-ganado-ovino.xlsx (accessed on 30 August 2018). (In Spanish)

26. Scott, E.M.; Fitzpatrick, J.L.; Nolan, A.M.; Reid, J.; Wiseman, M.L. Evaluation of welfare state based on interpretation of multiple indices. *Anim. Welf.* **2003**, *12*, 457–468.

27. R Core Team. *R: A Language and Environment for Statistical Computing*; R Foundation for Statistical Computing: Vienna, Austria, 2016.

28. Fleiss, J.L.; Levin, B.; Paik, M.C. The measurement of interrater agreement. In *Statistical Methods for Rates and Proportions*, 3th ed.; Balding, D., Cressie, N., Fisher, N., Johnstone, I., Kadane, J., Ryan, L., Scott, D., Smith, A., Teugels, J., Eds.; John Wiley & Sons: Hoboken, NJ, USA, 2003; pp. 598–626.

29. Mellor, D.J. Updating animal welfare thinking: Moving beyond the "Five Freedoms" towards "A life worth living". *Animals* **2016**, *6*, 21. [CrossRef] [PubMed]

30. Laytte, M.J. Caracterización del grado de innovación según género en sistemas de producción ovina en las comunas de Navidad y Litueche. In *Memoria de Título*; Universidad de Chile: Santiago, Chile, 2015. (In Spanish)

31. Sutherland, M.A. Painful husbandry procedures and methods of alleviation: A review. *Proc. New Zeal Soc. Anim. Prod.* **2011**, *71*, 178–202.

32. Guesgen, M.J.; Ngaio, J.; Beausoleil, N.; Minot, E.; Stewart, M.; Jones, G.; Stafford, K. The effects of age and sex on pain sensitivity in young lambs. *Appl. Anim. Behav. Sci.* **2011**, *135*, 51–56. [CrossRef]

33. Larrondo, C.; Gallo, C.; Calderón, J.; Bustamante, H.; Garay, L. Dolor e inflamación asociados al corte de cola en corderos. In Proceedings of the Xº Congreso Latinoamericano de Especialistas en Pequeños Rumiantes y Camélidos Sudamericanos, Punta Arenas, Chile, 2–4 May 2017. (In Spanish)

34. Lester, S.J.; Mellor, D.J.; Ward, R.N.; Holmes, R.J. Cortisol responses of young lambs to castration and tailing using different methods. *New Zeal. Vet. J.* **1991**, *39*, 134–138. [CrossRef] [PubMed]

35. Lester, S.J.; Mellor, D.J.; Holmes, R.J.; Ward, R.N.; Stafford, K.J. Behavioural and cortisol responses of lambs to castration and tailing using different methods. *New Zeal. Vet. J.* **1996**, *44*, 45–54. [CrossRef] [PubMed]

36. Wohlt, J.E.; Wright, T.D.; Sirois, V.S.; Kniffen, D.M.; Lelkes, L. Effect on docking on health, blood cells and metabolites and growth of Dorset lambs. *J. Anim. Sci.* **1982**, *54*, 23–28. [CrossRef] [PubMed]

37. Kent, J.E.; Molony, V.; Graham, J. The effect of different bloodless castrators and different tail docking methods on the responses of lambs to the combined burdizzo rubber ring method of castration. *Vet. J.* **2001**, *162*, 250–254. [CrossRef] [PubMed]

38. Atashi, H.; Izadifard, J. Effect of fat-tail docking on the reproductive performance in Ghezel and Mehraban sheep. *Iran. J. Appl. Anim. Sci.* **2016**, *6*, 645–647.

39. Orihuela, A.; Ungerfeld, R.; Fierros-García, A.; Pedernera, M.; Aguirre, V. Rams prefer tailed than docked ewes as sexual partners. *Reprod. Domest. Anim.* **2018**, 1–5. [CrossRef] [PubMed]

40. Waghorn, G.C.; Gregory, N.G.; Todd, S.E.; Wesselink, R. Dags in sheep; a look at faeces and reasons for dag formation. *Proc. New Zeal. Soc. Anim. Prod.* **1999**, *61*, 43–49.

41. Thomas, D.L.; Waldron, G.D.; Lowe, G.D.; Morrical, D.G.; Meyer, H.H.; High, R.A.; Berger, Y.M.; Clevenger, D.D.; Fogle, G.E.; Gottfredson, R.G.; et al. Length of docked tail and incidence or rectal prolapse in lambs. *J. Anim. Sci.* **2003**, *81*, 2725–2732. [CrossRef] [PubMed]

42. Fisher, M.W.; Gregory, N.G. Reconciling the differences between the length at which lambs' tails are commonly docked and animal welfare recommendations. *Proc. New Zeal. Soc. Anim. Prod.* **2007**, *67*, 32–38.

43. Kent, J.E.; Molony, V.; Jackson, R.E.; Hosie, B. Chronic inflammatory responses of lambs to rubber ring castration: Are there any effects of age or size of lamb at treatment? In *Farm Animal Welfare—Who Writes the Rules?* British Society of Animal Science: Cambridge, England, 1999; pp. 160–162.

44. Viñuela-Fernández, I.; Jones, E.; Welsh, E.M.; Fleetwood-Walker, S.M. Pain mechanisms and their implication for the management of pain in farm and companion animals. *Vet. J.* **2007**, *174*, 227–239. [CrossRef] [PubMed]

45. Phillips, C.J.C.; Wojciechowska, J.; Meng, J.; Cross, N. Perceptions of the importance of different welfare issues in livestock production. *Animal* **2009**, *3*, 1152–1166. [CrossRef] [PubMed]

46. Phillips, C.J.C.; Phillips, A.P. Attitudes of Australian sheep farmers to animal welfare. *J. Int. Farm Manag.* **2010**, *5*, 1–26.

47. Tamioso, P.; Bittencourt, P.; Maiolino, C. Attitudes of South Brazilian sheep farmers to animal welfare and sientence. *Ciencia Rural.* **2017**, *47*, 1–6. [CrossRef]

48. Turner, S.P.; Dywer, C.M. Welfare assessment in extensive animal production systems: Challenges and opportunities. *Anim. Welf.* **2007**, *16*, 189–192.

49. Doughty, A.K.; Coleman, G.J.; Hinch, G.N.; Doyle, R.E. Stakeholder perceptions of welfare and indicators for extensively managed sheep in Australia. *Animals* **2017**, *7*, 28. [CrossRef] [PubMed]

50. Molony, V.; Kent, J.E.; McKendrick, I.J. Validation of method for assessment of an acute pain in lambs. *Appl. Anim. Behav. Sci.* **2002**, *76*, 215–238. [CrossRef]

51. Goddard, P.; Waterhouse, T.; Dwyer, C.; Stott, A. The perception of the welfare of sheep in extensive systems. *Small Rum. Res.* **2006**, *62*, 215–225. [CrossRef]

52. Paton, M.W.; Buller, N.B.; Rose, I.R.; Ellis, T.M. Effect of the interval between shearing and dipping on the spread of Corynebacterium pseudotuberculosis infection in sheep. *Aust. Vet. J.* **2002**, *80*, 494–496. [CrossRef] [PubMed]

53. Carcangiu, V.; Vacca, G.M.; Parmeggiani, A.; Mura, M.C. The effect of shearing procedures on blood levels of growth hormone, cortisol and other stress haematochemical parameters in Sarda sheep. *Animal* **2008**, *2*, 606–612. [CrossRef] [PubMed]

54. Hudson, G.D. Behavioural principles of sheep handling. In *Livestock Handling and Transport*; Grandin, T., Ed.; CAB International: Wallingford, UK, 2014; pp. 193–216.

55. Lizarraga, I.; Chambers, J.P. Use of analgesic drugs for pain management in sheep. *New Zeal. Vet. J.* **2012**, *60*, 87–94. [CrossRef] [PubMed]

![animals logo] *animals*

MDPI

Article

Stakeholder Perceptions of the Challenges to Racehorse Welfare

Deborah Butler *, Mathilde Valenchon, Rachel Annan, Helen R. Whay and Siobhan Mullan

School of Veterinary Sciences, University of Bristol, Langford, North Somerset, Bristol BS40 5DU, UK;
mathilde.valenchon@bristol.ac.uk (M.V.); rachel.annan@bristol.ac.uk (R.A.); Bec.Whay@bristol.ac.uk (H.R.W.);
siobhan.mullan@bristol.ac.uk (S.M.)
* Correspondence: deborah.butler@bristol.ac.uk

Received: 3 May 2019; Accepted: 14 June 2019; Published: 17 June 2019

Simple Summary: British horseracing industry stakeholders were asked to discuss some of the challenges they perceived as having an effect on the welfare of racehorses in training. A shortage of racing staff was mentioned in six of the nine themes stakeholders identified as having an effect on racehorse welfare. Staff shortages were perceived as having an effect on welfare directly, through standards of care given to racehorses in training, and indirectly, through poor employee relations between racehorse trainers and staff, perceived as affecting attitudes and behaviour which, in turn, can affect the welfare of horses in training and potentially their performance.

Abstract: The purpose of this paper is to highlight some of the key challenges to racehorse welfare as perceived by racing industry stakeholders. The paper draws upon statements and transcripts from 10 focus group discussions with 42 participants who were taking part in a larger study investigating stakeholders' perceptions of racehorse welfare, which participants recognised as maintaining the physical and mental well-being of a performance animal. Analysis of the 68 statements participants identified as challenges produced nine themes. Among these, 26% (18 statements) of the challenges were health related, whilst 41% (28 statements) focused on the effect staff shortages were having on the racing industry. Staff shortages were perceived as affecting standards of racehorse care and the opportunity to develop a human–horse relationship. Poor employee relations due to a lack of recognition, communication and respect were perceived as having a detrimental effect on employee attitudes, behaviour and staff retention which, in turn, can have a sequential effect on the welfare and health of horses in training. Although the number of challenges produced is small (68), they emphasise the perceptions of stakeholders closely associated with the racing industry.

Keywords: racehorse welfare; staff shortages; horse–human relationship; standards of care; employee relations

---

1. Introduction

The British racing industry has been said to have been experiencing an ongoing labour shortage from at least the early 1970s [1–4], a shortage that has yet to be resolved. The aim of this paper is to highlight some of the effects staff shortages were perceived as having in racing yards and how a shortage of labour and poor employee relations may affect horse husbandry and potentially racehorse welfare.

The role of stable staff as carers is an area that has received relatively little academic attention even though research has shown that animal carers/stock people can have a major impact on the welfare of the animals in their care [5]. These attributes play an important part in working practices that promote positive interactions between humans and, as in this study, horses. Drawing on data collected during focus groups with racing industry stakeholders, the aim of this study is to provide an insight into one of the main challenges racing industry stakeholders identified, that of staff shortages, and the impact this is perceived as having on horse welfare. Maintaining a standard of care in the face of structural changes in the organisation of work in racehorse training yards is of increasing importance to the racehorse welfare debate and is an area that has become visible during discussions with racing industry stakeholders.

At present, there are 550 racing yards training on average 16,221 racehorses per month in Great Britain [6]. The industry is mainly based in rural areas and has three main training centres with communal gallops: Lambourn, Newmarket and Middleham. Training racehorses is a challenge in itself, a precarious profession in that, to use a well-worn epithet, 'you are only as good as your last winner'. Trainers' income is very variable and is mainly derived from training fees, owners and, if they are successful, a percentage of any prize money their horses might accrue. If the trainer has a bad season or trains horses of moderate ability, the amount of prize money received may be negligible. Poor results can result in owners moving their horses to another trainer, thus reducing income coming into the business.

Training racehorses is a relatively labour-intensive occupation. In 2018, there were 6734 registered employees, 4428 full-time and 2306 part-time, of which 3493 were male and 3241 were female [7]. Employees in the racing industry have to be registered with the British Horseracing Authority (BHA) by their trainer on the Register of Stable Employee Names [8], a procedure not seen in other sections of the equine industry. Stable staff have a trade organisation, The National Association of Racing Staff (NARS), 'the trade union for racing staff' [9] who negotiate with the National Trainers Federation (NTF), 'the voice of Britain's racehorse trainers' [10] over, for instance, pay and conditions, hours of work and other work related policies.

2. Staff Shortages and Changes in the Organisation of Work in the Racing Industry

At present, it has been estimated that the racing industry has a shortfall of approximately 500–1000 available stable staff. Whilst steps are being taken to improve domestic recruitment of stable staff, staff shortages may be may be further exacerbated should changes be made to British immigration policy which restrict the number of workers, as the racing industry has for many years employed migrant workers from 23 European Economic Areas (EEA) excluding the United Kingdom as well as non-EEA countries to supplement its shrinking workforce [7].

Staff shortages, however, are not a new phenomenon. In 1974, The Committee of Inquiry into the Manpower of the Racing Industry (CIMRI) was appointed by the Joint Racing Board (JRB) as 'a result of the growing concern throughout the racing industry about the current and future labour position' [1]. Although the most severe shortages of staff were found in Newmarket, where the highest proportion of horses were in training, there was a country-wide shortage of labour that was thought to be caused by an inability to attract and retain suitable stable staff. Faced with a shortage of staff, some trainers were beginning to employ girls, something of which the JRB [1] was aware. Women began working as stable staff in the mid-20th century when the racing industry was faced with a shortage of male labour; women's suitability for this work was couched in terms of their 'natural love of horses' [1].

Research by Filby [11] identified other significant changes that were having an effect on the organisation of work in racing yards. These included, for instance, the introduction in the late 1970s of a national minimum wage for racing staff, which meant trainers found themselves having to adapt their labour processes rather quickly once a national minimum wage had to be paid. Indentured apprenticeship, once the only entry route into racing for small, lithe young men with aspirations of being a jockey, was abolished in 1976, thus bringing to an end the superexploitation of indentured apprentices. The once ready supply of cheap labour (indentured apprentices) stopped (see Butler [12] for a more detailed explanation of indentured apprenticeship). Other factors played their part: a declining workforce lured by the opportunities available of alternative better-paid employment, an increase in the size and weight of the population and an increase in the number of horses in training [6,11]. In 1975, there were 11,491 horses in training; this has now risen to over 16,000 in 2018 [6,13] which, coupled with the introduction of Sunday racing in the 1990s [14], an increase in the fixture list from 1132 meetings in 2000 to 1508 in 2018 [6,13,15], a greater regulation in the hours stable staff can work and overtime payment, [9] has meant staff and trainers are constantly under pressure.

All of these changes meant that the historic practice and custom of 'doing your two [horses]' is no longer used as a standard measurement of work as set out in the Memorandum of Agreement between the National Trainers Federation and The National Association of Racing Staff [9]. The practice involved one member of staff having responsibility for their 'own two horses' which they would muck out in the morning, ride them on exercise and feed them at lunch time (12:30–1:00 p.m.). Two or three 'lots' (exercise routines) would be the norm. In some yards, staff were not to go back into the yard until 'evening stables', as it was viewed as the time horses would be able to rest during the day. Evening stables typically started at 3:30 p.m. and finished at 5:30–6:00 p.m., sometimes later. Staff would skip out their two, giving their horses a 'dressing over', that is, a thorough groom, possibly strapping them with a wisp made from plaited hay, then leave them tied up until they were told to let the horses down, that is, untie them. The head 'lad' would, during this time, check each horse over, asking the lad if he or she had noticed or sensed if anything was amiss. The trainer would often 'look round', typically starting at 5:15 p.m. when he or she would enter every box, check the horse over whilst it was being held by the lad (see [12] for more detail). Once the procedure had finished, the head lad could then feed the evening feed with each lad feeding their two horses. The weekly working week was set as 48 h per seven days, although most racing staff would have typically worked over that with no overtime paid.

The title of 'stable lad' (male and female) has now been changed to 'racing grooms' [16] and has given way to a different occupational hierarchy [1,2,11,12]. This has involved a disaggregation of work roles, a polarisation of skills and knowledge and the creation of different roles within the yard. For example, yards will typically employ yard staff who will only muck out and tidy the yard, 'rider outers', or 'work riders', who only ride out in the mornings, paid by the number of exercise lots they ride, together with full-time staff who will carry out a more traditional composite role within the yard and will have five or more horses assigned to them, although not always to ride out. Evening stables will typically last no more than two hours and involve staff skipping out five horses, 'setting them fair', that is, brushing off any sweat marks, picking out feet and removing and replacing horses' rugs. If staff are away racing or, as now happens in some yards, a certain percentage of staff get an afternoon off if they have worked a weekend, the remaining staff will 'work round', that is, carry out the basic husbandry routines such as skipping out, putting forage in, checking water and straitening rugs. The head lad or assistant trainer will move around the yard and check each horse's legs, check they have eaten up and generally appear healthy and settled.

There are very few biographies describing what life as a racing groom, a stable lad, is like, who perform similar roles to stockpeople in the livestock industry. It is therefore reasonable to assume that relationships similar to those reported in a number of livestock industries may exist in recreational and working horse populations. Whilst there are differences in the degree of human interaction that exists with management and husbandry practices both within livestock industries as well as between horse and livestock industries, the frequency and quality of interactions and the context in which they occur will determine the quality of the relationship [17].

Hemsworth et al. [5] discussed the topic of 'stockmanship' in farm animal welfare monitoring schemes, highlighting how, as a topic, it has received relatively little attention even though research has shown that stockpeople have a major impact on the welfare of their stock. As Seabrook [18] found, the welfare of farm animals is dependent upon the actions of stockpeople, who regularly handle, observe and monitor the animals in their charge. Research to date has focused upon the more obvious aspects of stockmanship: how the animals are handled and how they become fearful of people. However, stockmanship involves much more, and the relationships that can develop between people and animals can be quite subtle [19–21].

The use of horses, in principle, differs little from the use of other animals for food, transport or entertainment [22–24]. The racehorse is no exception. It is the central player in a complex relationship that revolves around, at a macrolevel, the betting industry and Thoroughbred horse breeding, and at a microlevel, the racehorse trainer, the racehorse owner, the jockey and the stable staff. All parties are reliant on the racehorse to provide their leisure, employment and financial security. Given their role within the racing field, the racehorse could be defined as a production animal, an 'ambiguous commodity' [25]. The racehorse as a specific breed, the Thoroughbred, was and still is subject to asymmetries of power where their genealogy, their working and reproductive life (if they have one) and ultimately their death is dominated by a political ecology of human dominance and exploitation in the same way livestock can be.

This study identifies the effect staff shortages were perceived as having on racehorse welfare, firstly, on standards of care and, secondly, on employee relations in some racing yards. It highlights the challenges in maintaining positive contacts between horse and human despite the trend in racing to increase the number of horses each staff member looks after, as has been seen to be the case in agriculture [26].

3. Materials and Methods

3.1. Participant Recruitment and Response

This study drew upon statements collected from an exercise that formed part of 10 focus group discussions with 42 participants who were taking part in a larger study investigating stakeholders' perceptions of racehorse welfare (see Butler et al. [27] for more details of the study). Participants were recruited to reflect the main stakeholder groups within the industry, which included equine veterinary surgeons; racehorse trainers; assistant trainers; stable staff; owners; human and animal nongovernmental organisations, for instance, Racing Welfare and World Horse Welfare, respectively; paraprofessionals such as equine physiotherapists; BHA veterinary officers and BHA racing yard inspectors. Potential participants were contacted through a variety of methods. One of the authors (D.B.) had worked and is still closely involved in the racing industry and was able to draw upon personal contacts for stable staff and ancillary racing stakeholder groups, who in turn asked their friends if they would like to attend the focus groups. Such an approach could best be described as a snowballing technique [28]. Other participants were contacted through their place of work via email. The focus groups were posted on Facebook by Racing Welfare, the charity which supports the workforce of British racing. Posters were also displayed by Racing Welfare in Newmarket and Middleham. Holding focus groups in the main racehorse training areas and centrally in Britain ensured that, theoretically, attitudes and perceptions of welfare could be gathered from a range of racing industry participants. To facilitate a more open discussion, two separate focus groups were held in each training centre, one for trainers and one for stable staff and ancillary racing personnel who had worked in racing. These were held in either local pubs or hotels. In addition, a combined focus group was held in the Midlands, as there are a plethora of racing yards in these areas; one in London for representatives from equine charities and BHA veterinary officers; another at a University Veterinary School for members of their Equine Veterinary Hospital who work with racehorse trainers and as specialists in the Equine Hospital; and the tenth was in the Cotswolds for BHA stable inspectors, BHA veterinary officers who work at the racecourse and paraprofessionals who work with racehorses. Racing's stakeholders are located across the length and breadth of Britain; focus groups were thus a more resource-efficient way of reaching a wider spread of participants when compared to carrying out face-to-face interviews.

3.2. Structure of Focus Group Discussion

Participants were given a participant information sheet and signed a consent form which informed them of their right to withdraw from the focus group. The study and consent process had been given ethical approval by the University of Bristol Faculty of Health Sciences Ethics Committee (R113851-101).

Each focus group was run by at least two members of the research team, including a trained facilitator (S.M.) for the first eight.

Participants worked together on three exercises. The first scenario-based exercise asked participants to imagine themselves as a racehorse in training. As a racehorse, participants were asked to identify within their group what important elements would form the minimum welfare standards they might be kept under, and conversely, what would contribute to the 'best life' they, as imaginary racehorses, might experience, where money and other factors were no object. These were written down on sticky notes by participants as they discussed the scenarios together and placed on a flip chart paper divided into two sections marked 'minimum welfare standards' (MWS) and 'best life' (BL).

This study drew upon data collected from the second exercise, in which participants were asked to individually identify the three main welfare challenges racehorses faced. The challenges were written on sticky notes and put up onto a large flip chart to facilitate a group discussion.

3.3. Data Analysis

Using thematic analysis, nine themes were identified from the 68 statements that were provided and ranked according to number of statements (Table 1).

Table 1. Number of statements per theme and associated frequency ranking.

Associated Freq. Ranking	Themes	No. of Statements	Statements/Theme (%)
1	Health	18	26%
2	Staff management and education	17	25%
3	Training/exercise and recovery	9	13%
4	Daily routine and monitoring	8	12%
5	Physical comfort/living environment	6	9%
6	Policy and procedures	3	4%
7	Turnout and social contact	3	4%
8	Owner/breeders	2	3%
9	Feeding	2	3%
Total		68	100%

Out of nine themes, health was perceived as the main challenge, although 11 statements relating to staff shortages were also associated with five other themes. These were training, exercise and recovery, daily routine and monitoring, physical comfort/living environment, turnout/social contact and feeding. In total, 41% (28 statements) of the statements were staff related.

Further thematic analysis by D.B. of the six themes which contained staff-related challenges identified two strands, standards of care and employee relations, areas which were perceived as impacting upon racehorses in training and thus on their welfare.

Once the two strands had been identified, audio recordings from the 10 focus groups were transcribed verbatim and analysed by D.B. Transcripts were read through and first coded manually to the nine themes initially identified. Sections of the transcripts assigned to each of the six themes were then recoded manually to the two identified strands of standards of care and employee relations.

4. Results

In terms of health, veterinary surgeons and trainers identified similar challenges such as 'soundness' and 'maintaining good health' and the avoidance of 'disease, flu/herpes and low-grade bacterial infection'. The veterinary surgeons outlined how, for a horse in training, factors such as 'veterinary aspects (pain, disease) [were] more important welfare issues than management aspects' where 'on-going health issues such as stomach ulcers and repetitive injuries cause lack of performance'.

The results shown below drew on some of the 68 statements participants perceived as challenges to racehorses in training together with an analysis of the transcripts from each focus group to illustrate participants' perceptions of staff shortages and the impact they see these having on welfare.

4.1. Standards of Care

Many of the factors participants identified were related to standards of care and how, if levels of horse husbandry are not maintained, horse welfare may be affected. Statements such as 'the staff-to-horse ratio' and 'lack of staff to look after horses in the yard when others [are] off racing'. However, as participants were aware, some trainers cannot afford to employ more staff to care for the horses they have in training:

> "In an ideal world you'd have one lad to every three, you now have one lad to every four or five. I think that's a reasonable number. In a yard that's struggling financially, your minimum is probably one lad to every eight (female racing staff)."

Trainers and staff alike were outlined as 'not knowing their horses', with participants highlighting 'how the lack of time spent out on exercise' was in some instances a challenge. Trainers acknowledged they had to adjust their general organisation of work as staff are perceived as not being available:

> "We've had to adjust our work routines to accommodate what's really happening out there and there's a massive staff shortage. You've got to have a way of adjusting your routines to use staff differently (racehorse trainer)."

As another trainer explained:

> "You do the warm up and cool down on an electric walker while your staff are out on another lot (racehorse trainer)."

Participants were acutely aware how a 'lack of knowledgeable experienced staff' and 'few experienced staff' can have an impact on standards of care and welfare and was something that was identified frequently:

> "I know there's nothing we can do about it, and that [staff experience] it's something in decline [and] that becomes a welfare issue. Things get missed, I've been in big yards where we've had enough staff, we've not had the one-to-three ratio, but we've had enough and things get missed. Someone doesn't pick a horse's feet out and then the next day it comes out and it's got an infection that's gone into its white line and it's lame (female participant)."

Participants felt that the lack of staff meant corners were bound to be cut:

> "If you're doing a line of six horses in the evening where you're running through them as fast as you can, because there's a million and one things to do in the hour, I feel like that's having a negative effect on the horses indirectly (female racing staff)."

As one participant outlined in a statement, there are 'poor standards of care around horse husbandry when the stable routine is compromised through lack of staff'. Other statements such as, 'standards of facilities and standard of stables', 'cleanliness of yard' and 'the level of health and welfare in a yard being affected by yard hygiene and the incidence of low-grade bacterial infections' were areas affected by a shortage of time available caused by a lack of staff to make sure routines were completed properly. Comments such as:

> "The yard and different areas like the feed room don't get swept, cleaned out properly and the same with mucking out. Horses cope with it, lying in shit, but I can't see as it's good for them (male racing staff)."

It was thought that, if a situation such as this were to be avoided, the experience of staff employed was seen as vital in maintaining welfare:

> "And the mix [of staff] as well I guess, in that you could have, you know 10 horses, three knowledgeable staff and two kids who are working in the evening after school, then that's great but if you've got five kids that are all working in the evening after school and nobody else, that's not good (male racing staff)."

The lack of experience was perceived as having a direct impact on horse welfare:

"But if you take it from a welfare point of view the issue is labour, there is not enough labour full stop in terms of skilled labour (male racing staff)."

Two written statements focused on how 'the lack of contact time with horses' and how 'the lack of contact time between horse and handler affected their ability to build a horse–human relationship'. These were areas participants identified as potentially limiting the horse–human relationship that many stable staff see as an important and integral part of their daily routine. As one participant illustrated:

"I'm a very strong believer in horses getting a really big kick out of their person that's in their life every day. There's proof of that in the way your horses react to you. I had a colt at XXXXX and he'd call to me every time he heard me calling in the yard. They get a buzz out of that and I feel like the staff retention situation is affecting the horses in that way as they [horses] don't get one-to-one (female racing staff)."

The horse-to-staff ratio was perceived as an area that could compromise welfare and participants were aware how staff shortages can affect the horse–human relationship:

"We put here the staff ratio possibly 3 horse to 1 staff max. So that you really have time to know your horse and develop a relationship and understand their likes, dislikes (female racing staff)."

4.2. Employee Relations

In terms of employee relations, two written statements thought 'bad and incompetent trainers' could be detrimental to welfare, as a former, experienced member of staff expressed:

"I think there's almost a hard-nosed attitude towards staff, and if you don't look after your staff they can't look after the horses and they're not going to want to do it for you, go that extra mile. We all know (female racing staff)."

Communication and trainers not listening was a bone of contention, highlighted four times, for instance, 'trainers not listening when they are told a horse is lame' as well as 'poor employment relations' where 'yards should be investigated when there is a high number of staff down as leaving'. As one member of staff explained:

"I stopped saying something to the trainer because I can protect the horse more by shutting up. I go steady and look after him/her a little bit (male racing staff)."

5. Discussion

The aim of this study was to highlight one of the key challenges to racehorse welfare as perceived by racing industry stakeholders, that of staff shortages and the organisation of work and its perceived effect on horse husbandry and thus on racehorse welfare. Nevertheless, what must be acknowledged is that the perceptions and practical reasoning of the stakeholders are based on their knowledge and experience, their taken-for-granted assumptions which help to define the racing field even if their assumptions may be contingent and arbitrary.

Within the larger study, participants perceived understanding of welfare focused around maintaining the physical and mental well-being of a horse living the 'best life' in training, which was strongly associated with racing performance [29]. Given stakeholders' understanding of welfare in this way, it is not surprising that health, that is keeping an animal free from illness and disease, should be identified as a challenge to welfare.

Little research has ever been carried out on racehorse training as a business, and the nature of racehorse training means it is labour intensive with wages of stable staff representing the largest expenditure item on a trainer's balance sheet [30]. As has been highlighted earlier in the paper, there is a shortage of staff entering and being retained within the racing industry [4]. Wages too, for experienced staff, are also higher [9]. Trainers may not also be able to afford to have a horse-to-staff ratio of two to three horses; this is now typically four to five horses to one member of staff with a reliance on part-time staff [30].

As was highlighted by participants, it was thought that 'standards of facilities and stables were being compromised by lack of staff', thus allowing for the potential spread of infection and disease. An increase in the number of horses to care for means that standards of care in a yard do, as was highlighted, slip, and one of the key aspects of disease control is decreasing exposure to pathogens, which at the most basic level includes the provision of a clean living environment for the horse. Many disease-causing pathogens can be transmitted indirectly through sharing of contaminated fomites [31], which could include horse equipment, clothing or bedding [32]. A lower number of staff means there are more horses per person to muck out and care for, which means less time can be spent in each box, especially when the morning routine must be completed in time to ride out.

As participants highlighted, when a yard has a shortage of staff, it means that less time could be given to each horse, thereby limiting the opportunity to build up a horse–human rapport, seen as an integral part of the horse–human relationship [33–36]. As Waiblinger et al. outlined [21], the human–animal relationship can be defined as the degree of relatedness between or distance between the animal and the human, that is, the mutual perception which develops and expresses itself in their mutual behaviour and experience with a particular human [36], as one participant illustrated, when 'her colt recognised her' as he heard her voice. She will have looked after him on a daily basis, knowing his habits and routine quite often from the first day the colt entered training. A positive human–animal relationship makes it possible for the stockperson, in this case, stable staff, to determine earlier deviations in the animal's behaviour, which might express the first disease symptoms [37,38]. A good relationship with the animal can therefore be associated with a better health status [39].

If a trainer, whether through affordability or yard reputation as a poor employer, is short of staff, the development of an established horse–human relationship requiring mutual recognition will be harder to establish. This in turn means that the feedback effect created through a close positive human–horse relationship cannot be established, which in turn can lead to an increase in fear and suspicion when, for instance, a horse is being handled, resulting in increased levels of anxiety for horse and handler [40,41]. Whilst some animals, such as the participant's colt, may also have generalised his experiences, this does not mean he was not able to differentiate between her and other members of staff [41].

If the human–animal relationship can be conceptualised in terms of inter-individual relationships [42], where the quality and frequency of interactions between the two individuals as well as the context in which they occur, can determine the quality of the relationship, the low human to high horse ratio identified by participants in some yards where trainers are short of staff may make building up a human–animal relationship difficult, as seen in with other livestock [43]. Studies into the behavioural effects of interactions between humans and horses has been a topic of investigation in recent years [17,35,36]. Many have speculated that a person's attitude and confidence level will affect the behavioural reactions of the horse [44–46]. In a similar vein, Hausberger et al. [3] highlighted that whilst there is no prescriptive method of being able to adapt and handle horses with different temperaments [39], only well-trained observational skills allied with advanced knowledge of horse behaviour can mean horses can be handled safely in terms of horse–human interactions [39,47]. If this is the case, the association participants made between the lack of experienced staff and potential detrimental effects on welfare may be substantiated. Participants were in agreement that a lack of knowledgeable experienced staff was a challenge and one which could have an impact on welfare and performance. Whilst literature on the role of stable staff in maintaining and promoting horse welfare is extremely limited both from a social and animal science perspective, as Hemsworth et al. [17] identified with recreational horse welfare, there are possible relationships between horse owner attributes and horse welfare outcomes. There are a small number of studies on recreational horse owners where one of the key factors in horse–human relationships and its link to the health and welfare of recreational horses may be owners' attitudes and their performance of husbandry and management practices [17,38].

Employee relations, that is, a company's efforts to manage relationships between employers and employees [48,49], were said to be poor by some participants, which can affect the retention of staff and thus the number of employees available in a yard. Working closely with any animals means staff will often make an extra effort, 'go that extra mile' as the participant highlighted. When being conscientious and empathetic is routinely not recognised and staff are constantly criticised, as is implied within the quote, attitudes and behaviour towards the horse can be affected [19]. For instance, one participant discussed the way in which staff 'protect' the horse they are riding by 'shutting up'. What is being referred to is the fact that the rider felt the trainer was not listening to the feedback the rider was giving him about how the horse did not seem to be moving very well on exercise. The reference to 'shutting up' means that when the trainer asked the rider again when the horse was exercised how he/she was moving, the rider will have just said 'fine, moving okay', although in the rider's opinion there was no improvement. As the trainer will think the horse is fine, the horse's exercise will reflect this. However, the rider will ignore the instructions from the trainer by going at a slower pace than the trainer has instructed the rider to do, thus 'protecting' the horse when the horse did not seem to be moving correctly. This scenario has implications both for the welfare of the horse and for employee relations. The goal of a racehorse trainer is to produce winners; having riders tell him or her a horse is lame is not what he or she wants to hear. Nevertheless, for staff members, their experience and knowledge are ignored and their concern for the well-being of the horse is ignored, creating a barrier to two-way communication and job satisfaction.

Ethical Approval

The study and consent process have been given ethical approval by the University of Bristol Faculty of Health Sciences Ethics Committee (R113851-101). Only adults (>18 years) were involved; all participants took part on a voluntary basis and could withdraw at any time. Information regarding the purpose, intent, motivation, funding body, potential use of data and methods of data collection were provided to participants prior to the beginning of the focus groups. All data were collected anonymously and it was not possible to identify participants in the raw research data.

Author Contributions: Conceptualization, H.R.W.; data curation, D.B., M.V. and R.A.; formal analysis, D.B., M.V. and R.A.; writing—original draft, D.B.; writing—review and editing, S.M.

Funding: The study was funded by The Racing Foundation grant number R113851-101.

Acknowledgments: Deborah Butler would like to thank all the racing industry participants who gave up their time to take part in the focus groups and Siobhan Mullan for her insightful guidance in the production of this paper. We would also like to particularly thank the British Horseracing Authority for their support and help in facilitating the study.

Conflicts of Interest: The authors declare no conflict of interests.

Ethical Approval: The study and consent process have been given ethical approval by the University of Bristol Faculty of Health Sciences Ethics Committee. Only adults (>18 years) were involved; all participants took part on a voluntary basis and could withdraw at any time. Information regarding the purpose, intent, motivation, funding body, potential use of data and methods of data collection were provided to participants prior to the beginning of the focus groups. All data were collected anonymously and it was not possible to identify participants in the raw research data.

References

1. Board, J.R. *Report of the Committee into the Manpower of Racing Industry*; The Jockey Club: London, UK, 2006.
2. Rock, G. Stable Lads are a Dying Breed. Available online: https://www.theguardian.com/observer/sport/story/0,6903,406236,00.html (accessed on 30 May 2019).
3. Stable and Stud Staff Commission. *Report of the Stable and Stud Staff Commission*; British Horseracing Board: London, UK, 2004; pp. 1–134.

4. Racing Post. Stable Staff Shortage Having Major Impact on Industry. Available online: https://www.racingpost.com/news/recruitment-and-lack-of-training-having-big-impact-on-industry/276532 (accessed on 12 April 2019).
5. Hemsworth, P.H.; Coleman, G.J. *Human-Livestock: The Stockperson and the Productivity and Welfare of Intensively Farmed Animals*, 2nd ed.; Cab International: Oxford, UK, 2010; pp. 1–189.
6. British Horseracing Authority. BHA Racing Data Pack. Available online: https://www.britishhorseracing.com/wp-content/uploads/2019/02/January-2019-Data-Pack.pdf (accessed on 2 April 2019).
7. British Horseracing Authority. *Migration Advisory Committee—EEA Workers in the UK Labour Market*; British Horseracing Authority: London, UK, 2017; pp. 1–10.
8. British Horseracing Authority. The Rules of Racing. Available online: http://rules.britishhorseracing.com/Orders-and-rules%26staticID=126786%26depth=3?zoom_highlight=employees (accessed on 2 April 2019).
9. National Association of Racing Staff. Available online: http://www.naors.co.uk/ (accessed on 3 April 2019).
10. National Trainers Federation. Available online: http://www.racehorsetrainers.org/ (accessed on 3 April 2019).
11. Filby, M.P. The Newmarket Racing Lad: Tradition and change in a marginal occupation. *Work Employ. Soc.* **1987**, *1*, 205–224. [CrossRef]
12. Butler, D. *Women, Horseracing and Gender: Becoming One of the 'Lads'*, 1st ed.; Routledge: Oxfordshire, UK, 2017; pp. 1–244.
13. House of Commons. Memorandum by The Epsom Trainers' Association (PPG 27). Available online: https://publications.parliament.uk/pa/cm200102/cmselect/cmtlgr/238/238ap12.htm (accessed on 3 April 2019).
14. Independent. Sports Listings: Sunday/Horse Racing. Available online: https://www.independent.co.uk/sport/sports-listings-sunday-horse-racing-1535192.html (accessed on 3 April 2019).
15. Weatherbys. *The Racing Industry Statistical Bureau Statistics*; Weatherbys Group Ltd.: Wellingborough, UK, 1999; pp. 1–55.
16. Racing Post. Available online: https://www.racingpost.com/news/end-for-lads-and-lasses-in-new-bha-initiative/278009 (accessed on 2 June 2019).
17. Hemsworth, L.M.; Jongman, E.; Coleman, G.J. Recreational horse welfare: The relationships between recreational horse owner attributes and recreational horse welfare. *Appl. Anim. Behav. Sci.* **2015**, *165*, 1–16. [CrossRef]
18. Seabrook, M.F. Stockpersonship in the 21st century. *J. R. Agric. Soc. Engl.* **2005**, *166*, 1–12.
19. Hemsworth, P.H.; Barnett, J.L.; Coleman, G.J. Improving productivity with better stock handling. *Proc. N. Z. Soc. Anim. Prod.* **1993**, *53*, 211–213. Available online: http://www.nzsap.org/system/files/proceedings/1993/ab93050.pdf (accessed on 2 April 2019).
20. Rushen, J.; Taylor, A.A.; de Passillé, A.A. Domestic animals' fear of humans and its effect on their welfare. *Appl. Anim. Behav. Sci.* **1999**, *65*, 285–303. [CrossRef]
21. Waiblinger, S.; Boivin, X.; Pederson, V.; Tosi, M.-V.; Janczak, E.; Visser, K.; Jones, R.B. Assessing the human-animal relationship in farmed animal species: A critical review. *Appl. Anim. Behav. Sci.* **2006**, *101*, 185–242. [CrossRef]
22. Midgley, M. *Animals and Why They Matter*; University of Georgia Press: Athens, GA, USA, 1998; pp. 7–155.
23. Regan, T. Animal rights, human wrongs. In *Ethics and Animals*, 1st ed.; Miller, H.B., Williams, W.H., Eds.; Humana: Clifton, NJ, USA, 1983; pp. 19–44.
24. McGreevy, P.; McClean, A. (Eds.) Ethical equitation. In *Equitation Science*, 1st ed.; Wiley Blackwell: Oxford, UK, 2010; pp. 1–305.
25. Wilkie, R. Ambiguous Commodities: Changing Perceptions and Status of Food Animals. Available online: http://www.oxfordhandbooks.com/view/10.1093/oxfordhb/9780199927142.001.0001/oxfordhb-9780199927142-e-16?print=pdf (accessed on 28 March 2019).
26. Boivin, X.; Lensink, B.J.; Veissier, I. *The farmer and the animal: A double mirror, In Human-Animal Relationship: Stockmanship and Housing in Organic Livestock Systems*; Hovi, M., Bouilhol, M., Eds.; University of Reading: Reading, UK, 2000; pp. 5–12.
27. Butler, D.; Valenchon, M.; Annan, R.; Whay, H.R.; Mullan, S. 'Living the best life' or 'one size fits all'. *Animals* **2019**, *9*, 134. [CrossRef]
28. May, T. *Social Research Issues, Methods and Process*, 2nd ed.; Open University Press: Buckingham, UK, 1997; pp. 5–222.

29. Racing Board Report 2017. Available online: http://www.hri.ie/uploadedFiles/HRI-Corporate/HRI_Corporate/Press_Office/Economic_Impact/HRI%20Report.pdf (accessed on 20 April 2019).

30. Racing Post. Members' Club. Available online: https://newspaper.racingpost.com/html5/reader/production/default.aspx?pubname=&edid=da43d273-f307-476e-904e-17eaffbd0395 (accessed on 30 May 2019).

31. Timoney, J.F.; Kumar, P. Early pathogenesis of equine *Streptococcus equi* infection (strangles). *Equine Vet. J.* **2008**, *40*, 637–642. [CrossRef]

32. Yarnell, K.; Le Bon, M.; Savova, M.; McGlennon, A.; Forsythe, S. Reducing exposure to pathogens I the horse: A preliminary study into the survival of bacteria on a range of equine beddingtypes. *J. Appl. Microbiol.* **2016**, *122*, 23–29. [CrossRef]

33. Estep, D.Q.; Hetts, S. Interactions, relationships and bonds: The conceptual basis for scientist-animal relations. In *The Inevitable Bond—Examining Scientist–Animal Interactions*; Davis, H., Balfour, A.D., Eds.; CAB International: Cambridge, UK, 1992; pp. 6–26.

34. Hausberger, M.; Roche, H.; Henry, S.E.; Visser, K. A review of the human–horse relationship. *Appl. Anim. Behav. Sci.* **2008**, *109*, 1–24. [CrossRef]

35. Ijichi, C.; Griffin, K.; Squibb, K.; Favier, R. Stanger danger? An investigation into the influence of human-bond on stress and behavior. *Appl. Anim. Behav. Sci.* **2018**, *206*, 59–63. [CrossRef]

36. Fureix, C.; Jego, P.; Sankey, C.; Hausberger, M. How horses (*Equus Caballus*) see the world: Humans as significant objects. *Anim. Cogn.* **2009**, *12*, 643–654. [CrossRef]

37. Henderson, A.J.Z. Don't fence me in: Managing psychological well being for elite performance horses. *J. Appl. Anim. Welf. Sci.* **2007**, *4*, 309–329. [CrossRef] [PubMed]

38. Minero, M.; Canali, E. Welfare issues of horses: An overview and practical recommendations. *Ital. J. Anim. Sci.* **2009**, *8*, 219–230. [CrossRef]

39. Sankey, C.; Richard-Yris, M.-A.; Henry, S.; Fureix, C.; Nassur, F.; Hausberger, M. Reinforcement as a mediator of the perception of humans by horses (*Equus Caballus*). *Anim. Cogn.* **2010**, *13*, 753–764. [CrossRef] [PubMed]

40. Smith, A.C.; Wilson, C.; McComb, K.; Proops, L. Domestic horses *(Equus Caballus)* prefer to approach humans displaying a submissive body posture rather than a dominant body posture. *Anim. Cogn.* **2018**, *21*, 307–312. [CrossRef] [PubMed]

41. Proops, L.; Grounds, K.; Smith, A.V.; McComb, K. Animals remember previous facial expressions that specific humans have exhibited. *Curr. Biol.* **2018**, *28*, 1428–1432. [CrossRef] [PubMed]

42. Payne, E.; DeAraugo, J.; Bennett, P.; McGreevy, P. Exploring the existence of dog-human and horse-human attachment bonds. *Behav. Process.* **2016**, *125*, 114–121. [CrossRef] [PubMed]

43. Boivin, X.; LeNeindre, P.; Chupin, J.M. Establishment of cattle–human relationships. *Appl. Anim. Behav. Sci.* **1992**, *32*, 325–335. [CrossRef]

44. Popescu, S.; Diugan, E.A. The relationship between behavioural and other welfare indicators in working horses. *J. Equine Vet. Sci.* **2013**, *33*, 1–12. [CrossRef]

45. Valenchon, M.; Lévy, F.; Neveux, C.; Lansade, L. Horses under an enrichment program showed better welfare, stronger relationships with humans and less fear. *J. Vet. Behav.* **2012**, *7*, 16. [CrossRef]

46. Hockenhull, J.; Young, T.J.; Redgate, S.E.; Birke, L. Exploring synchronicity in the heart rates of familiar and unfamiliar pairs of horses and humans undertaking an in-hand task. *Anthrozoos* **2015**, *28*, 501–511. [CrossRef]

47. Chamove, A.S.; Crawley-Hartrick, O.J.; Stafford, K.J. Horse reactions to human attitudes and behaviour. *Anthrozoos* **2002**, *15*, 323–333. [CrossRef]

48. Hrzone. What are Employee Relations? Available online: https://www.hrzone.com/hr-glossary/what-are-employee-relations (accessed on 17 April 2019).

49. Employee Relations: An Introduction. Available online: https://www.cipd.co.uk/knowledge/fundamentals/relations/employees/factsheet (accessed on 3 April 2019).

MDPI

St. Alban-Anlage 66

4052 Basel

Switzerland

Tel. +41 61 683 77 34

Fax +41 61 302 89 18

www.mdpi.com

Animals Editorial Office

E-mail: animals@mdpi.com

www.mdpi.com/journal/animals

www.ingramcontent.com/pod-product-compliance
Lightning Source LLC
Chambersburg PA
CBHW051838210326
41597CB00033B/5694